D0519276

ALIS
3010465

The Music Instinct

How Music Works and
Why We Can't Do Without It

BY THE SAME AUTHOR

Designing the Molecular World:
Chemistry at the Frontier

Made to Measure:
New Materials for the 21st Century

H_2O:
A Biography of Water

The Self-made Tapestry:
Pattern Formation in Nature

Bright Earth:
The Invention of Colour

Stories of the Invisible:
A Guided Tour of Molecules

The Ingredients:
A Guided Tour of the Elements

Critical Mass:
How One Thing Leads to Another

Elegant Solutions:
Ten Beautiful Experiments in Chemistry

The Devil's Doctor:
Paracelsus and the World of Renaissance Magic and Science

Nature's Patterns:
A Tapestry in Three Parts

Universe of Stone:
Chartres Cathedral and the Triumph of the Medieval Mind

The Sun and Moon Corrupted

The Music Instinct

How Music Works and Why We Can't Do Without It

PHILIP BALL

THE BODLEY HEAD
LONDON

Published by The Bodley Head 2010

2 4 6 8 10 9 7 5 3 1

First published in Great Britain in 2010 by
The Bodley Head
Random House, 20 Vauxhall Bridge Road,
London SW1V 2SA

www.bodleyhead.co.uk
www.rbooks.co.uk

Addresses for companies within The Random House Group Limited can be found at: www.randomhouse.co.uk/offices.htm

The Random House Group Limited Reg. No. 954009

A CIP catalogue record for this book is available from the British Library

ISBN 9781847920881

The Random House Group Limited supports The Forest Stewardship Council (FSC), the leading international forest certification organisation. All our titles that are printed on Greenpeace approved FSC certified paper carry the FSC logo. Our paper procurement policy can be found at www.rbooks.co.uk/environment

Typeset in Dante MT by Palimpsest Book Production Ltd., Grangemouth, Stirlingshire
Printed and bound in the UK by
CPI Mackays, Chatham ME5 8TD

Contents

Preface

'Must the majority be made "unmusical" so that a few may become more "musical"?' This question, in John Blacking's seminal 1973 book *How Musical Is Man?*, apparently sums up the status of music in Western culture: it is composed by a tiny minority and performed by a slightly larger one, and these are the people we call 'musicians'. But Blacking points to the contradiction inherent in the way that music is at the same time utterly pervasive in this culture: at supermarkets and airports, in movies and on television (every programme must have its theme tune), at important ceremonies, and now, on the private, portable soundscapes that snake ubiquitously from pocket to ear. '"My" society', says Blacking, 'claims that only a limited number of people are musical, yet it behaves as if all people possessed the basic capacity without which no musical tradition can exist – the capacity to listen to and distinguish patterns of sound.' He implies that this assumption goes still further: 'his' society presupposes shared ground in the way those patterns will be interpreted, understood, responded to.

And of course these assumptions are justified: we *do* have this capacity to hear music, and to develop a cultural consensus about how to respond to it. Yet we have, in the West at least, decided that these mental faculties are so commonplace that they are hardly worth noting, let alone celebrating or designating as 'musical' attributes. Blacking's experiences among African cultures in which music-making was far less rigidly apportioned between 'producers' and 'consumers' – where, indeed, those categories were sometimes meaningless – helped him to appreciate the oddness of this situation. Personally, I suspect that it might in any event be easy to overplay that schism, which, to the extent that it exists at all, may prove to be partly a transient aspect of the emergence of mass media. Before music could be recorded and broadcast, people made it themselves. And now it is increasingly easy and cheap to make and broadcast it yourself, huge numbers are doing so. Yet we still tend to ascribe primacy of musicality to the 'production' mode. In this book I hope to show why the 'capacity to listen to and distinguish patterns of sound', which we nearly all possess, is the essence of musicality. The book is about how this capacity arises. And I want to suggest that, while hearing great music played by great performers is an incomparable pleasure, this is not the only way to get enjoyment and satisfaction from music.

Because the question of how music does what it does is so phenomenally complicated and elusive, one could easily construct an illusion of cleverness by pointing out flaws in the answers offered so far. I hope it will be clear that this is not my intention. Everyone has strong opinions on these matters, and thank goodness for that. In a subject of this nature, ideas and views that differ from one's own should not be targets for demolition, but whetstones for sharpening one's own thoughts. And since it is likely that everyone will find something with which to disagree in this book, I hope that readers will feel the same way.

For helpful advice and discussion, providing material, and for general support or even just good intentions, I am grateful to Aniruddh Patel, Stefan Koelsch, Jason Warren, Isabelle Peretz, Glenn Schellenberg, Oliver Sacks and David Huron. I am once again much indebted to my agent Clare Alexander for her encouragement, insight and incomparable combination of experience, diplomacy and resolve. I am thankful to be in the safe and supportive editorial hands of Will Sulkin and Jörg Hensgen at Bodley Head. And I cherish the music that Julia and Mei Lan bring into our home.

I should like to dedicate this book to everyone with whom I have made music.

Philip Ball
London
November 2009

Author's Note

To listen to the musical examples illustrated in this book, please visit www.bodleyhead.co.uk/musicinstinct

1

Prelude

The Harmonious Universe

An Introduction

Fourteen billion miles away from Earth, Johann Sebastian Bach's music is heading towards new listeners. An alien civilization encountering the *Voyager 1* or *2* spacecraft, launched in 1977 and now drifting beyond our solar system, will discover a golden gramophone record on which they can listen to Glenn Gould playing the Prelude and Fugue in C from the second book of *The Well-Tempered Clavier*.

You couldn't fit much on a long-playing record in 1977, but there was no room for a more extensive record collection – the main mission of the spacecraft was to photograph and study the planets, not to serve as an interstellar mobile music library. All the same, offering extraterrestrials this glimpse of Bach's masterwork while denying them the rest of it seems almost an act of cruelty. On the other hand, one scientist feared that including Bach's entire *oeuvre* might come across as an act of cosmic boasting.

Recipients of the *Voyager* golden record will also be able to listen to the music of Mozart, Stravinsky and Beethoven, as well as Indonesian gamelan, songs of Solomon Islanders and Navajo Native Americans, and, delightfully, Blind Willie Johnson performing 'Dark Was the Night, Cold Was the Ground'. (They are denied the Beatles; apparently EMI couldn't see how to maintain copyright on other worlds.)

What are we thinking of, sending music to the stars? Why should we assume that intelligent life forms that may have no human attributes, perhaps not even a sense of hearing, could comprehend what happens if – following the pictorial instructions included with the discs – you spin the *Voyager* golden records and put the needle in the groove?

That question is, in a sense, what this book is all about. Why is the succession of sounds that we call music comprehensible? What do we mean when we say that we do (or don't) 'understand' it? Why does it seem to us to have meaning, as well as aesthetic and emotional content? And can we assume, as the *Voyager* scientists have done implicitly, that these aspects of music are communicable to others outside our own culture, or even our own species? *Is* music universal?

A glib argument for the universality of music would say that it is at root mathematical, as Pythagoras proposed in the sixth century BC, so that any advanced civilization could 'decode' it from the vibrations excited in a stylus. But that is far too simplistic. Music is not a natural phenomenon but a human construct. Despite claims to the contrary, no other species is known to create or respond to music as such. Music is ubiquitous in human culture. We know of societies without writing, and even without visual art – but none, it seems, lack some form of music.

But unlike the case of language, there is no generally agreed reason why that should be so. The evidence suggests it is an inevitable product of intelligence coupled to hearing, but if so, we lack an explanation for why this is.

It is deeply puzzling why these complex mixtures of acoustic frequencies and amplitudes make any sense to us, let alone why they move us to joy and tears. But little by little, that is becoming a less mysterious question. When we listen to music, even casually, our brains are working awfully hard, performing clever feats of filtering, ordering and prediction, automatically and unconsciously. No, music is not simply a kind of mathematics. It is the most remarkable blend of art and science, logic and emotion, physics and psychology, known to us. In this book I will explore what we do and don't know about how music works its magic.

Confectionery for the mind?

'Music is auditory cheesecake, an exquisite confection crafted to tickle the sensitive spots of at least six of our mental faculties,' claimed cognitive scientist Steven Pinker in his 1997 book *How the Mind Works*. He went on:

> Compared with language, vision, social reasoning, and physical know-how, music could vanish from our species and the rest of our lifestyle would be virtually unchanged. Music appears to be a pure pleasure technology, a cocktail of recreational drugs that we ingest through the ear to stimulate a mass of pleasure circuits at once.

These claims provoked predictable outrage. Imagine it, comparing Bach's B minor Mass to the Ecstasy pills of club culture! And by suggesting that music *could* vanish from our species, Pinker seemed to some to be implying that he wouldn't much mind if it did. So Pinker's remarks were interpreted as a challenge to prove that music has a fundamental evolutionary value, that it has somehow helped us to survive as a species, that we are genetically predisposed to be music-makers and music-lovers. It seemed as though the very dignity and value of music itself was at stake.

Pinker responded to all of this with understandable weariness. No one was suggesting, he said, that music could only be taken seriously as an art form if it could be shown to be somehow evolutionarily beneficial. There are plenty of aspects of human culture that clearly did not arise as adaptive behaviour and yet which are deeply important components of our lives. Literacy is one: an evolutionary psychologist who argues that writing is obviously adaptive because it preserves vital information in a way that can be reliably passed on to our offspring is making a hopeless case, because writing is simply too recent an innovation to have any dedicated genetic component. We can read and write because we have the requisite intrinsic skills – vision and pattern recognition, language, dexterity – and not because we have literacy genes.

Joseph Carroll, professor of English at the University of Missouri-St Louis, has offered a more substantial reply to Pinker. 'Art, music and literature are not merely the products of cognitive fluidity,' he says. 'They are important means by which we cultivate and regulate the complex cognitive machinery on which our more highly developed functions depend.' These arts aren't at all like stimulation of the taste buds – they embody emotions and ideas:

> They are forms of communication, and what they communicate are the qualities of experience. Someone deprived of such experience would have artificially imposed on him a deficiency similar to that which is imposed on autistic children through an innate neurological defect . . . a child deprived of all experience with art and literature would still have innate capacities for social interaction, but these capacities would remain brutishly latent. The architecture of his or her own inner life and that of other people would remain dully obscure. In the place of meaningful pattern in the organization of emotions and the structure of human needs and purposes, such a child would perhaps scarcely rise above the level of reactive impulse.

This is the classical argument of the ennobling nature of art, which goes back to Plato. The problem is that it is awfully difficult to prove. Carroll cites the example of the Smallweeds in Dickens' *Bleak House*, who 'discarded all amusements, discountenanced all storybooks, fairy tales, fictions, and fables, and banished all levities whatsoever'. The Smallweed children are, as a result, 'complete little men and women [who] have been observed to bear a likeness to old monkeys with something depressing on their minds'. But that is all so much literary invention; and more to the point, the absence of art in the lives of the Smallweed children is obviously a symptom of their general lack of love and nurture, not a cause of it. Do we have any real evidence that being deprived of music would impoverish our spirit and diminish our humanity?

In this book I suggest why, even though there is something to be said for both Pinker's and Carroll's positions, they both miss the point. While it is possible in principle to disprove Pinker's thesis (and while, as we'll see, there are already reasons to suspect that it is indeed wrong), it is a mistake to think that this would somehow establish the fundamental value of music. Neither do we need to show that Carroll is right – that exclusion of music leads to a brutish nature – to prove that we can't go without it. After all, the reverse is not true: beastliness and a finely honed musical aesthetic may coexist, as personified by Anthony Burgess' Alex in *A Clockwork Orange*, not to mention Hitler's notorious love of Wagner. It is misguided to think that music enriches us in the mechanical manner of a nutrient. The fact is that it is meaningless to imagine a culture that has no music, because music is an inevitable product of human intelligence, regardless of whether or not that arrives as a genetic inheritance. The human mind quite naturally possesses the mental apparatus for musicality, and it will make use of these tools whether we consciously will it or not. Music isn't something we as a species do by choice – it is ingrained in our auditory, cognitive and motor functions, and is implicit in the way we construct our sonic landscape. Even if Pinker is correct (as he may be) that music serves no adaptive purpose, *you could not eliminate it from our cultures without changing our brains.* Boethius seemed to understand this in the early sixth century AD: music, he said, 'is so naturally united with us that we cannot be free from it even if we so desired'.

For that very reason, Pinker is also wrong to suggest that music is simply hedonistic. (Besides, however much cheesecake or recreational drugs we consume, we do not thereby exercise our intellect or our humanity – quite the reverse, one might say.) Here is the surprising thing: *music does not have to be enjoyed.* That sounds terrible, but it is a fact. I don't mean simply that everyone does not enjoy every possible sort of music; that's obviously true. I mean that we don't just listen to music for pleasure. In some cultures that doesn't seem to be the primary function, and it has been debated whether an aesthetic response to music is at all universal. Of course, there are also many reasons for eating other than hunger – but it is far from clear that the main reason for music is enjoyment in the same way that the main reason for eating is survival.

Happily, though, on the whole we *do* enjoy music, and one of the major themes of this book is to find out why. But the pleasure may well be a consequence, not a cause, of music-making. Pinker's 'auditory cheesecake' is itself a side effect of our urge to find music in our auditory environment, although in fact the image of a guzzling diner is absurdly inappropriate to span the range from an indigenous tribesperson singing a ritual chant to a composer of the hard avant-garde plotting out his mathematical music in the 1970s. We have a music instinct as much as a language instinct. It might be genetically hard-wired, or it might not. Either way, we can't suppress it, let alone meaningfully talk of taking it away.

What's more, it no longer makes much sense to reduce this instinct to some primitive urging on the savannah, any more than it makes sense to 'explain' the minutiae of courtship, personal grooming, extra-marital affairs, romantic fiction and *Othello* on our urge to reproduce. Cultures elaborate basic instincts out of all recognition or proportion, even inverting what one might presume to be their biological origin (if such a thing exists). Does it really make any sense to apply Pinker's phrase, or indeed Carroll's argument, to John Cage's *4'33"*, or to Motörhead playing 'Overkill' at a volume close to the pain threshold?

Whose music?

Although my survey ranges across cultures, much of it will draw on Western music. This is only in part because it is the music with which I am (and probably most readers will be) most familiar. More importantly, it is the best-studied system of highly developed 'art music', and so provides the richest source so far of information about how music is processed. Yet by exploring non-Western music, I hope not only to avoid the common error (many composers have made it) of asserting universal relevance to culturally specific concepts, but also to bring to light those aspects of music that *do* seem to have some cross-cultural validity. I shall look in particular at some highly developed and sophisticated non-Western forms of music, such as Indian classical music and Indonesian gamelan.* I will also show that 'more sophisticated' does not by any means imply 'better', and that art music is no more highly developed in some respects than traditional or folk music. Indeed, I will not on the whole be attempting to make judgements about music in an artistic or aesthetic sense, although we will find some clear, objective indications of why certain types of music seem to be more satisfying and enriching than others. I hope this book might encourage you, as researching it encouraged me, to listen again to music that you previously dismissed as too boring, too complicated, too dry, too slushy, or just plain incomprehensible. I doubt that there is a single one of us whose musical horizons could not be broadened with a little more understanding of what the music is doing and why.

* In a recent study of the 'complexity' of musical styles, gamelan scored highest: an emphatic reminder that we should not associate the geographic ubiquity of Western music with superiority.

Music is not a luxury

The fact that we humans are bound to be musical come what may could be taken as advocacy for a laissez-faire attitude to music education. And it's true that children who don't venture near an instrument and never have a music lesson at school are as likely as anyone else to end up plugged into their iPods. But to neglect music education is to stunt development and to deny opportunity. If we don't teach our children to cook, they won't starve, but we can't expect them to take much delight in food, or to be able to discover how to distinguish good fare from bad. Music is the same. And no one needs to be taught *what* to cook, but only *how*.

Whether or not Joseph Carroll is right to assert that deprivation of music makes us lumpen brutes, there is no question that the provision of music enriches in countless ways. One of the most remarkable illustrations of that is the National System of Youth and Children's Orchestras of Venezuela (commonly known simply as El Sistema), which has offered music tuition to around 250,000 impoverished Venezuelan children. Its 200 youth orchestras have provided a haven from crime and drugs for young people in the barrios, and the flagship Simón Bolívar Youth Orchestra plays with a passion and musicality that is the envy of more 'developed' nations. In part, the social benefits of El Sistema no doubt stem from the mere fact that it supplies some degree of structure and security in lives that had precious little before – perhaps football or literacy schemes would have similar effects. But it seems hard to doubt that the music itself, generally taken from the European classical repertoire, has instilled a focus, curiosity and optimism in the young Venezuelan players. In contrast, music education in the West is often seen as elitist and irrelevant, a drudge that promises little in the way of either satisfaction or inspiration. At best, it is something that children do if they have the spare time and resources.

Yet music ought to be a central and indispensable facet of a rounded education. For one thing, we will see that it is a gymnasium for the mind. No other activity seems to use so many parts of the brain at once, nor to promote their integration (the tiresome, cod-psychological classification of people as 'left brain' or 'right brain' is demolished where music is concerned). The spurious 'Mozart Effect' aside (see Chapter 9), it is clearly established that music education has a positive effect on general intellect. It is also a potentially socializing activity, and one that most young people are interested in, often passionately. And sensitive music teaching (as opposed to hothousing of little virtuosos) will bring out one of music's most valuable attributes, which is the nurturing and education of emotion.

The case for musical education should not rest on its 'improving' qualities, however, even if these are real. The fact is that music no less than literacy gives access to endless wonders. To cultivate those avenues is to facilitate life-enhancing experience.

But what usually happens instead? Children stop singing and dancing, they get embarrassed about their piano lessons (if they're lucky enough to be offered them) and frustrated that they don't sound like the stars on MTV. As adults, they deny that they possess having any musicality (despite the extraordinary skills needed to listen to and appreciate just about any music), they jokingly attribute to themselves the rare clinical condition of tone-deafness. They probably do not know that there are cultures in the world where to say 'I'm not musical' would be meaningless, akin to saying 'I'm not alive'.

This book is about that too.

2

Overture
Why We Sing

What is music and where does it come from?

Hadn't I better explain what I am talking about, when I am talking about music? That would seem a sensible preliminary, but I am going to decline it. Just why I do so should become clear when I begin to explore shortly the different forms that music takes in different cultures, but let me say right away that there is no meaningful definition of music that doesn't exclude some or other aspect of it. The most notorious counter-example to most definitions is John Cage's silent opus *4'33"* (a work that is more accurately described as specifying that the performer play no notes). Some might, with reason, argue that this is not music but a conceptual artwork* – but that risks straying towards arid semantics. Such eccentricities aside, one definition of music that seeks to acknowledge its cultural and historical diversity, proposed by musicologist Ian Cross, runs as follows:

> Musics can be defined as those temporally patterned human activities, individual and social, that involve the production and perception of sound and have no evident and immediate efficacy or fixed consensual reference.

And of course you can at once choose your own favourite nitpick – for example, that this includes a person idly scuffing his shoes at the street corner. 'Evident and immediate efficacy' in particular raises enormous questions. But perhaps most significantly, you need only

* Philosopher Stephen Davies does just that in an illuminating and dogma-free discussion, while concluding that 'we should acknowledge the originality and importance of Cage's contribution to our understanding of music and of the philosophy of the arts'. *4'33"*, he says, 'challenges the boundary between noise and music', but 'is likely to include more of the former'.

start listening to Sam Cooke or Ravel to sense that straining for a definition of music is a vapid exercise that tells us nothing important about what it is or why we listen.

Such definitions tend to consider music as an acoustic phenomenon, with the result that they seek distinctions between musical and non-musical sound. 'Organized sound' isn't a bad working description of music, so long as you recognize that this definition would make no sense to some cultures, and that it excludes some types of music and includes sounds not generally considered to be music.* Yet whatever you think of the twentieth-century Western avant-garde, it served as a reminder that this sort of exercise in definition is futile. Music can be made from mistuned radio sets, from the incidental noises of a concert hall, from the hum of machinery. No one says you have to like it.

There's a good argument that music is better defined in sociological and cultural than in acoustic terms. It's a thing we do. It is universal only in the sense that every culture seems to have it. But about what music is, and what purpose it serves, we can make no other generalizations.

Some cultures make music by banging on drums, blocks, pieces of metal: they value rhythm (and perhaps timbre) above melody. In others, the main instrument is the human voice; in others, music is inseparable from dance. In some cultures, music is reserved for special occasions; in others, people create a more or less continual musical soundtrack to their lives. Some reserve a term like 'music' only for a subset of the apparently musical things they do. Some analyse music in obsessive detail; others are puzzled by any need to discuss it. Perhaps most significantly, there is no reason to suppose that all musics should share any particular trait in common – that music has universal features. As semiologist Jean Molino has put it, 'Nothing guarantees that all forms of human music contain a nucleus of common properties that would be invariant since the origination of music.'

*There's a good case for saying that a viable definition of music doesn't have to be all-inclusive, but should simply apply to the central instances of the art. Of course, that in itself provokes questions; but I sympathize with the idea that we need feel no great obligation to encompass extreme experiments in our definitions.

Music in the world

One of the strongest objections to Steven Pinker's characterization of music as 'auditory cheesecake' is its ethnocentricity: it implies that all people listen to music simply because they like to do so. That is probably not true even in Western culture, where a type of music might for example serve as a badge of affiliation to a subculture, or proclaim a regime's military might. And music may serve very specific social functions that do not obviously demand (and perhaps do not involve) aesthetic judgements. For the Kaluli people of Papua New Guinea, music allows communion with the dead. The music of the Venda people in South Africa helps to define social relations.

And music is not merely structured sound. The word for 'music' in the language of the Igbo people of Nigeria is synonymous with 'dance', and in Lesotho there is no distinction between dance and song. In parts of sub-Saharan Africa, music without a steady rhythm, to which one cannot therefore dance, is not considered music at all, but instead a form of lamentation.

Ethnomusicologists have documented an abundance of social functions for music: it expresses emotion, induces pleasure, accompanies dance, validates rituals and institutions, promotes social stability. The last of these is not merely a matter of 'bringing people together' – music can serve as a socially sanctioned outlet for negative or controversial behaviours too. In some African cultures, the 'protest song' is tolerated as a form of political dissent that could not be voiced in speech or text. And in Bali, musicians and dancers may enact socially disruptive emotions such as rage so as to discharge them publicly in a way that serves the community. In Senegal, the low-caste *griots* of the Wolof people play music to and dance for the nobles in an emotional way. The *griots* are considered expressive and excitable, the nobles cool and detached. These musical performances enable both groups to maintain the stereotypes, which may have nothing to do with the true nature of the individuals. The music stands proxy for the emotions of the nobles, to ward against apathy, without them having to break their mask and actually display such attributes.

Music can be a vehicle for communication, sometimes with exquisite precision. The 'talking drums' of African cultures are legendary, and may be used to convey quite specific information in intricate codes almost like Morse, which seem to be tied to the pitch structure of African tonal languages. Villagers might burst into laughter during a xylophone performance as the musician uses his instrument to make a joke about a particular member of the tribe. Everyone gets it – except, sometimes, the butt.

The music of the Sirionó Indians of Bolivia, meanwhile, seems extremely simple: each song is a short phrase containing just a handful of closely spaced pitches. The function of this music seems to be entertainment rather than ritual, but in some ways that objective works on a far deeper level than it does in the West. Each member of a tribe has a 'signature' tune that forms the basis of all he sings, and these songs are voiced in the morning and evening almost as a kind of conversation, a way of saying 'here I am again'. Here's a musical culture of which composer Paul Hindemith would surely have approved when he wrote that 'Music that has nothing else as its purpose should neither be written nor be used.'

In many of these cases, music serves a symbolic purpose. It has been tacitly agreed in advance what the 'meaning' of the music is, and no one is too bothered about judging how 'well' the music meets that goal – the mere fact of the performance guarantees it. Yet whether such a purpose involves any element of pleasure is a difficult question. It has sometimes been claimed that some cultures show no aesthetic response to music, but other ethnomusicologists consider this a prejudiced view: the absence of critics, fanzines and discussion groups doesn't mean the listeners don't assess and enjoy what they hear. Such disagreements may themselves be a consequence of imposing alien categories on the musical experience. It has been claimed that composers among the Basongye people of the Congo have no explicit intention of creating music that others will admire, partly because there is nothing to be judged: they deem music to be an intrinsic good, not something that might be 'good' or 'bad'. Some ethnologists say that art criticism in general has no place in African tribal culture, since it is taken for granted that what we call art is a positive activity that meets essential needs. Meanwhile, David McAllester, a pioneer in the study of music in pre-Columbian American societies, suggests that aesthetic judgements in Native Americans are commonly tied to function: people like particular songs because they are associated with enjoyable ceremonies. (He says that some people in these cultures also express preferences for songs that are more easily learnt.)

The Canadian ethnomusicologist Colin McPhee claimed in 1935 that Balinese music is utilitarian, 'not to be listened to in itself' and containing no emotion. Rather, he said, it is simply, like flowers or incense, a necessary component of ceremony, as though someone had said 'We need three hours of music here' in the same way that they might specify the necessary dimensions of the meeting hall. Anthropologist Margaret Mead later argued that this did not mean listeners derive no enjoyment from the performance, but she suggested that this pleasure came from the performance aspects itself – from 'the way in which music is played rather than the music'. No one would, in that sense, be saying 'Oh, I love this song'. Yet these responses may have been the result of the Western observers' distance from Balinese culture; certainly, modern listeners to gamelan seem to derive aesthetic satisfaction from it. The ethnomusicologist Marc Benamou cautions that aesthetic and emotional judgements about music can be tricky to compare across cultures: Javanese people don't necessarily recognize the same categories of musical affect (happy/sad, say) as those in the West.

In any event, the Basongye people consider music to be inseparable from good feelings – they say they make music in order to be happy, or to express that happiness: 'When you are content, you sing.' But that's a more complex statement than it sounds; it is as if the emotion defines music rather than stimulates it, for music is not something that can be made in anger: 'When you are angry, you shout.' What's more, any notion that Basongye music is thereby a spontaneous outburst of pure joy is complicated by the claim of one tribesman that 'When one shouts, he is not thinking; when he sings, he is thinking.' All this suggests that Basongye music has a subtle and sophisticated social function that can't easily be described by analogies with the West.

The Basongye readily admit that music can be utilitarian too: another reason for making it, they say, is to be paid. In some cultures it is a tradable commodity, a form of wealth. In New Guinea, tribal people might sell dances from village to village, bundled up with new forms of clothing and some bits and pieces of magic. The Navajo Native Americans can own songs and sell them to others. And this, at least, is hardly unfamiliar in the West.

Neither is the sacred aspect of music and song. A further reason for making music, say the Basongye, is that God (whom they call Efile Mukulu) tells them to do so. The Yirkalla aborigines of Arnhem Land in Australia hear sacred song words in the babbling of babies. To them, songs are never composed but only discovered: all songs exist already. Where music is used for ritual purposes, a concern with accuracy may become almost obsessive, for a ceremony that is conducted incorrectly loses its power. If a single error is made in the song accompanying a Navajo ritual, the whole thing has to be started again – a standard that even the most exacting of Western concert performers hardly feels a need to uphold.

The purely functional roles of music are also evident in its association with healing. The ancient Egyptians regarded music as 'physic for the soul', and the Hebrews used it to treat physical and mental disturbance – an early form of music therapy. The Greek philosopher Thales is said to have used music to cure a 'plague' of anxiety suffered by the Spartans. According to Plutarch, Thales' songs dispelled the affliction with concord and harmony, in an echo of the magical healing powers attributed to the music Orpheus sang while playing his lyre. That myth is also reflected in the Bible:

> Now the Spirit of the Lord had departed from Saul, and an evil spirit from the Lord tormented him . . . Whenever the spirit of God came upon Saul, David would take up his harp and play. Then relief would come to Saul; he would feel better, and the evil spirit would leave him.

In antiquity and the Middle Ages, music was deemed (at least by the intelligentsia) to have a primarily moral rather than an aesthetic, much less a hedonistic, purpose. It was performed not for enjoyment but to guide the soul. For Plato and Aristotle, this made music a tool that could either promote social harmony or, if improperly used, discord. (It's no coincidence that those are both musical terms.) For the early Christian writer Boethius in the sixth century, music was to be judged 'by means of reason and the senses', not by the heart, and this made it the province of the philosopher rather than the artist. None of this is to deny that these classical listeners took pleasure from music, but that pleasure was supposed to be a means, not an end. No wonder St Augustine worried that people listening to religious singing might be 'more moved by the singing than by what is sung'. Philosopher Roger Scruton argues that music still has the capacity to provide moral education:

> Through melody, harmony, and rhythm, we enter a world where others exist besides the self, a world that is full of feeling but also ordered, disciplined but free. That is why music is a character-forming force.

In this view, music has as part of its function an educative and socializing value, and I have some sympathy with that.*

Faced with all this diversity, ethnomusicologists have long tended to avoid any quest for universals in the forms and categories of music. Yet there do seem to be some overlaps and parallels between traditions. African music, for example, can be crudely divided into two types, divided by a boundary that lies along the southern edge of the Sahara. To the north, the music is primarily vocal and monophonic, supported by a drone or rhythmic accompaniment. There is a lot of improvisation and ornamentation of the vocal line, often using microtones. In sub-Saharan Africa, in contrast, music is usually performed in groups, being polyphonic and often harmonized and making use of complex, multilayered rhythmic patterns. And the modes of singing are quite different: full-throated in the south, nasal in the north. The musicologist Alan Lomax argued that these distinctions reflect cultural attitudes towards cooperation, sex, hierarchy and class, and that the two styles are in fact representative of twin progenitors of all musical traditions. He proposed that a form based on (mostly male) improvised solos, with free rhythm and complex, highly ornamented melodies, sprung up in eastern Siberia, while sub-Saharan Africa gave birth to a 'feminized', many-voiced, rhythmically regular style. From these twin roots Lomax claimed to discern the branching of ten families of musical styles throughout the world. Although few

* I have less sympathy for Scruton's suggestion that both music and morals are on the decline – see p.335.

ethnomusicologists now accept this idea, the basic traits that Lomax identifies can indeed be recognized in the musics of many disparate cultures.

The science of music cognition is starting to make the question of universals respectable again. Perhaps this is because it has tended to break down music into the simplest structural elements, such as pitch and tone and rhythm, the perception and organization of which would seem to be an essential part of any listener's ability to transform sound into music regardless of the function that it serves. This approach can only get so far, however, because even the issue of what we perceive is not simply a matter of auditory acoustics: just as emotional, social and cultural factors may promote selective hearing of language, so they impinge on music. A Western listener can hear just about any music on the radio and make some kind of assessment of it without knowing anything about the composer, performer, period or context of the music. This would be an alien notion to some preliterate societies such as the Basongye or the Flathead Native Americans, for whom a response to, and even a recognition of, music depends on context, on the reason why it is being played and heard. To ask, as music psychologists routinely do of their test subjects, how one feels about particular intervals or rhythms has no meaning for these people – those are not, to them, questions about music at all.

That's one reason why studies in music cognition have focused almost entirely on the music of large and generally industrialized cultures: it is not only logistically difficult but also potentially ambiguous to test the perception of and response to music among tribal societies. Highly developed musical traditions usually have a rather explicit set of rules for composition, performance and analysis – to put it in coldly scientific terms, we have a better view of what the relevant variables are. But given this lacuna, we have to wonder whether a cognitive science of music can really say anything universal about it as a human activity. Jean Molino doubts that there is much we can learn about the question of *why* we make music by studying the 'great' music of the European classical *oeuvre* – he asserts that music used for ritual and dance (even disco), as well as poetry, is more relevant to that question.

This is not to say that cognitive studies based on the Western tradition – which means most of those discussed in this book – need be hopelessly parochial. We'll see that there is no reason to think, for example, that Western scales and musical structures are somehow processed with mental apparatus uniquely designed for that purpose, any more than the English language has dedicated modules that could not be used for other languages. And by asking how Western listeners cope in cognitive terms with non-Western music, we can gain insight into the general mechanisms the human brain uses to organize sound. Besides, the Western musical tradition, while having no claim to primacy, is unquestionably one of the most sophisticated in the world, and worth exploring in its own right.

In fact, music cognitive studies are helping to dismantle the old prejudices of ethnomusicology. While even much of the early work in this field had the virtue of challenging centuries of presumed Western musical supremacy,* it nevertheless tended to assert an exceptionalism of the sort voiced by one of the founders of the field, Bruno Nettl, who in 1956 defined ethnomusicology as 'the science that deals with the music of peoples outside of Western civilization'. A modern definition characterizes it instead as 'the study of social and cultural aspects of music and dance in local and global contexts', which rightly implies that Western music is as much a part of the subject as any other form. (Popular culture has yet to catch up, which is why we have the absurd genre category of 'world music', where the world is one in which a West is curiously absent.) As they probe deeper into such questions as how music is linked to emotion, researchers have become struck by how important it is to turn the ethnomusicologist's lens on their own culture. For according to music psychologist John Sloboda, 'It is a curious paradox that we probably know . . . more about the different purposes for which music is used within certain non-Western societies than we do about how it is used in Western consumer societies.'

* It is a relief to be free of the culture that enabled the early philosopher of music Eduard Hanslick to state in 1891 that South Sea Islanders 'rattle with wooden staves and pieces of metal to the accompaniment of fearful howlings', producing 'no music at all'.

In reawakening interest in universals, music psychology is also revitalizing an old question that ethnomusicologists have skirted with understandable caution. Once you start to ask how our brains make sense of music, you can't avoid the issue of *why* they are able to do so. And that summons forth the mystery behind any survey of how music is used in different cultures today: how and why did music arise in the first place?

The first musicians

In 1866 the Linguistic Society of Paris decided it had heard enough dogmatic bickering and vapid speculation about the origins of language, and decided to ban from its meetings any papers on the topic.

If you survey modern discussions about the origins of music – not just a parallel question but most probably a related one – you might have to concede that the Linguistic Society of Paris knew what it was doing. As seems to be the case with any academic enquiry, the stridency with which points of view are asserted seems to bear an inverse relation to the quantity and quality of supporting evidence. And about the origins of music, we have almost no evidence whatsoever.

Music in human culture is certainly very ancient. Several flutes made of bone have been found from the Stone Age – that's to say, the Palaeolithic period, deep into the last ice age. The oldest candidate known so far is carved from the bone of a young cave bear and dates to around 44,000 years ago. It was found in 1995 in Slovenia, and has two holes with the suggestion of a third, and perhaps another on the opposite side. When blown at one end and fingered, it will produce a diverse range of pitches.

It's possible that this object isn't an instrument, but a bone punctured by the sharp teeth of some carnivore which gnawed it later. But the holes appear to be carefully made, with no cracking at the edges, and it would be rather surprising if they had been formed in just these places without shattering the bone or presenting too much of a mouthful for any chewer. Besides, there is no doubt that bone flutes *were* made in the Stone Age. Several unambiguous examples have been unearthed in the Swabian Jura in Germany dating from around 40,000 years ago, including one more or less complete and rather elegant flute made from a bird bone (Figure 2.1). These instruments show that humans had already by this time rather thoroughly integrated music into their everyday lives.

Yet why should our ancestors have wanted or needed music, especially during an ice age when merely surviving from one day to the next was hard enough?

Charles Darwin offered one of the earliest speculations about why humans first made music. He couldn't ignore the puzzle it seemed to pose for evolutionary explanations of human behaviour, and in his *Descent of Man* (1877) he wrote:

> As neither the enjoyment nor the capacity of producing musical notes are faculties of the least direct use to man in reference to his ordinary habits of life, they must be ranked amongst the most mysterious with which he is endowed. They are present, though in a very rude and as it appears almost latent condition, in men of all races, even the most savage.

In other words, Darwin viewed music-making as an evolved behaviour without obvious adaptive value. He was, however, familiar with other apparently useless adaptations, and he believed his evolutionary theory could explain them. He argued that music had nothing to do with *natural* selection ('survival of the fittest'), but could be

Figure 2.1 A bone flute discovered by Nicholas Conrad of the University of Tübingen and his co-workers in 2008, during excavations at Hohle Fels Cave in the Swabian Jura. It is thought to be about 40,000 years old. (Image: H. Jensen/University of Tübingen.)

accounted for by his parallel notion of *sexual* selection, in which organisms gain a reproductive advantage not by living longer but by having more success at mating. He considered the music of our ancestors to be a form of exhibitionist display or prowess akin to the mating 'songs' and 'dances' of some animals. This hypothesis can accommodate the fact that music is not just of no 'direct use' but is seemingly anti-adaptive: it takes time to learn an instrument and to sit around playing it, which early humans or hominids might be expected to have spent more productively in hunting and gathering. The effort pays off, however, if the musician's skill makes him more attractive. (We can, for these purposes, assume a 'him', since these sexual displays are, in animals, the preserve of the male.) Why, though, should an ability to make music be deemed sexy? One possible answer is that it displays coordination, determination, good hearing, and perhaps stamina (some cultures engage in very lengthy musical ritual), all of which are arguably features that a female might wish to find in her offspring. In this view, music is like the peacock's tail: an elaborate display that is useless, and indeed a hindrance, in itself but which sends out a signal of 'good genes'. The pioneering American behavioural neurologist Norman Geschwind believed that musical ability is a genuine predictor of male reproductive prowess, because (he argued) both are promoted by high levels of foetal testosterone. Although this hypothesis was developed before we knew much about the links between brain anatomy and musicality (there's still much we don't know, as you'll see), it is sometimes still asserted today in support of Darwin's sexual-selection origin of music.

And indeed this idea isn't without merit. But its modern adherents too often mistake the accumulation of ad hoc arguments for the collection of scientific evidence.* One researcher points out, for instance,

*One argument adduced by psychologists Vanessa Sluming and John Manning has at least the virtue of being amusing: they found in 2000 that, averaged over eleven concerts of classical music, there were significantly more women in the seats nearer the (predominantly male) orchestras than in the back rows – a genteel form, they implied, of the female hysteria that greeted the Beatles in concert. Sluming and Manning admitted that the hypothesis might, however, need to take into account how many of these women were pre-menopausal. I suspect they may have found this information hard to acquire at classical recitals.

that just about all complex, varied and interesting sounds produced by other animals are made for the purposes of courtship, so why not humans? This would seem equally to argue that every phrase we utter has the aim of attracting a mate, which I rather doubt was true even of Casanova. In any event, the claim is not even true as stated: monkeys and apes don't appear to use calls for sexual purposes. And 'primitive' songs are by no means the tribal equivalents of 'Let's Spend the Night Together': those of the Australian Aborigines, for example, express the singer's feelings as a member of the community.

If music really did stem from sexual selection, we might reasonably expect that musicians will have more children (or children who survive better). Do they? We have no idea, and no one seems too concerned to find out. Even more regrettably, supporters of the sexual-selection hypothesis seem to find it extraordinarily difficult to refrain from drawing a facile analogy with the libidinal excesses of rock stars – which is why I suggest we view this as the 'Hendrix theory' of music's origins. Yes, Jimi Hendrix had plenty of sexual conquests (though he sired few children) before his untimely death (thereby, in evolutionary terms, making the dangerous drugs and drink a price potentially worth paying) – but if there's one thing worse than theorizing by anecdote, it is theorizing by celebrity anecdote. For every such case, there is a counter-example. We don't know much about the sexual adventures of troubadours, but most Western music in the Middle Ages was practised by (supposedly) celibate monks. In some African societies, musicians are regarded as lazy and unreliable – poor marriage prospects, in other words. (Some seem to find those characteristics in themselves aphrodisiac, but *chacun à son goût* is no evolutionary theory either.)

Besides, if music is an adaptation via sexual selection, we might expect it to be developed to different degrees in men and women. There's no evidence that it is (even though there may be slight differences in how music is processed in the brain – see p. 249). We know of no other example of sexual selection that is manifested the same way in both sexes. This doesn't mean that music *can't* be unique in that respect, but it does warrant some scepticism about the idea.

There's no lack of alternative hypotheses for the origin of music. One key question is whether human music has any connection to the 'songs' that some animals, from birds to whales, produce. Some people might be happy to call these 'songs' music simply because they sound a bit like it; and it hardly seems important to dissuade them, provided we accept that birdsong does not become music merely because Olivier Messiaen transcribed it as such (and because countless earlier composers, including Beethoven, mimicked it acoustically in their compositions). But as will become increasingly clear, music is not just a series of pitches, and neither is it a sound designed to impart information. One could, in theory, encode any message in music, merely by assigning notes to letters. If you know the code, the Bible could be communicated that way. But it wouldn't be the Bible rendered in 'music', because it would be musically empty.

Most animal sound is of this encoding type: the sounds have designated meanings, serving as warning signals or mating calls or summonses to the young. What is striking and suggestive about bird and whale songs is that they are not obviously like this: they aren't mere screams or whoops, but consist of phrases with distinct pitch and rhythm patterns, which are permutated to produce complex sound signals sometimes many minutes or even hours long. It seems clear that these sequences don't in themselves encode semantic information: songbirds don't mean one thing when they repeat a phrase twice, and another with three repeats. In this sense, no animal makes lexically meaningful combinations of sound – sentences, if you like, which derive a new meaning from the combined meanings of the component parts.

This challenges any notion that animal song is like human language – but what about human music? As we'll see, the questions of whether music may have either a grammar or any semantic meaning are hotly contested, but neither has yet been shown to be an essential characteristic of music. Songbirds, which comprise nearly half of all known bird species, tend to create their songs by the shuffling and combination of short phrases. In this way they can create a huge repertoire – sometimes hundreds of songs, each one apparently remembered and repeatable – from a small inventory of basic fragments. Yet the songs don't each have a distinct meaning; rather, it seems that the aim is merely to create sensory diversity, to produce a 'new' song that will capture the attention of potential mates through novelty. (In an echo of Darwin's sexual-selection hypothesis, the females of some of the songbird species with the most complex songs, such as sedge warblers and starlings, may choose males with the most elaborate songs.) This is certainly paralleled in some music, particularly that of the Western Classical period, in which many stock phrases (such as arpeggios, turns and so forth) recur in different sequences.

There are, nonetheless, all sorts of objections to the idea that this makes birdsong 'like music'. For one thing, the sonic building blocks of birdsong contain none of the note-to-note implications of Mozart: there is no indication that specific pitches imply which others will follow. And there is no indication that birdsong has the hierarchical structure of human music, in which even the simplest tunes tend to have 'nested' phrase structure reminiscent of the way language works. (I will explore this notion, as well as the validity and pitfalls of comparisons with language, in Chapter 12.) Birdsong is basically just a string of little bursts of sound: no more than one thing after another.* Furthermore, animal 'song' is typically non-volitional, but is stimulated by seasonal and hormonal changes.

One can nevertheless argue that an ability to create and remember different arrangements of sonic units is a necessary precursor to both music and language. Even there, however, we must take care. Don't imagine that birds and humans share some proto-musical ancestor: the ability to learn vocal repertoires of 'song' seems to have evolved separately in primates and birds, and indeed has evolved at least twice among birds themselves as well as separately in bats, whales and seals.

No other primates sing, but chimpanzees produce a structured call known as a pant-hoot, which seems to differ from one creature to another: each has a signature tune. Compared with birds, chimps always sing essentially the same 'song'. But both sexes 'sing' in all of the primate species that show this behaviour (about one in ten), and they accompany that activity with ritualized movements such as drumming, stomping and shaking branches, resembling the behaviours that humans show in musical contexts.† (They don't, however, keep a steady beat.)

* Whale song *does* seem to have some element of hierarchical organization. That doesn't mean whale song is 'music' either, but it does bring it somewhat closer to that status. Some philosophers of music insist, however, that music can only be a human attribute by definition, since listening to it as such requires imagination.

† Linguist Tecumseh Fitch thinks that primate drumming, which is sometimes conducted in a spontaneous and playful manner, could be closely related to human percussive music-making. But we don't know whether ape drummers can 'get with a beat' – see pp. 225–227.

Perhaps the most interesting aspect of primate singing is that the pant-hoot doesn't connote anything in particular, or at least nothing with a guaranteed meaning. Instead, it seems to be an expressive sound, a vehicle for emotion. Chimps pant-hoot when they are excited, or perhaps merely to say 'here I am'. African apes also use their voices more for affective than for informative reasons, unlike the coded calls of many other animals. This has led some researchers to speculate about whether music stemmed from emotive rather than semantic signalling.

Might there have then been a time when vocalization contained a bit of both information *and* emotion? Of course, language clearly does so today, particularly in poetry (which also shares with music the properties of rhythm and metre). But before language crystallized into consensual forms, one might conceivably have achieved a lot of communication and interaction with a 'musilanguage' that verged on the semantic while employing the emotional arsenal of simple music. This ancestral merging of language and music is a popular theory, and it also boasts Darwin's imprimatur. He wrote:

> it appears probable that the progenitors of man, either the males or females or both sexes, before acquiring the power of expressing their mutual love in articulate language, endeavoured to charm each other with musical notes and rhythm.

But the idea is older still: Jean-Jacques Rousseau expressed something like it in the eighteenth century, suggesting that our ancestors may have used music-like vocalization to convey passions before they became able to express their thoughts.

Adherents of the musilanguage hypothesis point to the analogies that have been drawn between the structures of language and music. Although these are still being debated, both do use combinatorial syntax – the rule-based combination of basic acoustic building blocks – and intonational phrasing using pitch and rhythm. Steven Brown of the Karolinska Institute in Sweden suggests that one can identify a continuum of forms between language and music, progressing through heightened speech, poetry and operatic recitative to song, musical symbolism (where semantic meaning is conveyed through devices such as descending melodic contours to denote 'falling') and finally 'pure' instrumental music. 'Music and language have just too many important similarities for these to be chance occurrences alone,' claims Brown. He believes that a musilanguage could have served as a launching stage for both music and language if it contained three essential features: lexical tone (the use of pitch to convey semantic meaning), combinatorial formation of small phrases, and expressive phrasing principles, which add emphasis and connote emotion (for example, fast tempos to convey happiness, and slow to convey sadness).

A remnant of such musilanguage might be perceived in the so-called Auchmartin and Enermartin of some Ecuadorian tribes, which are forms of song-speech used respectively by strangers meeting on a jungle path and by groups of men to promote courage before battle. More generally, perhaps relics of musilanguage survive in tonal languages, and in the way songs and rhythmic poems are often used in preliterate cultures to encode important knowledge. Here the musicality aids memory: it's generally much easier to memorize poetry than text, and we typically find song lyrics easier to recall when we sing rather than say them. (We'll look later at some of the possible neurological reasons for this.)

A curious variant of the 'musical communication' hypothesis was put forward by the Hungarian-Dutch psychologist Géza Révész, a friend of Béla Bartók, who pointed out that a song-like voice has acoustic features that allow it to carry over greater distances than a speech-like voice. You might say that this makes the earliest music a kind of yodel – if you like, the 'Lonely Goatherd' theory of music's origins.

One of the most obvious features of music the world over is that it tends to be a group activity. Even when performed by a select few, music commonly happens in places and contexts in which it creates social cohesion, for example in religion and ritual or in dance and communal singing. One of the clearest descriptions of this role was given by the English social anthropologist Alfred Radcliffe-Brown in his study of dance among people of the Andaman Islands in the Bay of Bengal:

> The dance produces a condition in which the unity, harmony and concord of the community are at a maximum, and in which they are

> intensely felt by every member. It is to produce this condition, I would maintain, that is the primary social function of the dance . . . For the dance affords an opportunity for the direct action of the community upon the individual, and we have seen that it exercises in the individual those sentiments by which the social harmony is maintained.

This function has led some to suspect that we shouldn't be searching for music's origins in the benefits it might confer on individuals, but rather, in how it is advantageous to an entire society or culture (and thus *indirectly* the individual). This is, you might say, the 'New Seekers theory' of music's origins: 'I'd like to teach the world to sing'. The music psychologist Juan Roederer puts it a little more soberly:

> The role of music in superstitions or sexual rites, religion, ideological proselytism, and military arousal clearly demonstrates the value of music as a means of establishing behavioural coherency in masses of people. In the distant past this would indeed have had an important survival value, as an increasingly complex human environment demanded coherent, collective actions on the part of groups of human society.

The notion of 'group selection' as an evolutionary force, whereby behaviours are selected because they benefit a group, has had a controversial history, and remains so today. The question of how much advantage you get from helping those only distantly related to you, or not related at all but nonetheless sharing common goals, is a very subtle one. All the same, the theory that music's adaptive value lay in the way it brought communities together and promoted social cohesion enjoys wide support.* A 'social' element is found in primate calls, which seem to help members of a group locate one another. And music in tribal societies often has communal functions: for example, it's said that Venda tribespeople can tell from a song what the singer is doing. The men of the Mekranoti tribe of Amazon Indians devote several hours a day to group singing, especially very

* There is in fact nothing necessarily incompatible with an origin of music in social cohesion and one in sexual selection, for individuals that take the lead in social music-making gain status.

early in the morning. Everyone is expected to attend these communal sessions, in which the singing possibly keeps the groggy men awake so that they can be alert to attacks from enemy tribes. That would certainly give music-making a survival value – and indeed, it would be odd to devote this much time to an activity that did not have some adaptive benefit.

Rhythmic sounds provide a great vehicle for synchronization and coordination of activity – witness (say the theory's advocates) the ubiquity of 'work songs'. And even when nothing tangibly 'useful' comes of group participation in music, it promotes a lasting sense of togetherness. Again, contemporary parallels offer themselves with treacherous alacrity: look at the way adolescent subcultures establish their identity through allegiance and shared listening to specific modes of music. Japanese researcher Hajime Fukui has found that people have lower testosterone levels when listening to their favourite music, which he interprets as an indication of music's socializing function, promoting sexual self-control and lowering aggression. Those notions seem squarely contradicted by rave parties and mosh pits, of course, but it's questionable whether we can deduce much about music's origins from the antics of Western teenagers. More importantly, Fukui's findings fail to tell us whether changes in listeners' testosterone levels are prompted by music as such or by the fact that they are hearing their favourite kind.

Rather more convincingly, the almost universal use of music in communal ritual might be understood on the basis that its ability to arouse emotion and teeter on the brink of meaning, without any such meaning ever becoming apparent at the semantic level (and we'll look at that contentious claim later), seems to recommend it for expressing or representing numinous concepts. Stravinsky seems to have shared this view, saying that 'the profound meaning of music and its essential aim . . . is to promote a communion, a union of man with his fellow man and with the Supreme Being'.

Sexual display, group bonding, transmission of information: it all sounds rather, well, male. Another hypothesis aims to relocate the musical impulse in the maternal, by pointing out that infants are much more receptive to speech when conveyed in the singsong tones dubbed 'motherese', and that mothers of all cultures use this instinctively. (So do fathers and siblings, although an infant's exposure to this kind of communication would usually come mostly from the mother.) And babies seem to come equipped with the mental apparatus for discerning simple musical attributes: they can discriminate from the time of birth between upward and downward musical contours, and within two months they can detect a change in pitch of just a semitone.

If better mother-child communication – not just semantic but emotional – leads to better-adjusted and more cognitively able adults who do well in the world, then there is a potential selective advantage to a predisposition to musicality.* But one can also interpret at least some features of motherese in purely linguistic terms: it facilitates language learning by helping to emphasize contrasts between different vowels and consonants, for instance. And the descending pitch contours that characterize lullabies throughout the world are also found in speech typically used to soothe a child.

Furthermore, it's not easy to explain how traits shaped by one-on-one infant interactions find their way into adult social ritual. Musicologist Ellen Dissanayake suggests that the music-like sensitivities and competencies developed in the mother-infant interaction 'were found by evolving human groups to be emotionally affecting and functionally effective when used and when further shaped and elaborated in culturally created ceremonial rituals where they served a similar purpose – to attune or synchronize, emotionally conjoin, and enculturate the participants'. That seems something of a leap of faith, and surely invites the question of why often (although by no means universally) it is men who have traditionally engaged in the production of music. Even less comfortable is the suggestion that an origin of music in infanthood explains why (Western) popular songs use childish words such as 'baby' to express sentiment, an idea that tempts me to call this the 'Ronettes theory' of music's origins: 'Be My Baby'.

Might it be that to ask 'what is the origin of music?' is simply to ask the wrong question? Some palaeontologists and archaeologists consider that the transition from our ape-like ancestors to humans

*Some Freudian psychologists, such as Heinz Kohut, have stood this argument on its head, suggesting that music is therefore a form of infantile regression, albeit one that has been made socially and aesthetically acceptable.

– what they call hominization – involved the appearance of such a rich complex of traits, such as language, arithmetic, logic, society and self-consciousness, in such a short time that it makes little sense to consider them independent of one another. Either they are all part of the same basic phenomenon, or the emergence of one made all the others inevitable. In a similar vein, Jean Molino argues that, as there seems to be no universal definition of music, we can't reasonably say that something called music emerged in evolution, but only that certain human capacities and tendencies appeared which have gradually found expression in what we now consider types of music.

Pinker redux

To almost all of this theorizing about origins, one can only say: yes, it sounds plausible. You could be right there. Maybe it's unfair to point out (as one nonetheless must) that all these ideas are more or less impossible to prove. After all, no one has access to a Palaeolithic society. But it's unfortunately rather rare to hear hypotheses about music's origins asserted with anything less than firm conviction, and rather common (as I've hinted) to find them supported by cherry-picking from the vast and diverse ways in which music is made and used throughout the world – or worse still, by anecdotes from Western popular culture, in which music has surely become a more artificial, abstracted, and fashion-bound medium than in any other culture.

More troubling still is the sense that these speculations are being offered in desperate determination to prove that music has been hard-wired into our brains by evolution – to show, in other words, that Steven Pinker is wrong to portray music as a kind of aesthetic parasite. One can't help feeling that an evolutionary role for music is often seen as the only way it can be afforded its true dignity. Even worse is the implication that if we understand where music came from, we will understand what it is about. Whether or not Pinker is right about why we have music, he is right to say that the debate should not become a surrogate for arguing over the *value* of music. When evolutionary biology becomes the arbiter of artistic worth, we are in big trouble.

The fact is that at this point there are no compelling scientific arguments against Pinker's position. We will see later that there are a few tentative indications that music might indeed be a genuine, adaptive 'instinct', but that case is far from secure. My view is that we should approach this interesting question as one that is utterly irrelevant to our reasons for treasuring music, thereby jettisoning the emotive baggage with which it has been burdened. 'Auditory cheesecake' is a phrase carefully chosen for polemical effect, but it's doubtful that Pinker meant it as any kind of artistic or aesthetic judgement, and we'd do him a favour to assume he did not. As for William James, who regarded music as a 'mere incidental peculiarity of the nervous system', it seems that this unfortunate philosopher strained in vain to make much sense of music (he may have been one of the rare individuals who is genuinely tone-deaf), and so his dismissive comment is more an expression of bafflement than a profound insight.

Richard Dawkins has rightly said that a question does not become meaningful simply because it can be articulated, and I suspect that is true of Pinker's suggestion that music 'could vanish from our species and the rest of our lifestyle would be virtually unchanged'. Aniruddh Patel of The Neurosciences Institute in San Diego regards music as a transformative technology – one whose appearance so alters its host culture that there is no going back. 'The notion', he says,

> that something is either a product of biological adaptation or a frill ['cheesecake'] is based on a false dichotomy. Music may be a human invention, but if so, it resembles the ability to make and control fire: It is something we invented that transforms human life. Indeed, it is more remarkable than fire making in some ways, because not only is it a product of our brain's mental capacities, it also has the power to change our brain.

In this view, you might as well imagine that we could give up theatre or sport, which equally don't seem essential for our survival (though again, one can tell superficially plausible stories about why they might be adaptive). It is no surprise, then, that we know of no culture that lacks music.

I agree with Patel, but would go further. It's not only that music is too deeply embedded in our cultures to be extracted. It is too deeply embedded in our *brains*. Regardless of whether evolution has given our brains musical modules, it seems to have given us intrinsic proclivities for extracting music from the world. Music is a part of what we are and how we perceive the world. Let's now start to see why.

3

Staccato
The Atoms of Music

What are musical notes and how do we decide which to use?

'Organized sound' might strike you as a pretty neat definition of music. But the phrase was coined by the French-born avant-garde composer Edgar Varèse, who wrote music in the early twentieth century that many of his contemporaries would not have accepted as 'music' at all. And Varèse wasn't seeking a catchy general definition that could be applied to anything from Monteverdi to Leadbelly; rather, he used the description to *distinguish* his bold sonic explorations from conventional music. His compositions called for howling sirens, the ghostly electronic wail of the theremin, and electronically taped ambient noises: rumbling, scraping, jangling, honking and the churning of machines. He gave these works pseudoscientific names: *Intégrales*, *Ionisation*, *Density 21.5*. 'I decided', he said, 'to call my music "organized sound" and myself, not a musician, but a "worker in rhythms, frequencies and intensities".' If *that* was meant to apply to Mozart too, it would seem to make him something like a cross between a laboratory technician and an industrial labourer.

But Varèse didn't regard himself as an iconoclast. He traced his heritage to music's ancient practices, and professed admiration for the music of the Gothic Middle Ages. This seems appropriate in many ways, for the composers and musical scholars of antiquity shared his view of music as a kind of technical crafting of sound. Unlike the nineteenth-century Romantics, they would have had no problem with a discussion of music in terms of acoustic frequencies and intensities.

I suspect many people share the romantic sense that music is a product of numinous inspiration, and are apt to feel disheartened, even appalled, when it is fragmented and seemingly reduced to a matter of mere acoustics, of the physics and biology of sound and audition. If that seems at first to be the business of this chapter, I hope you will soon see that it isn't really so. But I don't intend to apologise for a digression into the mathematics, physics and physiology of acoustic science by justifying it as an unavoidable introduction to the raw materials of music. It is much more interesting than that.

I will admit, however, that while dissection is as necessary in musicology as it is in biology, it is likewise apt to leave us contemplating a pile of lifeless parts. A better metaphor than the anatomical is the geographical: music is a journey through musical space, a process that unfolds in time and the effect of which depends on how clearly we can see where we are and how well we recall where we have come from. Only the vista ahead is obscure to us; but our sense of journey depends implicitly on our anticipation of what it might contain. And just as a journey is not made from trees and rocks and sky, music is not a series of acoustic facts; in fact, it is *not acoustic at all*. I can't emphasize this enough. It is fine to call music 'organized sound', so long as we recognize that this organization is not wholly, perhaps not even primarily, determined by the composer or performer.

Nonetheless, this chapter is not merely about the cold facts of the 'sound' that Varèse sought, in his idiosyncratic way, to organize. It is about how nature and culture interact to produce the diverse palettes of notes that most traditions draw on in creating their sonic art. There is very little that is preordained in this palette – contrary to common belief, it is not determined by nature. We are free to choose the notes of music, and that's what makes the choices interesting. They crystallized in Western music into the notes of the modern piano, which repeat in octave cycles with twelve notes each. Not everyone accedes to that arrangement. The maverick American composer Harry Partch (1901–74), searching for a system better adapted to the nuances of the human voice, devised a microtonal scale of 43 pitch steps per octave (he also experimented with 29-, 37- and 41-note scales). The music Partch wrote was played on special instruments that he designed and built himself, bearing exotic names such as the chromelodeon, the bloboy and the zymo-xyl. It sounds frighteningly experimental, but Partch's music is actually not as formidable or jarring as one might expect, especially if you've had any exposure to gamelan and South East Asian percussion orchestras.

The point is not necessarily *which* notes we choose, but the fact that we choose them at all. Painters understand that notion. They have in principle an infinite range of colours at their disposal, especially in modern times with the chromatic explosion of synthetic chemistry. And yet painters don't use all the colours at once, and indeed many have used a remarkably restrictive selection. Mondrian limited himself mostly to the three primaries red, yellow and blue to fill his black-ruled grids, and Kasimir Malevich worked with similar self-imposed restrictions. For Yves Klein, one colour was enough; Franz Kline's art was typically black on white. There was nothing new in this: the Impressionists rejected tertiary colours, and the Greeks and Romans tended to use just red, yellow, black and white. Why? It's impossible to generalize, but both in antiquity and modernity it seems likely that the limited palette aided clarity and comprehensibility, and helped to focus attention on the components that mattered: shape and form. That is perhaps even more true in music, because the notes carry a heavy cognitive burden. To make sense of music, we need to be able to see how they are *related* – which are the most important, say, and which are peripheral. We need to understand them not just as stepping stones in musical space, but as a family group with a hierarchy of roles.

Making waves

Most of the music in the world is made from notes. And to a good approximation, we hear each note as a *pitch* with a specific acoustic frequency. Notes sounded in succession create a *melody*, while notes sounded simultaneously produce *harmony*. The subjective quality of the note – crudely, what instrument it sounds like – is its *timbre*. The duration and timing of notes, meanwhile, define the *rhythm* of music. With these ingredients, musicians compile 'global' structures: songs and symphonies, jingles and operas, compositions that typically belong to a certain form, style and genre.

Most music is experienced as vibrations in air. That's to say, some object is struck, plucked or blown, or driven by oscillating electromagnetic fields, so that it vibrates at specific frequencies. These motions then induce sympathetic vibrations in the surrounding air that radiate away from the source like ripples in water. Unlike waves on the sea surface, the vibrations of air are not undulations in height; they are changes in the air's density. At the 'peaks' of sound waves, the air is compressed to greater density than it would be in a soundless space; at the 'troughs', the air is less dense (rarefied) (Figure 3.1). The same is true of acoustic waves passing through other substances, such as water or wood: the vibrations are waves of material density. Earpieces attached to portable electronic devices sit in contact with the tissues of the ear and transmit vibrations to them directly.

The perceived pitch of a sound becomes higher as the frequency of the acoustic vibrations increases. A frequency of 440 vibrations per second – the 'concert pitch' to which Western pitched instruments are conventionally tuned – corresponds to the A note above middle C. Scientists use units of hertz (Hz) to denote the number of vibrations per second: concert A has a frequency of 440 Hz. We can hear frequencies down to 20 Hz, below which we feel rather than hear them. Frequencies just below the lower threshold of hearing are called

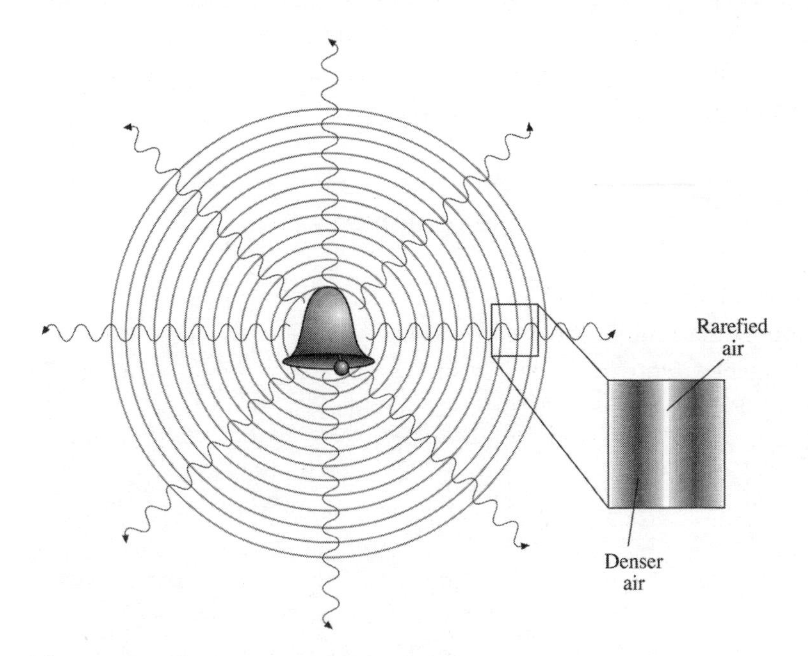

Figure 3.1 Sound waves in air are waves of greater or lesser air density.

infrasound, and are produced by some natural processes such as surf, earthquakes and storms. They seem to stimulate strange psychological responses, particularly feelings of unease, revulsion, anxiety and awe, and have been credited with inducing 'supernatural' experiences. Infrasound has been used for its unsettling effect in some contemporary music, for example in the soundtrack to the French shocker movie *Irréversible*. As a way of using music to induce emotion, that somehow seems like cheating.

The upper frequency limit of human hearing is typically about 20,000 Hz, but tends to be lower in older people because our ear's sound sensor cells stiffen with age. Higher frequencies than this (ultrasound) are inaudible to humans, but are detected by many other animals: bats use them for echolocation. Between these extremes, human hearing typically spans a gamut of about ten octaves. On the 88-note piano, the lowest A growls at 27.5 Hz, and the top C is a shrill 4,186 Hz. We find it harder to make out a single well-defined pitch for notes at either extreme of the piano than for those in the middle – which is why most music is played there, and of course why it is the middle of the keyboard at all. The human male speaking voice has a typical frequency of around 110 Hz, and the female voice an octave higher, 220 Hz; so when men and women sing together 'in unison', they are actually singing in octave harmony.

How we hear

Converting acoustic vibrations in the air to a nerve signal that is sent off for processing in the brain is the job of the cochlea, a spiral chamber of bone in the inner ear that looks like a tiny snail shell (Figure 3.2). Housed within it is a long membrane sheet called the basilar membrane, covered with sound-sensitive 'hair cells', so called because of the little tuft-like protrusions that poke above their surface. These cells are like mechanical switches: when the 'hairs' are set waving by acoustic vibrations in the fluid that fills the cochlea, the movement pulls open tiny pores in the cell walls, letting in electrically charged metal atoms from salt in the surrounding fluid that change the electrical state of the cell. This produces nerve signals that surge along nerve fibres leading to the brain.

Different hair cells respond to different sound frequencies, and they are arranged over the basilar membrane in a manner remarkably similar to the way piano strings are laid out on the soundboard: the basilar membrane resonates to low-frequency sounds at one end and progressively higher frequencies as we move along it.

All this is the easy part of auditory cognition – it converts sound into electrical signals, rather as a microphone does. Our perception of the sound depends on how those signals are processed. One of the first steps is to decode the pitches, which is done surprisingly directly: each part of the basilar membrane seems to have a dedicated set of

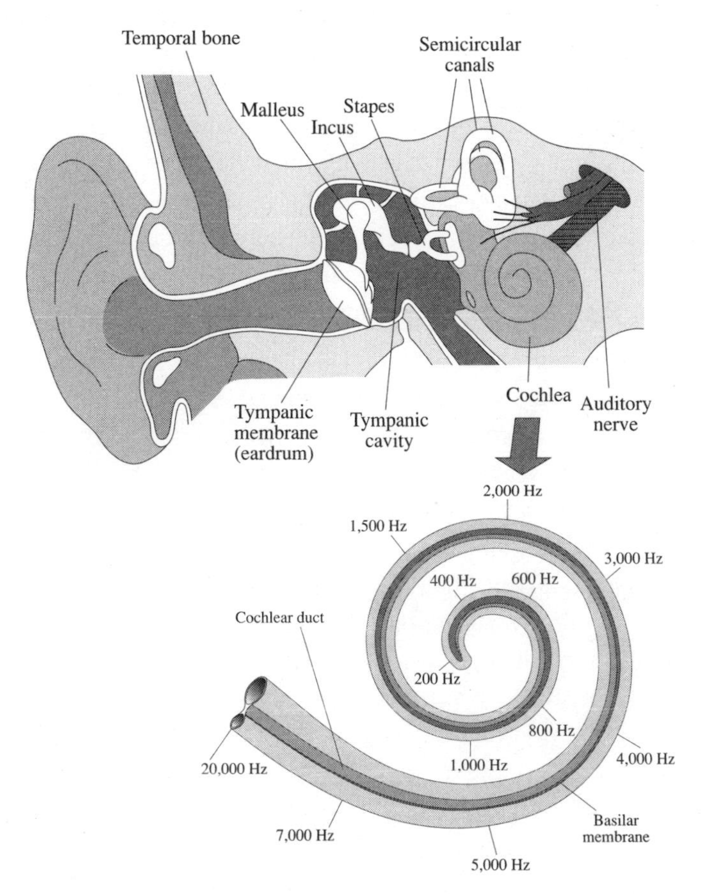

Figure 3.2 The anatomy of the ear.

neurons in the brain to pick up the activity. These pitch-selective neurons sit in the primary auditory cortex, the part of the brain where pitch is processed. It's very unusual for a perceptual stimulus to have this one-to-one mapping in the brain – we have no comparable neurons that, for example, respond to specific tastes, smells or colours.

How music is written

The ability to read music is sometimes regarded with awe, as though it offers privileged insights into the world of music from which non-initiates are excluded. The fact that many popular and jazz musicians could not read a note – Erroll Garner and Buddy Rich, for example, virtuosos both – only adds to this notion: if *they* couldn't do it, it must surely be a gargantuan feat.

Well, it isn't. The plain truth is that many self-taught musicians simply never felt the need to learn this skill. But it is not hard at all, at least to the extent of becoming able to name a note. A facility for reading complex music fast enough to play it on sight is indeed a technical skill that requires practice, as is the ability to 'hear' in one's head what is written on the page. But you will not need to do that in order to understand and benefit from the musical quotations reproduced in this book. And in any case, they can all be heard online at www.bodleyhead.co.uk/musicinstinct.

Reading music is perhaps most easily explained with reference to the piano, on which there is a transparent relationship between written notes and the piano keys – every written note corresponds to a specific key. The musical stave – the five horizontal lines on which notes are suspended – is just a representation of a piano stood on its side, with the high notes uppermost. Of course there are a lot more notes on the piano than there are lines on a stave, but the two staves in written music cover the notes most commonly used (Figure 3.3) – one typically played with the left hand, the other with the right. For notes outside this range, the staves are extended locally with extra lines.

Each of the white-note keys of the piano is assigned a place on these staves, successively either between or on the stave lines. The

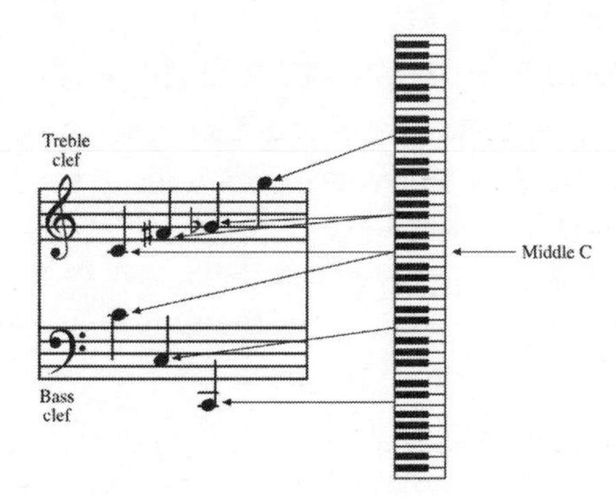

Figure 3.3 The musical clefs and staves. Notes higher on the stave indicate notes 'higher' on the piano. Every note either sits on a stave line or between them.

ornate 'clef' signs that appear at the start of the stave – the baroque treble clef on the upper stave, the coiled bass clef on the lower – are merely symbols to tell you where the correspondence between stave lines and piano keys is anchored. The treble clef looks a little like a fancy 'g', and its lower body circles the stave line corresponding to the note G above middle C. The bass clef is a stylized F, and its two dots straddle the stave line corresponding to the F below middle C (Figure 3.3). Other types of clef are sometimes used that define different 'origins' on the staves, but we needn't be concerned with them.

What about the black notes on the piano? These are indicated by so-called accidentals – the sharp and flat signs ♯ and ♭ – next to a note of a particular pitch, indicating that this note should be played either on the black note above (sharp) or below (flat). There is some redundancy in this system, because it means that the black notes have more than one 'identity'. That above F, for example, is simultaneously F♯ and G♭ (Figure 3.3). And for white notes that do not have a black note immediately adjacent in a certain direction, a sharp or a flat becomes another white note: B♯, for instance, is the same as C. We will see shortly where this notation for accidentals came from, and also how the 'multiple identities' of the various notes have a complex history and a subtle role to play in how we conceptualize music.

The notes go in a repeating cycle as they ascend. Rising from concert A, for example, we find the white notes B, C, D, E, F, G – and then A again. This second A is an octave above the first: eight notes higher as we rise up the musical scale. We'll see below precisely what that means. Each particular note can be given a unique label by numbering those in each octave, starting from the lowest A on the piano keyboard (AO). This makes middle C equivalent to C4, and concert pitch A4.

The blobs denoting notes take on different appearances in a musical score (Figure 3.4) – sometimes filled in (♩), sometimes empty (𝅗𝅥), some with wavy tails attached, some linked by bars or accompanied by dots or so on. These markings indicate the note's duration: whether they last a whole beat, or several beats, or a fraction of a beat. And there are other objects arrayed on the staves. Some (such as 𝄽 and 𝄾) denote *rests*, which are beats or fractions of a beat during which no notes sound. Arching lines might designate *ties*, which show that a note is to be prolonged over both of the note durations at each end; or they might tell the musician about patterns of phrasing, meaning how notes are to be articulated in groups.

There are rhythmic codes here too. The numbers at the start of a stave are the *time signature*, which indicates how many notes of a particular duration should appear in each of the *bars* denoted by vertical lines. In essence, they tell us whether we count the rhythm (more properly, the *metre* – see Chapter 7) in groups of two, three, four or more beats. Music scores are littered with other signs too, indicating such matters as dynamics (where to get louder or softer), accents, trills and so forth. This isn't the place for a comprehensive account of what music notation means; Figure 3.4 shows the main symbols that will suffice to understand most of the musical extracts I quote. To keep matters simple, I have generally stripped them down to their bare essentials rather than following the composers' scores rigorously. I hope that will not trouble purists.

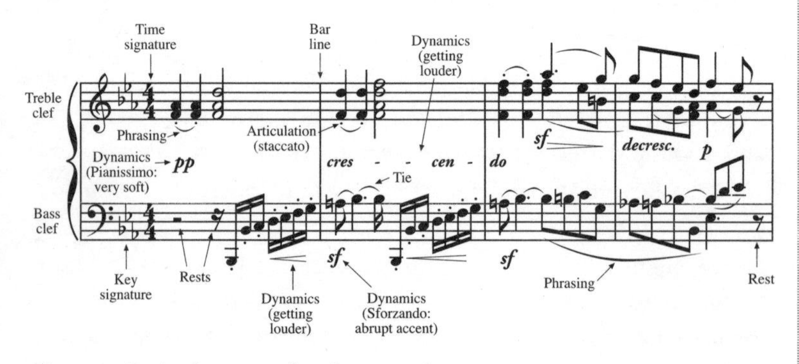

Figure 3.4 Basic elements of written music.

Arranging the sonic staircase

In principle, the relationship between the pitch and acoustic frequency of a musical note seems simple: the higher the frequency, the higher the pitch. But a striking aspect of music in nearly every culture is that its notes are *discrete*. There are an infinite number of pitches within the frequency range audible to humans, since the difference between two frequencies can be made as small as you like. Although there comes a point at which we can no longer distinguish between two very closely spaced tones, just as there are limits to the resolution with which we can discriminate objects visually, there are consequently very many 'notes' that we could arrange into music. Why do we use only a subset of them, and how is that subset selected?

Nature seems to have imposed a basic division of this continuous pitch scale: the octave. If you play any note on the piano and then play the same note an octave higher, you will hear it as a 'higher version' of the original note. This is so commonplace that it seems barely remarkable. Anyone who has ever picked out a simple melody on a piano with tentative prods of a finger will have discovered that the same melody can be played, with the same pattern of key strokes, an octave higher or lower. The octave is itself enshrined on the piano in the shape and arrangement of the keys: novice pianists learn to recognize the L shape of Cs and Fs, and the cyclic clustering of two and three black notes.

But octave equivalence is deeply strange. It is a perceptual experience unique to music: there is no analogous repeat pattern in visual perception, or in taste.* In what way is middle C 'like' the C above

* It is often remarked that the two extremes of the visible spectrum of light – red and violet – match up, so that the spectrum seems to come full circle. There is no obvious reason why this should be so: the frequencies of light corresponding to the two colours are unrelated. The circularity of colour space is, however, a genuine perceptual phenomenon, which has been tremendously useful to colour theorists in permitting them to construct closed 'colour wheels' for purposes of classification. It is an artificial construction, however: the 'colour' just as visible light turns into infrared does not really look identical to the colour as violet becomes ultraviolet. And in any case, there is no repetitive cyclicity here as there is in pitch perception – colour has, at best, only 'one octave'. Nonetheless, a semi-mystical conviction that colour and sound must be related wave-determined phenomena was what led Isaac Newton to postulate seven divisions of the rainbow, by analogy to the seven notes of the musical scale. This arbitrary Newtonian scheme of colour categories is still taught as objective fact today.

or below it – what quality remains unchanged? Most people would say 'they sound the same' – but what does that mean? Clearly they are not the same. Perhaps one might say that the two notes sound good or pleasing together, but that only generates more questions.

Pythagoras is attributed with the discovery of how octave pitches are related to one another. The apocryphal story has it that the Greek philosopher once walked into a blacksmiths' forge where notes rang out as the smiths' hammers struck their anvils, and he noticed that there was a mathematical relationship between the relative pitches and the size (the masses) of the anvils that produced them. When he went on to investigate the sounds made by plucking taut wires or strings, Pythagoras is said to have found that tones which sound harmonious together, such as octaves, have simple ratios of their frequencies. The octave is the simplest of all: a note an octave above another has double the frequency of the first. This is equivalent to saying that the higher note has *half* the wavelength of the lower.

We can visualize the relationship in terms of the lengths of the plucked string. If you halve the length of the vibrating string by putting your finger on its midpoint, you double the frequency and generate a note an octave higher. This is easily done on a guitar: the fret that produces an octave higher than the open string is located exactly halfway between the contacts of the string at the head and the bridge (Figure 3.5*a*).

The next higher octave has a frequency that is doubled again, making it $2 \times 2 = 4$ times that of the original note. The frequency of A6, say, is $4 \times 440 = 1{,}760$ Hz. You will get such a double-octave jump on a guitar string by pressing down three-quarters of the way along the string and plucking the upper quarter. And so it goes on: each successive octave entails another doubling of the original frequency.

Just about every musical system that we know of is based on a division of pitch space into octaves: it seems to be a fundamental characteristic of human pitch perception. We'll see later the likely reason why.

But what about the notes in between the octaves? Again, legend credits Pythagoras with an explanation for how these are chosen in Western scales, although in fact this knowledge was surely older. If

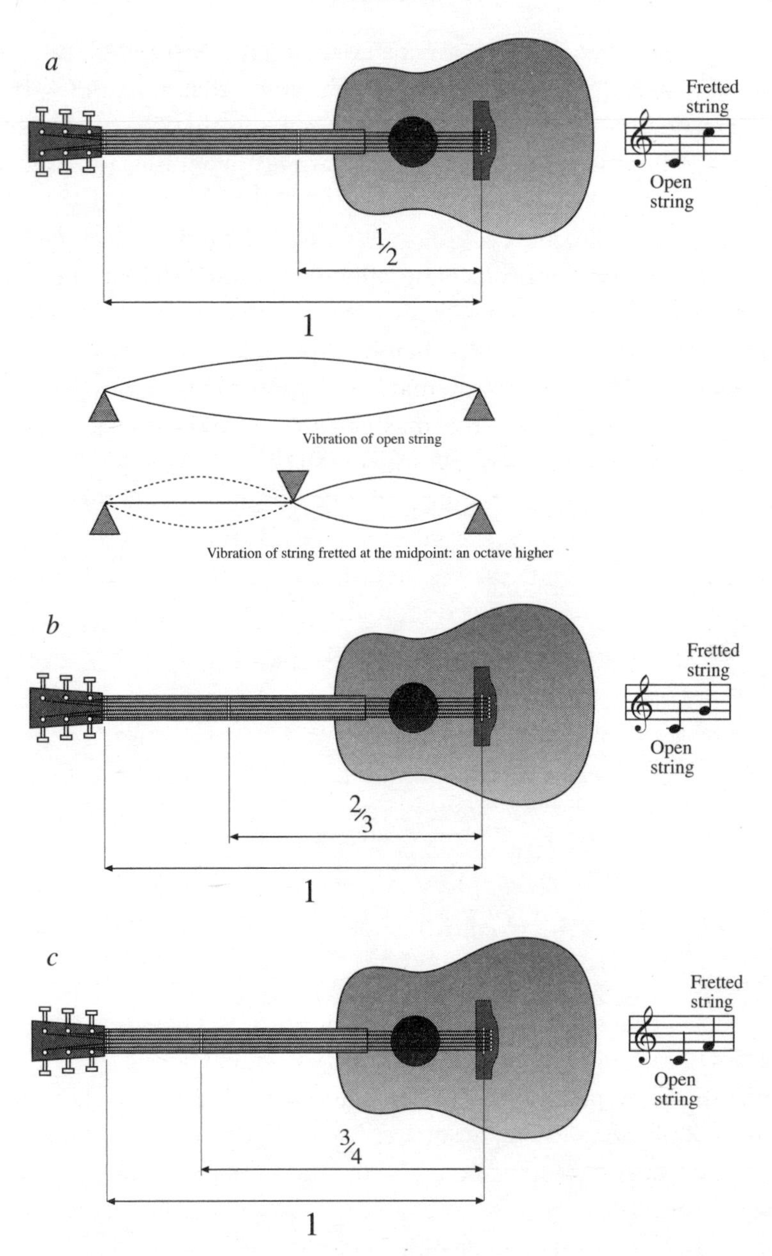

Figure 3.5 An octave can be produced on a plucked string by fretting it halfway along its length (*a*). The two notes have wavelengths in the ratio 2:1, and frequencies in the ratio 1:2. Fretting a third of the way along produces a note a perfect fifth above the open string (*b*), and fretting a quarter of the way along produces a perfect fourth (*c*).

you put your finger a third of the way along a vibrating string and pluck the longer segment – now two-thirds as long as the open string – you get a note raised by less than an octave (Figure 3.5*b*). Most people will agree that this note fits pleasingly with that of the open string – it sounds harmonious. The note is the fifth of a major scale that begins on the open string; in *do-re-mi* terminology it is the note *so*. If the open string is tuned to middle C, this new note is the G above. The distance, or *interval*, between these two notes is called a perfect fifth (see Box: *Scales and intervals*).

Scales and intervals

The set of pitches from which a tradition constructs its music is called a *scale*. As we've seen, the notes of a scale have discrete pitches in the continuous space of all possible pitches. We can think of the smoothly varying pitch of, say, the rising wail of a siren as a kind of ascending slope of frequency, while a scale is more like a staircase or ladder.

Western music uses scales inherited from Greek tradition, called diatonic scales ('diatonic' means 'progressing through tones'), in which there are seven tones between each octave. Just about all of Western music between the late Renaissance and the early twentieth century was based on two general classes of diatonic scale: the major and minor. The major scale is the sequence picked out, in the key of C, by all the white notes (Figure 3.6). There are several different minor scales.

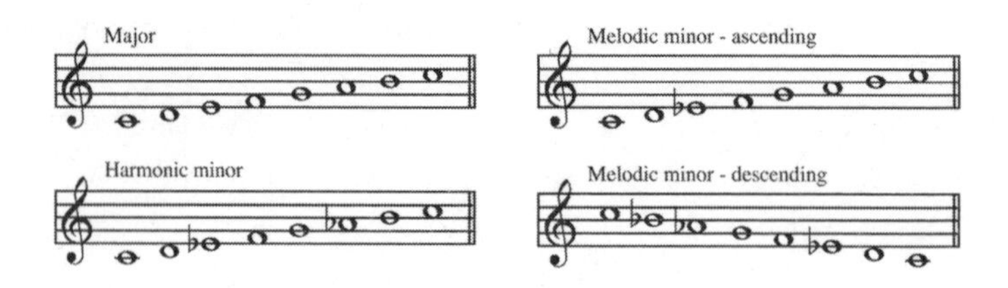

Figure 3.6 The major and minor diatonic scales.

Each note of a diatonic scale has a technical name. I will explain and occasionally use these later, but for the present purpose we need only recognize the first: the note on which the scale starts, which is called the *tonic*. In the key of C, for example, C is the tonic. I will often refer to the other notes of the scale not by their technical names but by their ordering in the scale: the second note (D in the scale of C major) is **2**, the third is **3** and so on. Likewise, I'll sometimes call the tonic **1**. That way we can avoid either having to use technical terms or having to specify a key.

As well as the seven notes of the diatonic scales, there are five others in one octave span – for example, all the black notes scattered among the C major scale. These notes lie outside the scale, but most tonal Western music makes occasional use of them. The scale that includes all twelve notes is called a *chromatic* scale, and when tonal music deviates from the tones of the diatonic scale, it is said to be chromatic.

A pitch step between one note and that immediately above – B to C, say, or F to F♯ – is called a *semitone*, while a step of two semitones (F to G, or C to D, say) is called a *tone*. This is a somewhat confusing terminology, because a 'tone' can also refer simply to any pitched musical sound – but I hope the distinctions will be clear from the context.

Any two notes are separated by an *interval*, which is defined by the corresponding number of steps in the scale. Thus the interval between the tonic note and the fifth note of the scale – C to G, say – is called a fifth (Figure 3.7). For technical reasons it is called a *perfect* fifth. Aside from the fourth and fifth, other intervals come in two different variants, depending on whether they involve a note from the major scale or the note a semitone below, which is often a note from a minor scale. An example of a major-third interval is the step from C to E, while the corresponding minor third is C to E♭ (Figure 3.7). The only interval not covered within this terminology is that between the tonic and the sharpened **4** or, equivalently, the flattened **5** – from C to F♯ or G♭, for example. This is sometimes called an augmented fourth or flattened fifth interval, but more commonly it is known as the tritone interval, because it is reached by three pitch steps of a whole tone: C→D→E→F♯. Intervals can also span more than an octave: that between C4 and D5,

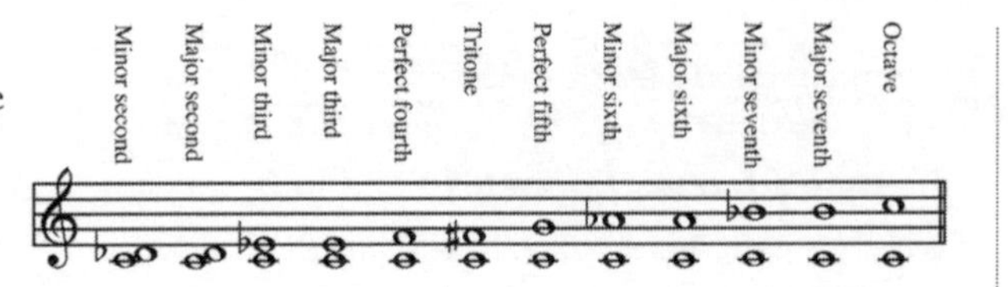

Figure 3.7 The intervals of the diatonic scales.

for example, is an octave plus a second, or nine scale degrees, and is called a (major) ninth. Arguably one could denote this step as **1→9**; but the **9** here is also the second note of the scale starting an octave higher, and so is more commonly written as **2'**, the prime denoting the start of a new octave. The octave interval is, in this notation, **1→1'**.

Any specific interval always spans a set number of semitone steps between the lower and upper notes: a major third corresponds to four semitone steps, a minor third to three, say.

So it is all really just a matter of counting. Where things get a little complex is that intervals aren't always defined with reference to the tonic note of the key in which they appear. Consider the interval E to G, say. This is a minor third: G is four semitones above E, and is also the third note of the E minor scale. But this doesn't mean that the E→G step is somehow confined to music in E minor. It is also a step between the **3** and **5** of the C major scale, for example, and between the **7** and **2'** of F major.

Thus an interval of a fifth separates pitches whose frequencies are in the ratio of 3:2. Having decided that this interval is pleasing to the ear, the followers of Pythagoras deduced from this a general principle: pitches whose frequencies are related by simple ratios sound 'good', which is to say, they are considered consonant. (The real relationship between 'consonance' and what we perceive is more complex, as we'll see.) For the consonant perfect fourth – C to F, say – the pitches are in the frequency ratio 4:3 (Figure 3.5*c*). The frequency of F is four-thirds that of the C below it.

These three simple frequency ratios – 2:1, 3:2 and 4:3 – give us three notes related to the original one: an octave, a perfect fifth and a fourth above, or **1'**, **5** and **4**. In the key of C, these are C', G and F. And so we have the beginnings of a scale, a set of notes that seem to fit together harmoniously and which we can arrange into music. One

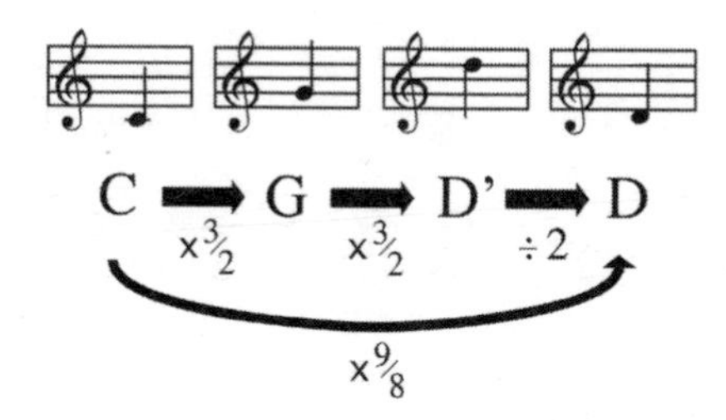

Figure 3.8 Getting from C to D in steps of fifths and octaves. The resulting frequency ratio is 9/8.

might then explore the notes that follow from other simple ratios, such as 5:4 and 6:5. We'll come to that shortly. But it isn't how the Greeks proceeded. They recognized that just these three intervals alone can provide a basis for generating other scale notes, because one can apply the same mathematical transformations on the 'new' notes **5** and **4**. For clarity, let's stay with the specific versions of these notes in C – that is, G and F. We can raise G by a perfect fifth by increasing its frequency in the ratio 3:2 (multiplying by 3/2). Relative to the original C, this new note has a frequency greater by a factor 3/2 × 3/2, or 9/4, and it corresponds to the note D'. Now if we fold this new note back into the octave span C→C' by dropping it an octave (halving its frequency), we find that we have a note 9/8 times the frequency of the tonic C, corresponding to the D above (Figure 3.8).

We can get to this D another way from just steps of fourths and fifths: by stepping up a perfect fifth to G, and then down a perfect fourth. I won't go through the maths, but that too brings us to a note with a frequency 9/8 that of the tonic. Consistency! So we have a scale C, D, F, G, C'. But there are big gaps between D and F, and between G and C'. We can fill them by taking step sizes equal to that between C and D, or between F and G, which are both frequency increments of 9/8. Applying this to D gives us an E with a frequency ratio 81/64 that of C, and applying it to G gives us an A at 27/16 the frequency of C. A further such step up from A gives B, with a ratio of 243/128. An equivalent way of getting these extra notes is simply to raise the tonic progressively in steps of a fifth – from C to G, G to D', D' to A', A' to E" and E" to B" (Figure 3.9) – and then to fold these notes back into the original octave span.

And there is our major scale. Closer inspection reveals that it has a curious, uneven pattern of pitch steps. The first two degrees, **1→2** and **2→3** are as wide as the steps **4→5** and **5→6**, and all are equal to frequency increments of 9/8. But the steps **3→4** and **7→1'** (E to F and B to C', say)

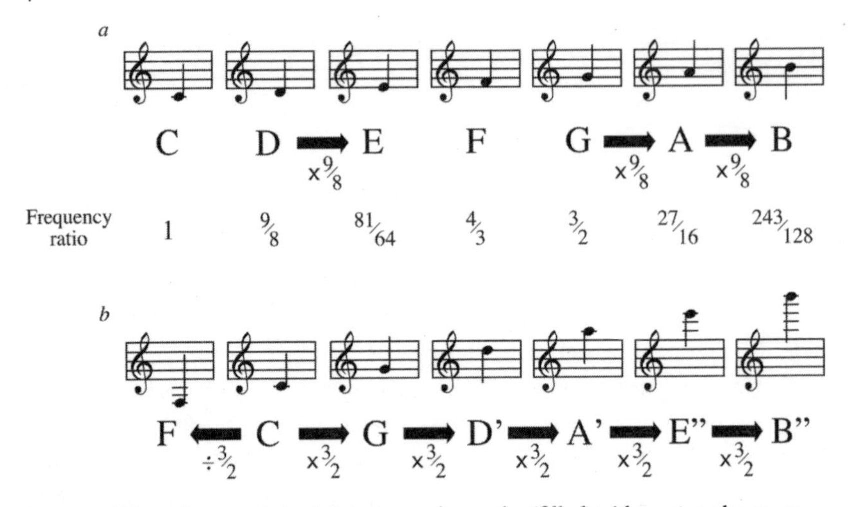

Figure 3.9 The other notes in the major scale can be 'filled in' by using the same whole-tone step of a 9/8 frequency increment (*a*). Equivalently, all the notes can be obtained through a repeated upwards step of a fifth (or downward to reach F), followed by 'folding' the notes back into a single octave span (*b*). The result is the Pythagorean tuning.

are smaller, equal to a factor 256/243. On the modern piano keyboard, these two types of step are the whole tone and semitone respectively.

The scale formed this way is called a Pythagorean scale. It seems to stem from a mathematically attractive way of iterating the harmonious interval of a perfect fifth – a kind of hierarchy of the simple 3/2 ratio. In fact, it turns out that all the note frequencies can be related to that of the tonic by factors of 3 and 2: the ratios for *3* and 7, say, are $3^4/2^6$: 1 and $3^5/2^7$: 1. So even if the maths ends up a little hair-raising, there is a sound, logical basis to it all, derived from simple proportion. To the Pythagoreans, for whom proportion and number constituted the fundamental ingredients of the universe, this seemed to place music on a solid mathematical footing, suggesting that music was itself a branch of maths with a structure that was embedded in nature.

Keys and tonics

The term 'tonal music' has come to be almost synonymous for many Western music-lovers with music that 'has a tune' – music you can hum in the street. Why this is so, and to what extent it is true, is

Figure 3.10
Key signatures.

one of the themes I will explore later. But for now it is enough to say that all 'tonal' really means here is that the music has a *tonic* – a pitch, or rather a pitch class (because of the perceptual equivalence of octaves), that in some sense roots the music, providing an anchor point that organizes all the others. The notion of a tonic is actually a far more complicated thing than is implied in much conventional music theory, but it is generally the same as saying that the music is written in a particular key. In other words, its notes are likely to be mostly drawn from those in a scale (major or minor) starting on the tonic note.

The key of a piece of music is indicated by the *key signature* at the start of the score, which shows which notes are to be sharpened or flattened (Figure 3.10). Novices rejoice at piano pieces written in C major, because they are unlikely to contain many black notes: the scale contains no sharps or flats. The key of F sharp, meanwhile, includes all five of the black notes in the octave span. Some keys are even worse than that: D sharp minor, which Bach was thoughtless enough to use for one of his most elegant fugues in *The Well-Tempered Clavier*, forces the pianist to break the habit of a lifetime and play an F (here an E♯) whenever an E appears on the stave.

Unless you have absolute pitch (the ability to identify musical pitches purely from hearing them), a composition should sound pretty much identical on modern instruments regardless of the key it is played in. Moving a piece to a different key is called *transposition*, and skilled musicians can do it instantly as they read the music. But many people insist that different keys have different 'characters' that are distinguishable even without absolute pitch: that Grieg's Piano Concerto would sound very different if moved from its 'native' A minor to, say, F minor. They are probably right, but for complex reasons explored in Chapter 10.

Modes

We don't know what ancient Greek music sounded like, because there are no examples of it in written or notated form, nor has it survived in oral tradition. Much of it was probably improvised anyway, within certain rules and conventions. So we are forced largely to guess at its basis from the accounts of writers such as Plato and Aristotle, who were generally more concerned with writing about music as a philosophical and ethical exercise than with providing a technical primer on its practice.

All the same, the very word music stems from this tradition: it is 'Music', inspired by the Muses. It seems Greek music was predominantly a vocal form, consisting of sung verse accompanied by instruments such as the lyre or the plucked kithara (the root of 'guitar'). In fact, Plato considered music in which the lyre and flute played alone and not as the accompaniment of dance or song to be 'exceedingly coarse and tasteless'. The melodies seem to have had a very limited pitch range, since the instruments generally span only an octave, from one E (as we'd now define it) to the next. Poetry intended for singing with lyre accompaniment was 'lyric', the origin of our term for the words of songs. In fact, just about all poetry was set to music; while Aristotle refers to a form of verse that uses language alone, he says that it doesn't even have a name.

Greek music would probably not sound so very strange to us, since its 'scales' (an anachronistic term here) seem to have been composed of notes rather close to those we use today. Indeed, something like the diatonic scale may be very ancient: it has been suggested, based on a decoding of a love song from around 1400 BC written on a clay tablet, that even the Sumerians used something of this sort.

As far as we can tell, the Greeks made no real use of harmony – of two or more different notes sounded simultaneously (although they would surely have overlapped when played on plucked insruments such as lyres). Instead their music was monophonic, in which a single voice sang a melody that was duplicated on an instrument. Purely instrumental music seems to have been a rarity. All this makes it the more striking that the Pythagoreans seem so concerned about 'harmony' – but here they are referring to a notion of orderly relationships between entities, such as the whole-number ratios of frequencies.

The oldest Greek treatise specifically on music is the *Harmonics* of Aristoxenus, a philosopher who studied within the Pythagorean tradition and became a pupil of Aristotle in the fourth century BC. From this text we learn that the musical system was based on the interval of a perfect fourth. The organizational unit was called a tetrachord, a sequence of four notes in which the highest and lowest were separated by a fourth and the inner notes were somewhat flexibly tuned: despite the Pythagoreans' carefully worked-out scheme, it seems that these inner notes were tuned by ear, not by maths. These tetrachords were combined in various ways to make 'scales' spanning an octave, which were called *modes*.

It's not clear what the various modes really meant to the Greeks: whether they were truly different types of 'scale', or the same 'scale' transposed to different 'keys', or perhaps simply music that has particular qualities. It may be that the names of the various modes actually referred to different things at different times. Some musicologists suspect that modes were not simply collections of notes from which the Greeks composed freely, but had associated with them melodic motifs, prefabricated units that were combined to make songs. A system like that was certainly used for Byzantine hymns, and it is also a feature of some other musical traditions, notably the ragas of India.

At any event, by the time Ptolemy (an authority on music as well as astronomy) wrote his own *Harmonia* in the second century AD, there were seven modes, and they do then seem to have acquired something of the status of scales (Figure 3.11). The most common

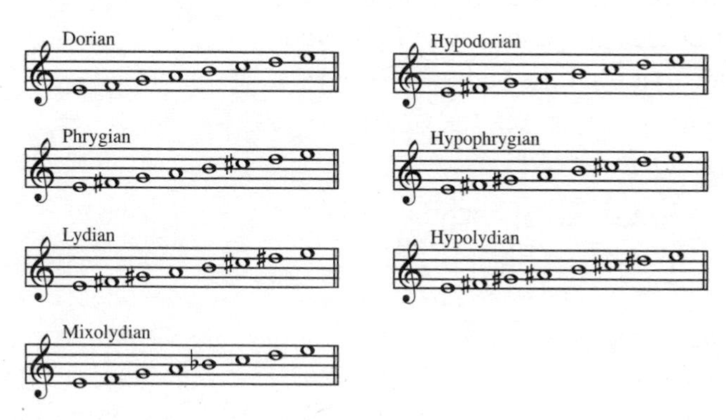

Figure 3.11 The Greek modes, as defined by Ptolemy in the second century AD.

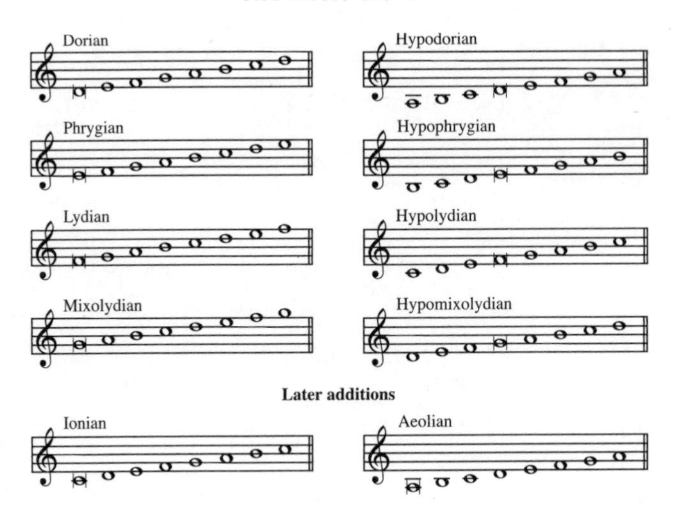

Figure 3.12 The medieval modes – authentic and plagal – and some of those added at later times. The parallel vertical lines denote the 'final'.

was the Dorian, which is similar to a Western minor scale. Ptolemy's Phrygian is also a 'minor' mode, while his Lydian is equivalent to the major scale. These modes were inherited in the West, with some modifications, in the Middle Ages, when they began to be notated for liturgical singing. Unfortunately, owing partly to misinterpretations by medieval scholars, the modes of medieval music used in church liturgy used different notes from their namesakes in ancient Greece (Figure 3.12). The Greek Mixolydian, for example, can be considered to run from B to B' on the white notes, whereas the medieval Mixolydian is the white-note scale from G to G'. And later music theorists introduced new modes to which they gave antique names: the Ionian introduced by the Swiss music theorist Heinrich Glarean in the sixteenth century is basically the same as the major scale, while Glarean's Aeolian is one of the modern minor scales.

In the diatonic scales used today, each scale is rooted in a special note, the tonic. Medieval modes didn't exactly have a tonic, but they had somewhat equivalent anchoring notes. One was called a 'final': in so-called authentic modes, the series of notes began (somewhat confusingly) on the final, whereas in plagal modes they began a perfect fourth below the final (Figure 3.12). Most chants had simple formulaic melodies called psalm tones, in which most of the text would be sung on a single 'reciting tone' called the *tenor*, from the

Latin *tenere*, to hold. There were also specific psalm tones associated with each mode. So not all notes in the mode had equal status: the final and tenor were special, just as some notes (such as the tonic, third and fifth) are privileged in diatonic tonal music.

To help medieval monks remember their notes, an eleventh-century Italian cleric named Guido of Arezzo came up with a handy mnemonic: the notes C, D, E, F, G, A, he pointed out, began successive phrases in the well-known hymn 'Ut queant laxis', with syllables *ut-re-mi-fa-sol-la*. With the addition of the seventh note *ti* and the replacement of *ut* by *do*, we move from Gregorian chant to Julie Andrews and her troupe of Austrian children, singing the standard designations of the major scale.

Guido's six-note series, called a hexachord, contains notes all separated by a whole tone except the middle two (E, F), which are a semitone apart. Guido proposed a musical scheme in which all the notes then in use formed a series of overlapping hexachords, starting on G, C and F. Melodies in medieval chant had a pitch range that spanned nearly three octaves, from a low G to a high E. The lowest G was designated by the Greek symbol for gamma, and as the first note of its hexachord it was called *gamma ut*, which is the root of our word 'gamut'.

To preserve the pitch steps of the hexachord, one of the 'white notes' must be altered when the series starts on an F. Here the hexachord is F, G, A, B, C, D. To keep the third step a semitone and the fourth a whole tone, the B must be brought down a semitone: in modern terms, it must be *flattened* to B♭. Therefore, new symbols were introduced to distinguish between the flattened B of a hexachord on F, and the 'natural' B of a hexachord on G (the hexachord on C doesn't reach B). The former was denoted by a rounded 'b' sign and known as a 'soft B' (*B molle*), the latter by a square one (♮) and called a 'hard B' (*B durum*). These are the origins of our modern symbols for a flat and a natural (♭ and ♮). The sharp sign ♯ is also related to the latter. Thus the medieval system of hexachord gave rise to the first *accidental* – the modification of a note to fit within a different 'scale'. From the modern perspective of key structures, we can say that the hexachord on F is a transposition of the hexachords on C or G to a new key.

Change of key

However 'natural' the Pythagorean scale might appear, it suffers from a serious problem. Actually, from two related problems. First, we started from the premise that harmonious sounds come from the combination of pitches with simple ratios of frequencies. But we have somehow ended up with a scale containing pitch ratios of the likes of 81/64 and 243/128 – not exactly simple! Second, if the 'big' steps in the Pythagorean scale correspond to a whole tone and the small steps to a semitone, two small steps should equal a big one. But they don't. Increasing a pitch by two semitones here means multiplying its frequency by $(256/243)^2$, which is a fearsome fraction and not equal to the whole-tone ratio of 9/8. It is rather close, however, so perhaps we shouldn't worry? But we must. It is because

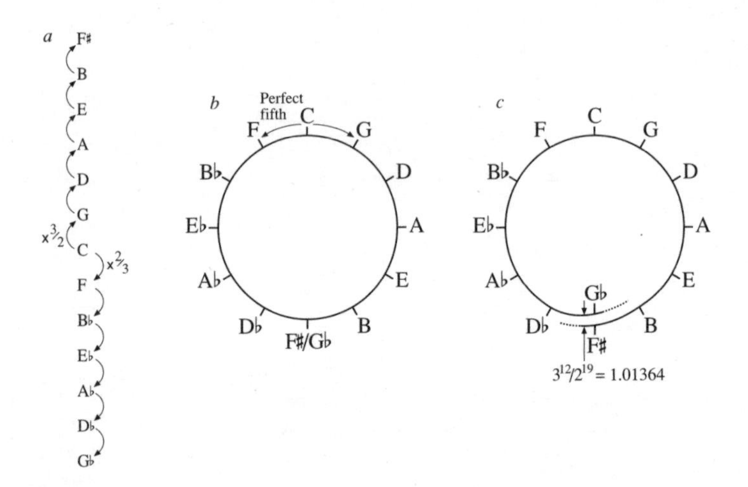

Figure 3.13 All the chromatic notes in Western scales can be reached through a repeated cycling of fifths (*a*). Eventually, the upward cycle brings us to F♯, and the downward cycle to G♭. On a modern piano these notes are equivalent – they both correspond to the same key. This means that the 'cycle of fifths' closes (*b*). But in the Pythagorean system, where each step corresponds strictly to a 'mathematical' perfect fifth (with frequency ratio 3/2), F♯ and G♭ do not coincide: they have frequencies differing by a factor called the Pythagorean comma, and equal to about 1.01364. This means that the 'cycle' does not close, but is instead a spiral (*c*). However far we continue cycling in fifths, we never find a point of closure, but just keep adding new notes.

of this mismatch that the Pythagorean system doesn't give us a small set of notes from which to make music – in fact, it gives us an infinite number.

How does that happen? We saw that one way of constructing the Pythagorean scale is by a succession of upward movements of the original (tonic) note by a perfect fifth. The one exception here is the fourth note of the scale, F in the key of C, which we reach by a *downward* step of a perfect fifth from the tonic (a frequency ratio of 2/3).

If we continue this 'cycling' of fifths, we alight on notes that aren't in the major scale, but are instead chromatic. A fifth up from B takes us to F♯. A fifth down from F takes us to B♭ (frequency ratio $2^4/3^2$: 1, or 16/9), and then successively to E♭, A♭, D♭ and G♭ (Figure 3.13*a*).

But now comes the crisis. On the piano keyboard, F♯ and G♭ are equivalent notes: they are both sounded on the same key. In other words, F♯ and G♭ represent the point where the so-called cycle of fifths starting at C closes in on itself and indeed becomes a cycle (Figure 3.13*b*). But in the Pythagorean system of tuning, the notes F♯ and G♭ are *not* the same. The F♯ reached by upward cycling has a frequency $3^6/2^9$ or 729/512 times that of C, while the frequency of G♭ reached by downward cycling exceeds that of C by a factor $2^{10}/3^6$ or 1024/729. These differ by a small amount: a factor of about 1.01364, or to be more precise, by the ratio $3^{12}/2^{19}$, called the Pythagorean comma.

So for Pythagorean tuning, the cycle of fifths doesn't close. We can continue this sequence of stepping up and down by a fifth and *never* find the upper and lower scales coinciding exactly. For example, from F♯ we go up to C♯, then G♯. These notes are identical on a modern piano to D♭ and A♭ respectively, but in the Pythagorean system they differ by the Pythagorean comma. The 'cycle' is in fact a spiral (Figure 3.13*c*).

This needn't be a big problem if all music is written in just one or a small number of keys, since the problematic notes that lie outside those keys are hardly ever encountered. In ancient Greece and medieval Europe there was in fact no system of keys at all like that used in Western music today (see Box: Keys and tonics). Instead, the musical palette was varied by employing a range of different *modes*, which we can think (somewhat crudely) of as scales that all use the same notes but simply start in different places. By the Middle Ages, the Greek modal system had crystallized into a series of seven-note 'scales' in which the notes were basically equivalent to the white notes on the piano; one mode (the Dorian) started on D, say, another (the Phrygian) on E, and so on (see Box: *Modes*). Thus, each different mode had a different sequence of step heights in its 'scale'. The Hypolydian sounds like our major scale; the Dorian is very close to some of our minor scales. Others sound distinctly 'odd' to our ears now when we hear them as scales.

So-called accidentals – sharps and flats – were initially added to this modal system in an ad hoc manner, starting with B♭ and soon followed by F♯. In effect this enabled the process of shifting the root note, or as we would now say, *transposing* the key, so that the same melody could be played or sung on a different starting note. If you play the rising set of white notes C, D, E, F, G on a piano and then move the same pattern up to begin on F, it doesn't sound the same when you use just the white notes: F, G, A, B, C. The wrong note is the fourth, B, because whereas the initial sequence consisted of the pitch steps t-t-s-t, the transposed one has a sequence t-t-t-s. To preserve the same melody, we need to drop the B by a semitone, making it a B♭ (Figure 3.14). Adding accidentals to the original scale also makes possible the musical process of *modulation*, in which a melody moves smoothly from one key to another. The Christmas carol 'We Three Kings', for example, modulates from E minor in the verses to G major in the chorus. Modulation became central to Western classical music from the Baroque era. It is a key part of the organisational scheme in the sonata form, say, where the initial themes or 'subjects' generally occur in different (but related) keys.

Transposition and modulation make it necessary to introduce a new scale. If this happens via a single step (up or down) in the cycle of fifths, only one new note is needed. The major scale of G has only one note (F♯) that doesn't also appear in the major scale of C, and so does the F major scale (B♭). With those additions, all the notes in the Pythagorean G and F major scales have the same pitch ratios as the corresponding

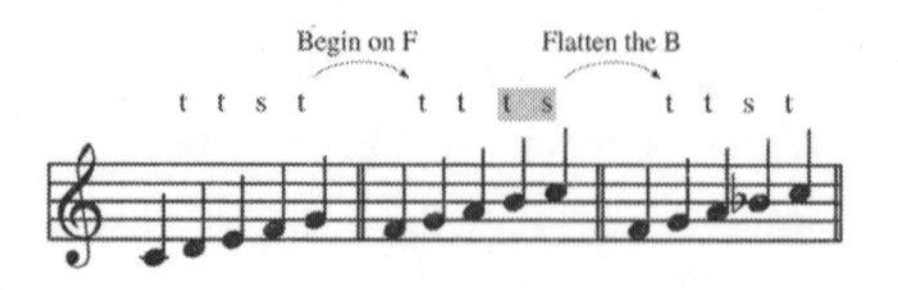

Figure 3.14 The ascending scale on C (left) has an altered sequence of pitch steps (tone/semitone) when moved up (transposed) to begin on F (middle). The original sequence is restored by dropping the fourth note (B) by a semitone, turning it into B♭ (right).

notes in C major. That's pretty easy to accommodate. But towards the Renaissance, composers began to experiment with new keys – or rather, one should say, to use more accidentals, for the notion of a key as such was still in gestation. By the fourteenth century, some of these experiments in chromaticism look by today's standards to be remarkably free. Musicians sometimes introduced accidentals not according to rigorous notions of scale, but for so-called *causa pulchritudinis* – they simply thought the melody sounded more beautiful that way. Some of the accidentals became well-established formulas, to the extent that they weren't even included in written scores: performers were expected to know the rules, or might merely flatten or sharpen notes just because they felt like it. In compositions such as the madrigals of the Italian Carlo Gesualdo (*c.*1560–1613), chromaticism was carried to such mannerist extremes that it sounds decidedly odd to us now.

New notes demanded by transposition and chromaticism can be progressively added to the repertoire using the same Pythagorean principles with which we generated the scales of G and F above. That is, we can cycle upwards in steps of a perfect fifth to create extra sharps, and downwards by the same steps to produce flats. Transposing the major scale from G to D picks up one new note: C♯. And stepping from D to A, we add G♯; and so on (Figure 3.15). But again, there is no end to this proliferation of notes, because the sequence does not close in on itself when we reach the keys of F♯ and G♭ – every single note in these two scales has a different frequency ratio to its counterpart in the other scale. And the mismatch continues as we progress to the keys of C♯, G♯ and so on.

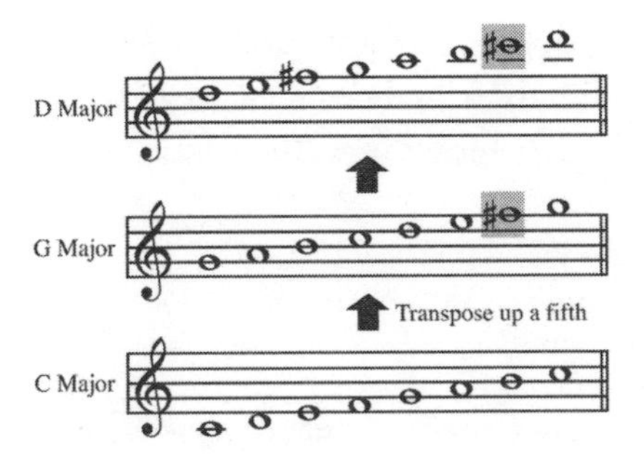

Figure 3.15 Each transposition of the scale upwards by a perfect fifth adds one new note, here highlighted in grey.

This means, in short, that there is no correct way to tune a piano using Pythagorean tuning. If we tuned the notes based on the upwards cycle of fifths to F♯, we'd find that some of them will sound out of tune whenever we play in one of the keys reached by downward transposition: the steps between successive notes wouldn't be quite right. It's not just F♯ and G♭ that fail to match – *all* of the sharps have different frequencies from the equivalent flats. So transposition and modulation play havoc with the tuning: the further we go from the 'root' key, the worse the music sounds. What is needed to avoid this problem is a system of tuning that is 'closed' rather than open under transposition, so that F♯ and G♭ become the same note.

Tuning up

The first alternative tuning systems were motivated by a different perceived defect in the Pythagorean scheme. The Pythagoreans deemed there to be an intrinsic 'goodness' in the juxtaposition of notes whose frequencies are related by a simple ratio. In the Middle Ages only octaves, fifths and fourths were regarded as consonances of this sort. But by the fifteenth century, music had become increasingly polyphonic: it tended to employ several melodic voices at once, which threw up new harmonic combinations of notes. In particular, the intervals of a third (C to E, say) and a sixth (C to A) became permissible and were deemed relatively consonant. But the Pythagorean ratio for a third (81/64) was far from simple. To restore the mathematical simplicity that was meant to characterize a consonance, this ratio was modified very slightly to 80/64, which simplifies to 5/4. Likewise, the sixth was altered from 27/16 to 25/15, or 5/3.

This system was formalized in 1558 by Gioseffo Zarlino, the choirmaster of St Mark's Church in Venice, who proposed the following sequence of frequency ratios (shown here for C):

Note	C	D	E	F	G	A	B	C'
Frequency	1	9/8	5/4	4/3	3/2	5/3	15/8	2

Zarlino's system became known as 'just intonation'. Although it keeps the ratios simple, it introduces a new complication. By shifting the third and sixth notes of the scale slightly, it creates two different 'whole tone' steps: one in which the two successive notes differ in frequency by a factor 9/8 (here C→D, F→G and A→B), and the other where this difference is 10/9 (D→E, G→A). Moreover, neither of these is equal to two semitone steps, for the semitone difference (E→F, B→C') is 16/15. In short, it is a bit of a fudge.

Moreover, just intonation does nothing to solve the problem of transposition. In fact, it makes it worse, because each transposition of a fifth introduces *two* new notes instead of one. For example, transposing the scale from C to G produces not just the new note F♯ but also an A that differs from the one in the key of C. The frequency difference is the mismatch between the two types of whole-tone step, 9/8 and 10/9, and is equal to 1.0125, called the syntonic comma.

Somewhat remarkably, music theorists of the period elected to live with this problem, even though it meant that instruments that sounded fixed pitches, such as organs and harpsichords, had to have extra notes for the different tunings. (String players could accommodate the differences by finely adjusting where they pressed on the string, even though that placed formidable demands on their accuracy and dexterity.) A keyboard designed by the French mathematician Marin Mersenne in the 1630s had thirty-one notes within an octave, including fifteen between the span F to A (Figure 3.16*a*). It sounds virtually unplayable, although allegedly the virtuosic Joseph Haydn performed on such an instrument in the Netherlands. Other systems of tuning proposed over the ages have an even more absurd proliferation of finely graded notes: the Bosanquet harmonium, an instrument with an unconventional tuning scheme built in the 1870s for the scientist and music theorist Robert Holford Bosanquet, had 84 notes per octave (Figure 3.16*b*).

Another popular scheme for fixing the infelicities of Pythagorean tuning was meantone temperament, introduced in the early fifteenth century. This takes the same approach of constructing notes from a chain of perfect fifths, but gets around the fundamental problem – that no number of perfect-fifth steps will ever fit exactly into a whole number of octave steps – by making the frequency ratio of fifths just a little smaller than the 'ideal' value of 3:2. There's no unique way to do this, but the most common, called quarter-comma meantone and introduced in 1523 by the Florentine music theorist Pietro Aron, reduces each perfect fifth by a quarter of a syntonic comma. Four steps of this reduced fifth arrive at a major third two octaves higher. However, it's

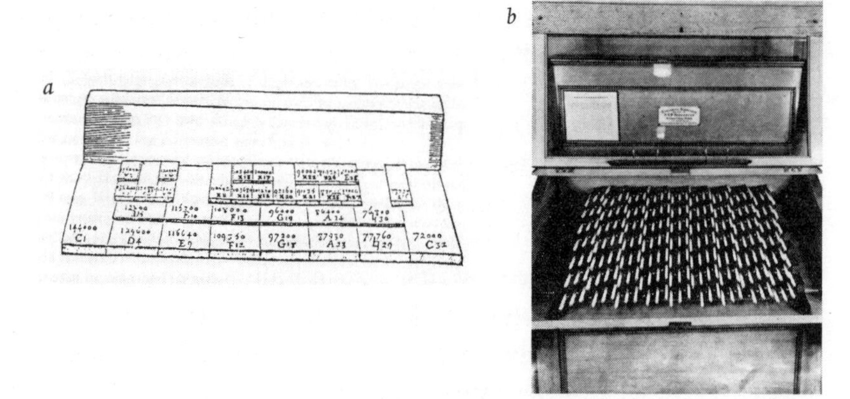

Figure 3.16 The keyboard devised by Marin Mersenne (*a*) and Robert Bosanquet's harmonium (*b*) are designed to accommodate more than twelve notes per octave, as is demanded by unequal temperaments.

not possible to consistently adjust all the fifths this way: one of them ends up significantly too 'wide', and is called the wolf interval (because of its sharp, 'howling' sound). Additionally, there are still more than twelve notes per octave: the sharps and flats fail to coincide.

The solution to all these difficulties was clear enough in principle. If the octave is simply divided up into twelve equal chromatic steps, each of a semitone, then a sequence of identical pitch steps, such as the major-scale sequence of t-t-s-t-t-t-s, can be begun from any note and always sound the same. The question is, what should that basic interval be? Cleaving still to the idea that it has to embody some relatively simple frequency ratio, Galileo's father Vincenzo Galilei, who studied under Zarlino, proposed in 1581 that an increment of a semitone should correspond to a frequency increase by a factor 18/17. Twelve such steps increase the initial frequency by about 1.9855, which is close to the ideal factor of 2. But this would mean that the intervals of an octave, fifth and fourth fall just short of their ideal values – they would be slightly flat. This seemed too insulting to classical ideas about harmony for the system to catch on.

There is only one exact solution to the problem, and it was discovered nearly simultaneously in two places: China and Holland. The only factor that, multiplied by itself twelve times (one for each semitone step), gives a precise doubling of frequency is the twelfth root of 2, written $\sqrt[12]{2}$ or $2^{1/12}$. And this definition of a semitone turns out to give more 'ideal' values for a fourth and a fifth than does Vincenzo Galilei's 18/17: the fourth is a factor of about 1.3348 greater than the tonic (compared with the Pythagorean 4/3 or 1.3333), and the fifth is 1.4983 (as opposed to 3/2 or 1.5).

This is the system known as equal temperament. It was first published in 1584 by the Chinese scholar Chu Tsai-Yü, a prince of the Ming dynasty. But the Flemish mathematician Simon Stevin put forward the same idea a year later. Some have claimed that Stevin knew of Chu's idea, which is not as unlikely as it sounds – from 1580, a trade fair was held every two years at the Portuguese colony of Macao on the Pearl River estuary, and here Westerners and Chinese exchanged ideas as well as goods. But a direct link from Chu to Stevin seems unlikely. In any event, the equal-tempered system was promoted in the early seventeenth century by Mersenne, his predilection for elaborate just-tempered instruments notwithstanding.

There is, however, a philosophical defect in the equal-tempered system, for $2^{1/12}$ is not just a difficult number but an *irrational* one, meaning that it cannot be expressed exactly as any fraction of whole numbers. Where now, then, are the harmonious mathematics of the Pythagorean concept of music? Stevin had no such qualms. What is so special about the Pythagorean fifth, he asked, if it leads to a semitone based on the formidable ratio 256/243? To a mathematical mind, $2^{1/12}$ is a perfectly elegant number – and if anyone thinks otherwise, said Stevin, that is their problem.

Plenty of people disagreed, and still do, insisting that the damage wrought on 'proper' harmony by the equal-tempered system sounds uncouth once you have heard the earlier alternatives.* Hermann von Helmholtz, a German physiologist who pioneered the understanding of auditory acoustics in the nineteenth century, claimed that equal temperament is 'unpleasant to uncorrupted ears'. And certainly, the differences between the systems are not always negligible: the major-third interval is particularly problematic, that of the equal-tempered scale being sharp relative to just intonation by almost one per cent, which is easily audible to most people. As we will see, however, claims

* In his spirited and enjoyable *How Equal Temperament Ruined Harmony*, musicologist Ross Duffin considers the deficiencies of equal temperament so self-evident that he feels no need to adduce a single perceptual test in support of the idea. The only evidence he provides is an anecdote about people's responses to experiments with tone generators, which makes no sense at all in acoustic terms.

about which system of tuning is 'best' lack any fundamental basis, and are much more a matter of what one has become accustomed to.

Several other forms of 'circular' tuning were designed from the sixteenth century onwards to close the cycle of fifths and thus allow modulation to any key without introducing notably out-of-tune notes. Typically these compromise the intervals of fifths so that they are not all of equal size, generally by narrowing those in white-note keys and widening them in black-note keys. The German music theorist Andreas Werckmeister proposed several schemes like this in the late seventeenth century, which he called *wohl temperiert* ('well tempered'). There has been much debate about whether J. S. Bach's *The Well-Tempered Clavier*, a collection of preludes and fugues in all twenty-four major and minor keys, was composed for this type of well temperament or for equal temperament. Either way, they were designed to showcase the advantages of a closed tuning system that allowed the composer to use all keys equally. No advertisement has ever been more astonishingly scored, but even this did not secure the general adoption of equal temperament until well into the nineteenth century.

It remains the case, however, that it is mathematically impossible to find any intonation scheme that can be modulated to any key while retaining simple frequency ratios for intervals such as the perfect fifth and fourth and the octave. Acoustic scientist William Sethares has proposed an ingenious solution that takes advantage of the arbitrary tuning capabilities of electronic instruments. He has devised an algorithm called Adaptun that enables an electronic keyboard to adapt its intonation in real time, note by note, so as to find the 'ideal' intervals at any point in the music. This doesn't require any knowledge of the key at any point in the music – the program simply finds the best 'local tuning' for each combination of notes. You can hear the results of Adaptun on Sethares' website and CDs at http://eceserv0.ece.wisc.edu/~sethares/.

The one-note chord

Clearly, then, attempts to ground the Western musical scales in pure mathematics – and thus to establish it as the 'natural' basis for music – quickly run into problems. There is, however, a rather different way to try to build these scales out of acoustic physics. As with Pythagoras' 'harmonious ratios', it has been cited for centuries as evidence of the superiority of the diatonic scale – and with equal spuriousness.

A key is characterized by two musical structures: a scale* and a triad chord composed of the scale notes **1**, **3** and **5**: C-E-G in C major. Sounded together in any combination or permutation, these three notes sound harmonious: they seem to 'fit'.

The triad chord is so familiar a piece of musical apparatus that it is easy to overlook why it is actually a bit odd in terms of the Pythagorean scheme. This infers that the most 'consonant' intervals are those in which the component notes have the simplest ratios of frequency. That accords with the presence of G – the perfect fifth – in the C major chord, and also with the fact that the octave C' may be added to create the full major chord: C-E-G-C' (with a doubled initial C, these are the notes that begin 'On Top of Old Smokey'). But on that basis we might expect F (with ratio 4:3 to C) to be included, not E (with ratio 81:64 in Pythagorean tuning). Why is the major third preferred here over the perfect fourth?

It is now that I must confess to having hidden from you so far a vitally important piece of information. I had good reason to do that, for it greatly complicates the picture we have built up so far of musical notes as pitches with specific frequencies. The truth is that you might have never heard notes of that kind in your life. Making them is rather hard, and only really became possible with the advent of electronic sound production. Sounds that contain a single acoustic frequency are almost unknown in the natural world, and no traditional instrument produces them (although some come fairly close).

If you pluck a guitar string, bow a violin, hit a piano key or blow into a flute or a trumpet, you don't just set the air vibrating at a single frequency. The resonating string or body generates a complex mixture of frequencies, all of which are generally whole-number multiples of the lowest one. An oscillating string, for example, doesn't just support a wave with a wavelength equal to the string's length, but also waves of half that wavelength (and thus twice the frequency), a third (three times the frequency), a quarter and so on. All that matters is that a whole number of wavelengths has to fit within the string's length (Figure 3.17). The lowest frequency is called the *fundamental*, and it is

* There are, as I've said, actually several different types of minor scale for any key.

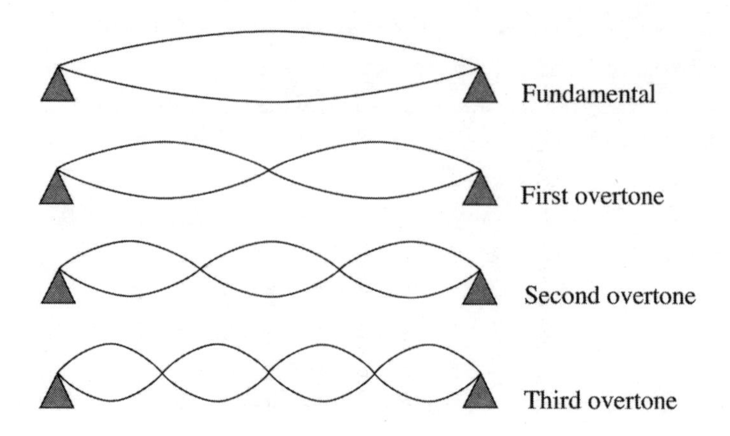

Figure 3.17 Harmonics: a fundamental, and overtones with wavelengths reduced by factors of 1/2, 1/3, 1/4 . . . The corresponding frequencies are greater by factors of 2, 3, 4 . . .

usually the loudest: most of the acoustic energy of the sound is channelled into this pitch. The higher pitches are called *overtones* or *harmonics*. (Rather confusingly, the fundamental is actually the first harmonic, but the tone with twice the fundamental frequency is the first overtone.)

Typically, an instrument produces dozens of overtones (Figure 3.18). But not all are easily detectable, since they get progressively weaker as they get higher in frequency. You can find some of the harmonics of a piano note by playing it with the sustain pedal depressed to free up all the other strings, and listening for which of the higher strings vibrates in sympathy. Notes that are combinations of a fundamental and overtones are called *complex tones*. The precise mixture of overtones is one of the factors that determines the characteristic sound or timbre of an instrument. The series of pitches in a complex tone is called the *harmonic series* (Figure 3.19*a*)

The presence of overtones means that just about every note we hear from musical instruments, including the human voice, is really a chord. Yet this is not how we hear it. Our ear and brain conspire to collapse all the harmonics into a single perceived tone.

This helps to explain why two notes an octave apart sound in some sense 'the same'. The first overtone is an octave above the fundamental, since an octave interval comes from a doubling of the frequency. Our brains create the sensation of sameness apparently as an evolutionary

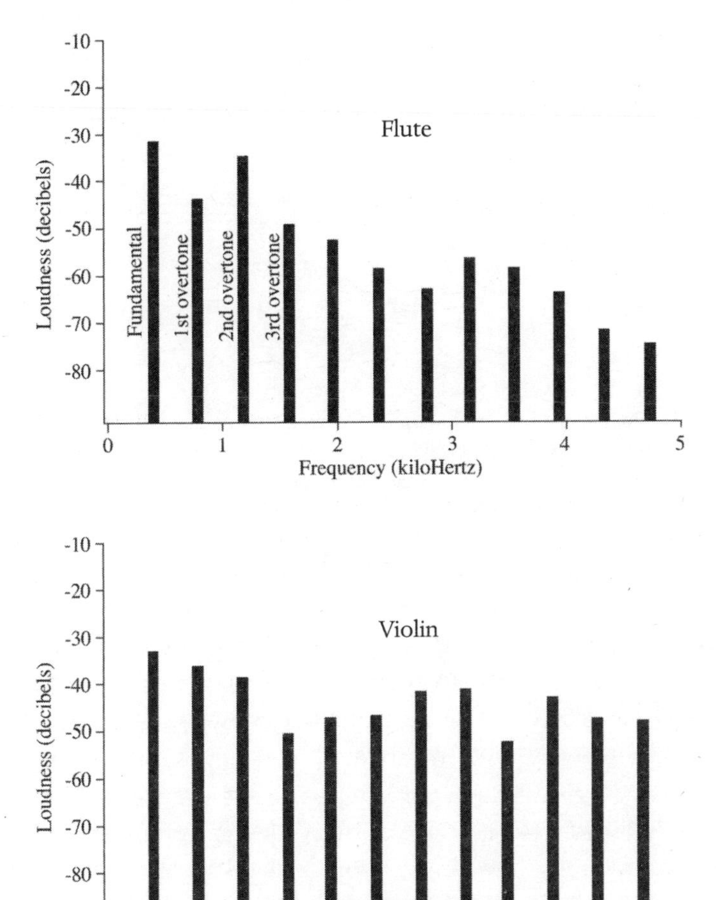

Figure 3.18 Typical overtone frequency spectra.

response to the fact that most natural sounds (not just those of musical instruments) are harmonically complex, with the first overtone – the doubled frequency – typically the strongest. The brain classifies the first overtone as having the same 'flavour' as the fundamental so that they can easily blend into a single perceived signal: it takes some effort and practice to 'hear' the octave overtone, although we'd notice a change in the quality of the sound if it was removed. This helps us to interpret the two frequencies as most probably coming from the same source, rather than setting us to look in vain for two separate sources.

Figure 3.19 (*a*) The harmonic series for a fundamental (*f*) of C. The numbers indicate the overtones – add 1 for the respective ordinal numbers of harmonics. A minus sign here indicates a slight detuning relative to the actual note annotated. (*b*) The major triad in C. All of the notes in this chord are represented in the first four overtones.

So there is apparently a 'natural' basis for building musical scales around the octave: it is a division of the continuum of acoustic frequencies that is grounded in the neurology of audition. And indeed just about all musical systems in the world are structured around the equivalence of octaves; the only well-documented exceptions are in some groups of Australian aboriginals.

The second overtone is an octave plus a fifth above the fundamental: for a C4 fundamental, say, it is a G5. Thus, it might seem reasonable that the interval of a fifth should also sound 'well fitting' when the two notes are played together. And indeed the fifth features in most musical scales. But not in all, as we'll see: even by the second overtone, the imprint of the harmonic series in the structure of musical scales seems to be fading.

The third overtone is yet another octave, which might be expected to reinforce the tendency for octave equivalence. And the fourth is a major third higher – for a C4 fundamental, it is an E6. Note that this is *not* the major third of the Pythagorean scale; its frequency ratio with respect to the nearest tonic is 5:4, the value in just intonation.* Then comes another fifth (in this case, G6). So the first six members of the harmonic series are all members of the major triad (Figure 3.19*b*).

* It's not hard to see why this is. As overtone frequencies are simple multiples of the fundamental, the fourth overtone will be a factor of five greater. When the resulting pitch is dropped two octaves, by dividing the frequency by $2^2 = 4$ to bring it back within the octave span above the fundamental, we end up with a frequency 5/4 that of the fundamental.

The brain converts the 'chord' produced by these overtones into a perception not of several simultaneous notes but of timbre: the blend of harmonics gives a characteristic sonic quality to what seems to be a single pitch. And yet, when two separate notes really are played at the same time, for example by pressing down two keys on the piano, we *do* hear them both, even if they are notes of the major triad. It seems that the distinct series of overtones attached to each of the two notes is enough to label them as having different sources: our brains are able to prevent the two fundamental notes from 'fusing' into one voice, because it can identify two sets of harmonics. All the same, we'll see later that this mental separation of sources is a precarious thing, and that musicians have to use careful strategies to prevent inadvertent fusion of harmonious notes.

The brain is tenaciously insistent on 'binding' overtones into a single perceived pitch. If one of the harmonics is detuned by electronic manipulation of the sound, our cognitive system at first tries desperately to resolve the discrepant harmonic series by searching around for a new fundamental frequency that 'fits', even though none exists: what we hear is not an out-of-tune harmonic, but a different single pitch. Only when the detuning is too large for any such compromise to be found does the brain admit defeat and register the 'bad' harmonic as a distinct tone. Some percussive instruments naturally produce overtones whose frequencies are not perfect multiples of the fundamental – most bells, for example, are like this. Then the brain can't find a simple scheme for blending all the overtones into a single pitch, and the result is that the tone has an ambiguous pitch: we can't quite tell which note is being played. Such sounds are said to be *inharmonic*. Debussy sought to emulate this ambiguity of pitch in the bell-like sounds of his piano composition 'Engulfed Cathedral', by sounding two groups of two notes, a C and D in the 'dissonant' major-second

Figure 3.20 In 'Engulfed Cathedral', Debussy uses dissonance to mimic the inharmonic sound of bells.

interval, an octave apart (Figure 3.20). The ill-fitting pairs of harmonic series here create a somewhat inharmonic effect.

The harmonic series continues beyond the triadic harmonies of the first six members. The seventh might come as a surprise, because it is a so-called flattened or minor seventh: for C, it is a B♭ (or rather, a note very close to it, depending on what system of intonation you are using). This doesn't even feature in the major scale, and yet here it seems to be asserting a kind of natural, if rather distant, relationship to the tonic. If you throw a B♭ into a melody in C major, it sounds out of place – or not exactly that, but it sounds as though something is about to happen. The tune hovers on the brink of moving to the key of F, because B♭ appears in the F major scale. The minor-seventh degree of a scale is thus a common launching point for modulating the key.

The harmonic series provides an alternative to the cycle of fifths for building up a scale 'mathematically' from a fundamental or tonic pitch. In this case what it seems to give us is a 'natural' basis for the major triad of Western music. This was pointed out in the eighteenth century by the French composer and theorist Jean-Philippe Rameau, whose 1722 work *Treatise on Harmony* used the mathematics of the harmonic series to derive fundamental laws of musical composition based on what he saw as the natural relationships between notes.

But does the major triad really have a perceptually privileged status – do we really hear it as something naturally harmonious – because of the way it is embedded within complex tones? That isn't clear at all. Although the well-trained and attentive ear of a musician can make out perhaps six or seven distinct overtones in a complex tone, generally the overtones become rapidly weaker after the first one or two, and so it isn't clear that the major third will register strongly enough in most of the sounds we hear to be perceived as a 'special' interval in relation to a tonic note. I will return to this question when I consider consonance and dissonance in Chapter 6.

Flimsier still are the attempts that have been made to derive the entire Western scales from the harmonic series. Many music theorists and composers have fallen into this trap, by arguing that increasingly higher overtones of the series coincide with other notes of the diatonic scales. The eighth overtone is two octaves plus a major second above the fundamental: for a C4 fundamental, it is a D6. And above the ninth overtone we get chromatic notes, such as (staying with a C fundamental) G♭, A♭

and C♯ (Figure 3.19). But these pitches aren't really the same as those in the diatonic scales. They are often significantly 'out of tune' relative to any of the standard systems of intonation, mere approximations to their 'true' values. It is entirely unsurprising that 'notes' like this should start to appear sufficiently high up in the harmonic series, since the mathematics of the series dictates that the overtones will get ever closer together in pitch the higher we go: we're bound to find some more or less close matches to diatonic or chromatic notes if we look high enough in the sequence. Not fully comprehending this, Arnold Schoenberg felt that the 'chromaticism' in the higher overtones implied that intervals commonly considered dissonant were actually as 'naturally consonant' as those that conform to the norms of tonal harmony in the early part of the series. Olivier Messiaen also sought 'complex consonances' in the higher overtones. But even if we put aside the 'mistunings' of these high harmonics, the fact is that we don't really hear them identifiably in complex tones anyway. It's a common habit of musical iconoclasts who seek 'theoretical' justifications for their experiments (Harry Partch was one such) to use abstract reasoning that takes no account of how music is actually *heard*. Of course, this need not invalidate the experiments themselves, which may sometimes fortuitously hit on interesting new ways to organize sound.*

Even the perceptual equivalence of octaves, which does plausibly have its root in the harmonic series, has its limits. It doesn't mean that, musically speaking, a note in a melody can be substituted by the same note in a different octave without perturbing the way we hear it. It seems intuitively obvious enough that a sudden plummeting or soaring of a melody to a different octave range will sound odd (we will see later precisely why this is), but it is worse than that. When music psychologist Diana Deutsch played people familiar tunes such as 'Yankee Doodle' in which all the notes were of the right *pitch class* (for example, a D appeared wherever there was a D in the melody) but were each selected from any one of three octaves at random, the listeners couldn't even recognize the tune.

There are other good reasons why we shouldn't place too much faith

* The same may be said in the visual arts, as for example when the neo-Impressionists Paul Signac and Jean Seurat used poorly digested optical theory to motivate their experiments in colour. Interestingly, the scientific ideas of Hermann von Helmholtz were appropriated in this way both by visual artists and by musicians.

in claims, like those made by Leonard Bernstein, that the harmonic series provides a natural basis for the whole of Western melodic and harmonic theory. For example, there are two very notable absences from this series. One is the interval of a fourth – according to Pythagorean theory, the most 'consonant' interval after the octave and the fifth. Even by the twentieth harmonic for a complex tone on C, we still have not landed on an F. The other absence is the minor third.*

Actually the minor third is not obviously present in the Pythagorean scale either, or at least not unless we use that scale to construct modes – for the minor tonality is related to the mode that starts on the second degree of the scale, or the white-note keys from D to D. Here the minor-third interval – D to F – involves pitches related by a factor 32/27. In Zarlino's just intonation this ratio was simplified to 6/5, which is a perfectly reasonable but ultimately arbitrary choice. So when, for example, Paul Hindemith explained the 'naturalness' of the minor third in terms of this latter ratio, he seemed unaware that it was nothing more than a reasonable compromise for the sake of keeping the frequency ratio simple.

Breaking the cycle

The claims for a natural foundation of diatonic scales in mathematical and acoustic principles have occasionally been comically absurd. So convinced was the German Jesuit scholar Athanasius Kircher in the seventeenth century that the major scale was a natural phenomenon that he claimed to hear it in birdsong and even in the calls of South American sloths.† And even Charles Darwin, in *Descent of Man*, reports

* Some texts blithely claim that both the fourth and the minor third are there in the series – for the interval between the second and third overtones (G and C) is a fourth, while that between the fourth and fifth overtones (E and G) is a minor third. But these intervals are heard in relation to the fundamental of C – that's to say, we hear the third overtone as an octave C, not as the fourth note of a scale on G. The fact is that the **4** and minor **3** degrees of the diatonic scales are simply not present in the series.

† It is easy to laugh at such claims, but I still find it hard to suppress my astonishment at hearing a wild bird sing what sounds like a rather catchy jazz riff on the blues scale: listen to the slowed-down version of the veery's song at www.whybirdssing.com. It's coincidence, of course, but a striking one. And in fairness, Darwin too fell for this idea, citing Helmholtz in support of the belief that animal tone production will converge on the notes of 'human musical scales' – whatever those are supposed to be!

respectfully the claim by one Rev. S. Lockwood that the 'songs' of a species of American mouse 'keep to the key of B [flat] (two flats) and strictly in a major key'.

When ethnomusicologist Norman Cazden suggested (somewhat optimistically) in 1945 that 'the naïve view that by some occult process mathematical ratios are consciously transferred to musical perception has been rejected', he was drawing on his ability to see beyond the Carnegie Hall. For some of the strongest arguments for rejecting this idea come from looking at the scale systems used in non-Western cultures. That much was clear to some even before the discipline of ethnomusicology began to take shape in the mid-twentieth century. In his seminal text on the physics and physiology of music *On the Sensation of Tone* (1877), Hermann von Helmholtz wrote that 'Just as little as the Gothic painted arch, should our diatonic major scale be regarded as a natural product . . . In selecting the particular degrees of pitch, deviations of national taste become immediately apparent. The number of scales used by different nations is by no means small.'

If we are biologically predisposed to favour intervals whose frequency ratios closely approximate simple fractions, we'd expect to find those intervals recurring in most if not all musical traditions, since it is hard to see why any culture would build its music on a system of notes that were deemed unpleasant to listen to. But we don't find this at all. The best we can say is that the octave division is probably something of a musical universal, and just possibly the fifth too, which is found in Indian and Chinese music and even in the music of relatively isolated tribal people in Oceania.

But there is at least one sophisticated musical system that largely ignores the perfect fifth. This is the gamelan tradition of Indonesia. Javanese gamelan uses two main scale systems, called the *pélog* and *sléndro* scales. The *pélog* scale contains seven notes within an octave, as in the Western diatonic scales, but these are tuned very differently (Figure 3.21). None of the intervals corresponds to a fifth, and indeed none is based on a simple frequency ratio with the first note of the scale. In a performance using the *pélog* scale, only five of the seven notes are used, defining a particular mode of the scale. The *sléndro* scale is even more unusual: it has five notes, each of which is separated from the other by the same pitch step. In other words, the *sléndro* scale simply divides up the octave range into five equal chunks.

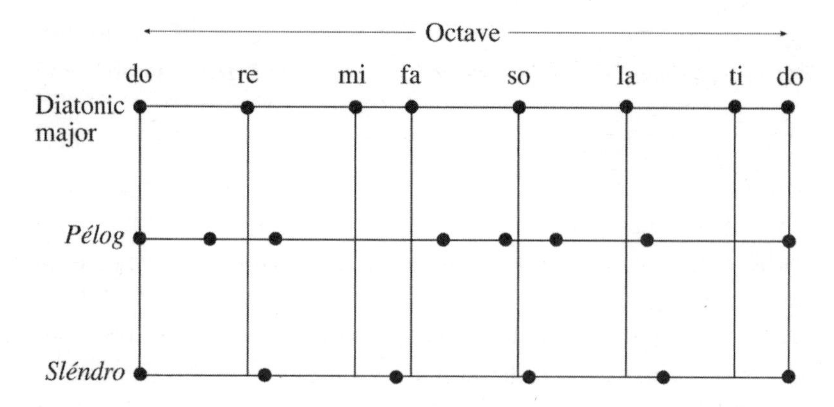

Figure 3.21 The Javanese *pélog* and *sléndro* scales, compared with the Western diatonic major scale.

Even this description doesn't really capture what Javanese gamelan scales are about. They do not really have fixed pitches at all: Figure 3.21 gives only an approximation. There is no standard tuning scheme: each gamelan ensemble tends to employ an individualized tuning in which the pitch steps are customized within fairly narrow limits, so it is said that 'there are as many scales as there are gamelans'. One possible reason for this flexibility of tuning, and for the apparent indifference to the perfect fourth and fifth intervals common in many other cultures, is that gamelan ensembles use many percussive instruments such as xylophones that have inharmonic overtones, so that the pitches produced are inherently ambiguous anyway. The result is that a traditional gamelan piece will sound different for every ensemble. This needn't seem odd from a Western perspective, however, since of course one could say the same of performances of any composition in the classical repertoire by different orchestras. It's just that we're not used to the idea of the very *notes* themselves being different. According to gamelan specialists Judith and Alton Becker, 'The specific size of intervals in gamelan music is of roughly the same structural importance as voice quality in the West: we recognize the same song sung in many voice qualities; a Javanese recognizes the same piece with many different intervallic structures as long as the contours remain the same.' That's the key to a gamelan composition: it is defined in terms of patterns of steps between degrees of the scales (for example, 'up one, down one, down another'), and not in terms of absolute pitch values.

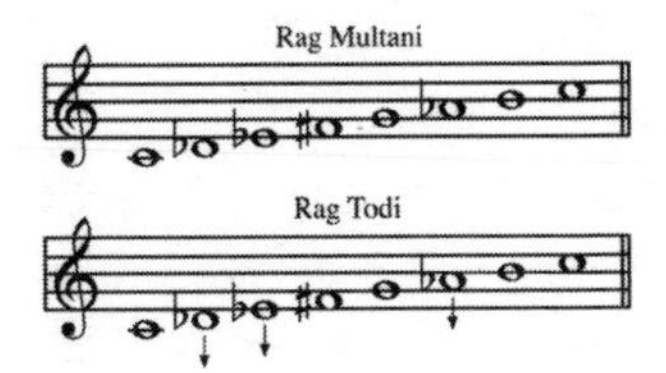

Figure 3.22 Two of the modes of north Indian music. The arrows indicate tunings slightly below those of the Western intonations notated here.

This is a demonstration that even pitched music need not be defined in terms of pitch as such.

Another highly evolved musical tradition, that of the Indian subcontinent,* also uses non-diatonic scales, although they include perfect fifths. The scale system of north Indian music is much richer than that of the West: it employs thirty-two different scales (called *thats*), each of seven notes per octave, drawn from a kind of 'chromatic' palette (the *sa-grama* scale) of twenty-two possible pitches per octave. All of these are based on pitches fairly close to the notes of the diatonic scale, but each of them except the 'tonic' (*Sa*, typically used as a drone) and fifth (*Pa*) can be sharpened or flattened in a given scale, giving them frequency ratios relative to the *Sa* that are quite different from those of Western scales (Figure 3.22). Particular scales are associated with specific musical forms, called ragas. Although analogies like this are always dangerous, it is rather like the way different modes were combined with specific melodic motifs in Greek and Byzantine music, albeit with a much more exquisite range of intonation and a strong dose of virtuosic improvisation.

Name that note

So how do different cultures decide on which notes to use? Are there any universals at all in musical scales? Aside from a reliance on the octave, what they mostly have in common is found not in the frequency relations of notes but in their number and distribution. Most music systems create melodies from between four and twelve distinct notes in the octave span, generally drawn from a larger subset of candidate notes. Music that uses more than the twelve pitch divisions of the

*There is in fact no single tradition of Indian music, but many, just as one can't really talk about a single tradition of European music.

Western chromatic scale is often said to be 'microtonal', but in most such traditions this doesn't mean that a composition includes many more distinct notes: it just means that there is a wider choice of candidate notes for a scale, as with the twenty-two pitches of the Indian *sa-grama*. Arab-Persian music similarly draws on pitch divisions smaller than the semitone, but it's not clear just how many there are in the basic octave range. Estimates vary between about fifteen and twenty-four, but it is hard to decide whether all of these are true 'notes' in the same sense as those of the Western chromatic scale, or just variations of a smaller subset. Here and in other allegedly 'microtonal' systems, smaller-than-semitone or variable pitch steps might be used as embellishments of a relatively small number of basic tones: the microtones aren't well-defined or basic scale notes, but modifications that are more like the bent notes of the blues.

Truly microtonal music, in which many distinct, finely graded pitch classes are employed within a single composition, usually tends to be made by unorthodox musicians emerging from the Western tradition, such as Partch. Another American iconoclast, Charles Ives, called for a microtonal nuance within a diatonic setting in one of his *114 Songs* (1992) to reflect the transcendental meaning of the words: in the phrase 'So near is God to man', the singer was to sing an A on the word 'near' as minimally flat as could be perceived. It sounds as much a prescription for a psychological experiment as it does a musical device.

These eccentricities notwithstanding, why are musical scales usually restricted to between four and twelve notes? It's not hard to guess. With fewer than that, we'd have insufficient building blocks from which to create music of any melodic complexity (although that is not by any means the only way to sustain musical interest). With many more, the brain could not keep track of all the distinctions. In principle we can tell the difference between at least twenty, and perhaps as many as 300, different pitch steps within a single semitone. But these fine gradations are mostly useless for music. This isn't merely a question of our limited capacity to retain so much pitch information; it's a matter of our ability to *organize* all the pitches. As we'll see, to comprehend most types of music we must discern a hierarchy of status between the notes of the scale, which in turn depends on being able to intuit the probabilities with which the different notes occur. With too many notes, we can't (subconsciously) collect and analyse the data needed to make such a ranking.

For that reason, we possess a mental facility for cutting down on the cognitive demands made by a diverse range of pitches and pitch classes. Without it, a tune played on a mistuned instrument would be gibberish, since the 'notes' wouldn't correspond to the ones we know. Yet the tuning has to be truly dire for a performance to lose coherence: we can cope with an awful lot of poor tuning. Partly this is because we can use other cues, such as rhythm, to recognize familiar melodies, but it is also because we learn through exposure to assign all pitches to a small set of categories. People who listen primarily to Western music have mental 'boxes' metaphorically labelled 'major second', 'major third' and so on. Of course, many people don't even know what these terms mean, but they come to recognize the pitch relationships between the various notes of a scale. Our cognitive faculties will 'place' in each box any pitch close enough to that box's 'ideal' interval size, just as we divide up the continuous colour spectrum into chunks labelled 'blue', 'red' and so on. It's not a case of being oblivious to the fine differences in pitch. We *hear* a slight mistuning of a major third, say – but we classify it as such and not as an entirely new and unknown pitch class. We know what it is 'supposed' to sound like.

This has been demonstrated by playing to people harmonic intervals that are increased in small microtonal steps – say, gradually changing a minor to a major third. Listeners experience the change as sudden, not gradual: what is heard at one point as a marginally sharp minor third becomes, with a further small sharpening, a somewhat flat major third. This seems to be an intrinsic aspect of the way we process sound; the same rather abrupt transition is found, for example, when two distinct syllables, such as 'da' and 'ba', are manipulated electronically so as to mutate one incrementally into the other. There's a visual analogy with optical illusions such as the Necker cube, which flick back and forth between one interpretation or the other but won't rest in between (Figure 3.23). Our brains won't tolerate ambiguity.

Incidentally, musicians perform a little differently in these tests of pitch classification: they more readily identify the deviations of intervals from ideal values, because musical training makes people more

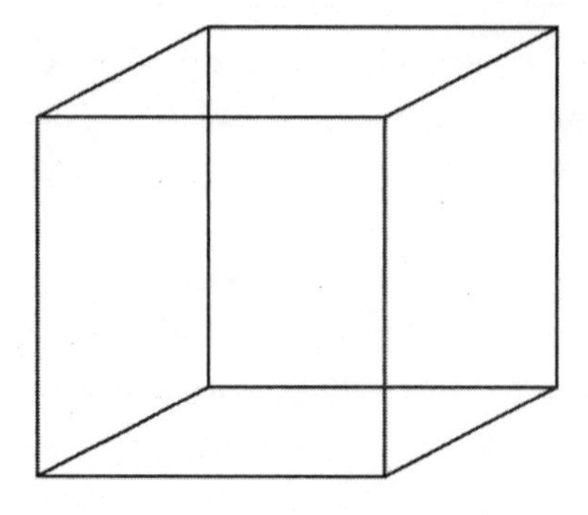

Figure 3.23 The Necker cube can be interpreted visually in either of two ways. Our minds may flip back and forth between these interpretations, but always unambiguously select one or the other.

perceptive of mistuning. But at the same time it seems to make the 'boxing' even stronger: the musician is more likely to recognize when an interval is out of tune, but also more likely to assign steps near the crossover point unambiguously to one box or another.

All this challenges one of the unspoken assumptions in arguments over intonation: that we can actually hear the differences. While some musical purists screw up their noses at the barbarities of equal temperament, many casual listeners would not even notice any difference from other systems. As ethnomusicologist Bruno Nettl put it, what we hear in music 'is conditioned not only by what sound is actually produced, but also by what sound one's ear is attuned to and expects'. This is true not only of pitch but of rhythm and other musical structures too.

For most of us, these 'pitch boxes' are defined only in relative terms: we learn to identify only the frequency relationships between notes, not the frequencies themselves. Even quite young children soon learn that an interval of a major third is one of the characteristic pitch steps of Western music: it's a familiar sound, and different from a major second or perfect fifth. But for some people, a C followed by an E is not simply a major third, but is precisely that: a C and an E. They can name the notes they hear, if they have learnt enough music theory to do so. Such people are said to have absolute pitch, and they constitute around one in 10,000 of the Western population. The ability is still not fully understood, and it has acquired a powerful mystique. Many people believe that an individual with absolute pitch is blessed with near-magical musical ability.

This is simply not the case. Absolute pitch doesn't seem to correlate with enhanced abilities for any other music-processing task – it doesn't, in other words, make you any more likely to be a gifted musician than the rest of the population. It is, in short, *not a musical endowment*. It is much more useful to be able to judge relative pitch – the pitch-step size between two notes – than absolute pitch.*

Confusingly, musicians *are* considerably more likely on average to have absolute pitch than are non-musicians. But there is no contradiction in this. Absolute pitch seems to be acquired at least partly by learning in childhood, and so it can result from greater exposure to music. In other words, it is in many cases an effect and not a cause of being musical. The role of learning is supported by the fact that people with absolute pitch are faster at identifying more common pitches (such as C and G) than less common ones (such as G♯). It's also possible that people found to possess absolute pitch are encouraged to become musicians, on the false belief that they have a special musical gift.

Absolute pitch is, however, literally mind-expanding: musicians who possess it have an enlargement in a brain area associated with speech processing. This suggests that it may be linked to an enhanced ability to detect verbal cues in the pitch variations of speech. That idea is supported by the fact that absolute pitch seems to be far more prevalent in cultures that have tonal languages. One study found that about fifty per cent of the first-year students at the Central Conservatory of Music in Beijing had absolute pitch, compared with ten per cent of their counterparts in the Eastman School of Music in Rochester, New York. These differences might not be all due to early learning – there seems to be a genetic basis for a predisposition to acquire absolute pitch, and it may be that this genetic component differs between East Asian and American populations. Even so, absolute pitch needs to be cultivated in youth: if you don't have it by the time you're an adult, you're very unlikely to be able to develop it by practice. There's no strong reason to feel despondent at that, unless you fancy using it for a party trick: absolute pitch won't make

* Musicologist David Huron suggests that, if there was once some genetic advantage conferred by absolute pitch, the advent of music-making may have *weakened* that because of the way it trades much more on relative pitch, so that absolute pitch is now a very rare ability.

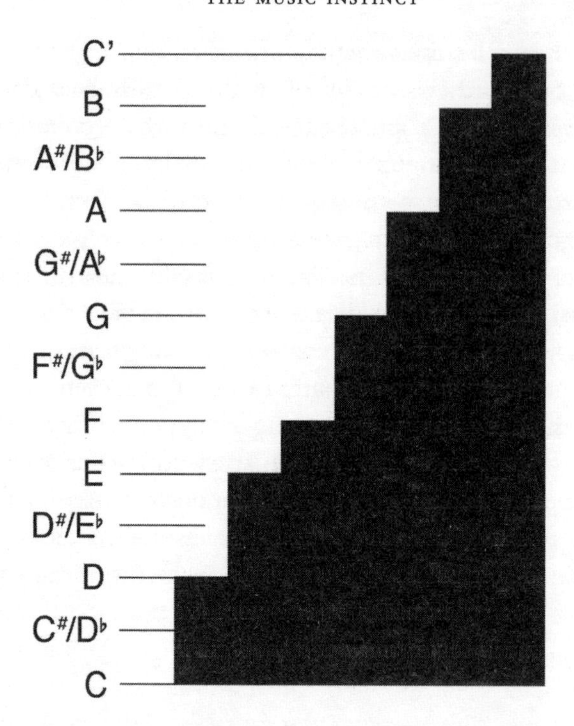

Figure 3.24 Pitch staircase of the major scale.

you better at playing or appreciating music, and can even be a hindrance, for those who possess it can suffer agonies when a familiar piece of music is transposed or a piece is played far from the standard tuning of concert A.*

Big steps and little steps

As well as having only a fairly small number of notes per octave, most scale systems throughout the world share another common feature: almost invariably they have unequal pitch steps between successive notes. (The Javanese *sléndro* scale is a rare exception.) For the diatonic

* All the same, absolute pitch can be useful to musicians. For example, when playing brass instruments, for which intonation is a matter of feedback between muscle and ear rather than just placing your fingers in the right spot, players with absolute pitch find it easier to negotiate the unfamiliar interval leaps common in atonal or chromatic music – they know from the outset what note they are aiming for. The same advantage is enjoyed by singers too.

major scale, the pitch steps have the sequence (tonic) t-t-s-t-t-t-s (octave) (Figure 3.24). The various modes of medieval music, as well as their presumed predecessors in ancient Greek music, have other sequences, but all contain an analogous irregularity, making the ascent to the octave occur on an erratic staircase.

Why not spread out the pitch steps as evenly as possible, to avoid any gaps or bunching in pitch space? The most likely explanation seems to be that the irregular steps provide a listener with reference points against which to judge where the 'tonal centre' of a melody lies – as Westerners would say, which key the piece is in. In short, the uneven steps provide a means of distinguishing one staircase from another. Imagine a scale with perfectly equal steps (Figure 3.25*a*). Transposing this scale to a new key would mean starting the staircase on a different step. But the new scale would ascend in just the same way – it would merely start and finish at different heights. Now, a melody is like a stroll up and down these staircases, progressing from one step to another. If you hear a

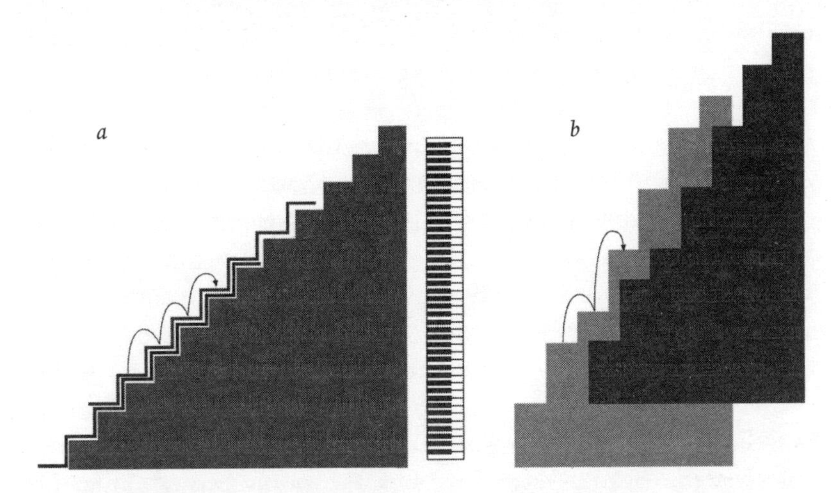

Figure 3.25 Scales in which every pitch step is the same can be superimposed on one another upon transposition: they are all 'the same pitch staircase', but starting on different notes. This means that a melody on one staircase fits on to any other overlapping staircase too (*a*). This lack of bearing about 'where the music sits' is comparable to what a piano player would experience if trying to play a keyboard in which there is a black note between *every* two white notes: the keyboard is homogeneous. In contrast, the unequal pitch steps of the diatonic scales mean that scales of different keys cannot be exactly superimposed (*b*). In this case, a melody that fits one staircase will not fit (most) others. That enables us quickly to deduce (or at least make a good guess at) what key a melody is in.

melodic fragment – a rising series of four steps, say – there is no way of figuring out on which of the possible staircases it belongs. You can't, in other words, tell where the staircase 'starts' (what the tonic is) from such an excerpt. This is precisely the case for music based on a chromatic scale, where the steps *are* all equal. So chromaticism is a way of creating ambiguity about the key. We'll come back to why this can be a musically effective thing to do.

In contrast, staircases of unequal step sizes can't be exactly superimposed one on the other by this sort of transposition: where in the original key there was a small step, one might find a large one in the new key (Figure 3.25*b*). This means that it becomes possible quite quickly to intuit the tonic root of a melody: we can get our 'key bearing' just from a snatch of tune.

If, for example, you hear the three-note fragment of melody E, F, G (the first and third 'Falling down' in 'London Bridge is Falling Down') it's likely that already you'll have subconsciously guessed that the piece is in C major.* Why? Because you will have heard the pitch steps semitone-tone. Where do we find that in the diatonic scale? It implies that the fragment must correspond either to the scale degrees **3, 4, 5** or **7, 1', 2'**. The former places the tonic, **1**, two whole tones below the first note of the sequence, that is, on C. The latter would make the tonic F, but we're more likely to choose C because the fragment begins and ends on notes that are more important and thus more 'stable' in the key of C – the third and fifth – than in F.

Of course, our guess may be quite wrong. The fragment could, for example, be in a *minor* key, in which case it would seem to be D minor or, for a certain definition of the minor scale, A minor. Or it could be in other keys with modified (sharpened or flattened) versions of the scale notes – there's no fundamental reason why a composer couldn't find a way to make this fragment 'work' in, say, the keys of G major or F♯ minor. But those are much less likely. (I will specify later exactly what I mean by less likely.) So on the basis of just three notes we have a good chance of finding the right tonal centre. In fact, we can and often do even better than this, although for other reasons that we'll see later.

Infants just six to nine months old appear to attend more closely to

*I don't mean that you'll say 'Oh yes, C major' (unless you have absolute pitch). I mean that you'll hear the melodic fragment as being in some sense rooted in the tonic note two whole-tone steps below the first note of the fragment.

scales with uneven than with even steps, suggesting that they have already learnt by then which sort of scale to prefer. (You might wonder how we can know anything about the 'musical' preferences of babies, but there are well-established techniques for measuring such things, for example by looking at how infants turn towards or away from sources of different sounds and sound patterns.) It has been claimed that this greater sensitivity of babies to scales with uneven rather than equal pitch steps persists even when the scales are newly minted inventions rather than conventional diatonic scales. If this is true, it is deeply puzzling: it would be altogether remarkable if the human brain were hard-wired to look for tonal centres in a series of pitches, and thus to be receptive to any mechanism that assists this. Indeed, I can think of no reason why this should be so; even if they were true, none of the speculations about an adaptive function of music seems to call for such a thing. It is possible, perhaps, that infants are simply able to discern similarities to the kinds of scales that, by the age of six months, they might have heard many times before. But for now this observation is merely baffling.

You might reasonably ask why it matters in the first place whether we have a tonal bearing – a sense of key, say – as we listen to music. To the glib answer that no one likes to feel lost, you could say that surely plenty of people enjoy being 'lost in music'. Why can't we just relax and let the music do its thing? The real answer is central to a great deal of what follows later in this book. For the fact is that music simply cannot do its thing if we are genuinely lost. Experiencing music is an active affair, no matter how idly we are listening. *If it wasn't, we would not be hearing music at all.* And a part of this activity, perhaps even the major part, requires a level of understanding about what the music is doing, what its rules are. If your initial response to this is 'But I know nothing about the rules of music!', all I will say for now is: with respect, you are wrong. You know a lot.

Scales with five or seven notes seem especially widespread: the former are called pentatonic, exemplified by the Chinese scale in which the pitch relations correspond to those of the piano's five black notes.*

* The common practice in the West of calling this *the* Chinese scale is, however, quite wrong. Chinese music is generally pentatonic in essence, yet not only are supplementary tones often added, but also there are several modes – somewhat akin to Greek modes – in each of which the pentatonic scale starts on a different note. Furthermore, there are many regional variants of these scales with quite different structures, some of which are defined over more than a single octave span.

This may be because they permit a particularly simple way of interconverting scales with different tonic notes (that is, modulating the key): it turns out that only scales with five or seven notes per octave can be interconverted by changing a single note. Recall that this is how the cycle of fifths works in the diatonic system: each scale in the cycle can be reached from the previous one by altering a single note by a semitone. C goes to G by sharpening the F, or to F by flattening the B. This makes it relatively easy to perceive a relationship between the original and modulated scale.

Music theorist Gerald Balzano has shown that a scale of seven notes chosen from twelve, like the diatonic scales, has several other significant properties that different choices of dividing the octave don't generally share. For example, transpositions never fully overlap: you can make a distinct scale, while preserving all the interval step sizes, starting on each of the twelve notes. And if we consider all possible pairings of notes, then we can obtain all possible intervals (minor seconds, major seconds, minor thirds and so on) up to the octave, and the number of intervals of one type is never the same as the number of another type. There is, in this sense, *maximum variety* of interval types. Balzano points out that this has nothing to do with the actual *intonation* of the notes: it is a mathematical property of any seven-from-twelve scale. This leads him to suspect that it may be these characteristics, and not the tunings of the notes themselves, that are the 'perceptually important' features which encouraged the adoption of diatonic scales in the West.

Scaling up

Western tonal music has not universally used the diatonic scales. As we have already seen, in earlier times it was modal. Although diatonic scales began to replace modal scales in most classical music after the Renaissance, the modes continued to be used in much folk music, and are still widespread today, often being preferred in rock and pop music. Whereas there are plenty of pop songs that use the major scale (which can also be considered to be the Ionian mode), such as the Beatles' 'I Wanna Hold Your Hand' or the Who's 'The Kids Are Alright', others

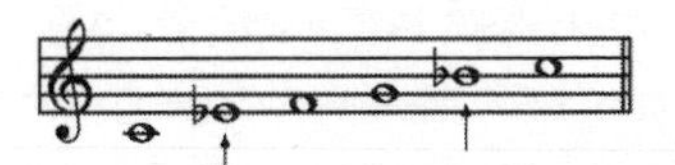

Figure 3.26 The pentatonic blues scale. The arrows show the 'blue' notes, whose pitch is ambiguous.

commonly use modes such as the Mixolydian ('Rebel Rebel' by David Bowie, 'The Last Time' by the Rolling Stones), the Dorian (Steppenwolf's 'Born To Be Wild', Pink Floyd's 'Another Brick in the Wall'), and the Aeolian (Blue Oyster Cult's 'Don't Fear the Reaper', Nirvana's 'Smells Like Teen Spirit'). The ubiquitous minor third of rock music (T. Rex's 'Twentieth Century Boy' and Led Zeppelin's 'When the Levee Breaks' are classic examples) has its bastard origins here, not in the diatonic minor scales.

The fundamental pitch soundscape for most rock and jazz is often called the blues scale, which is basically also a modal form but without any strict definition. The simplest version is pentatonic (Figure 3.26), but any attempt to pin the blues scale to a set of fixed pitch classes arguably omits its most important and expressive feature: the pitch ambiguity of the so-called blue notes, the flattened third and seventh. The third in particular is manipulated with gusto, being commonly raised to a pitch that hovers midway between a major and minor third. This is easy enough to do vocally – it is audible in performances by just about any great jazz and blues singers, particularly Billie Holiday – and is also possible on brass instruments such as the trumpet and saxophone where tuning can be adjusted by breath control. The clearest exploitation of this 'blue third', however, is probably to be heard in blues guitar music, particular the electrified urban blues that came out of Chicago, in which pitches can be readily bent by sliding the string across the fret. Even on the fixed-pitch piano, blue thirds are often given ambiguity by 'crushing' together the major and minor third – somewhat like the ephemeral grace notes used in Western classical music for centuries, but with less grace and more primal force, as though demanding the impossible from the hapless instrument.

The blue notes may have originated from attempts of black slaves to match the pentatonic African scales to the Western diatonic tradition they encountered in the New World. There is no major third or seventh in the former scheme, and so the Africans, singing the spirituals from which blues and jazz were born, may have 'fished' for something approximating these scale degrees, creating notes that were inherently variable and unstable. On the other hand, a 'neutral third' somewhere between the minor and the major is not uncommon in various non-Western cultures, such as Thailand. And the folk music of other Western traditions often mixes major and minor intervals quite freely, especially those around the 'blue' notes of jazz. According to the Australian composer Percy Grainger in the early twentieth century, 'the folk-scales in which their so-called "modal" melodies move are not finally fixed as are our art-scales, but abound with quickly alternating major and minor thirds, sharp and flat sevenths, and (more rarely) major and minor sixths.' There is absolutely nothing 'wrong' about such ambiguities, contrary to the way some early musicologists dismissed the blue notes as clumsy, degraded forms of diatonics.* Jazz historian Andre Hodeir suggests that, as jazz matured, its musicians became quite deliberate in their manipulation of the blue thirds and sevenths, 'depending on how open or how disguised an allusion to the major scale is desired'.

The second key feature of the blues scale is somewhat related: the addition of an augmented fourth interval, such as an F♯ in the key of C. In the Western tonal tradition this has long been considered one of the most uncomfortable dissonances, and was widely shunned (although its real 'meaning' in music theory is often misunderstood, as we'll see later). But in the blues scale the augmented fourth is not really meant to be heard as such: it is always heading somewhere else, either up to the fifth or down to the fourth. It functions more like a kind of 'bent' fifth, introducing a delicious wonkiness that can be heard in Duke Ellington's incomparable title track for *Anatomy of a Murder*. In Charles Mingus' 'Goodbye Pork Pie Hat' it is the augmented fourth that carries all the languid melancholy. Such deftness is rarer in rock music: the most famous use of the augmented fourth here, in Deep Purple's 'Smoke on the Water', sounds clumsy and forced in comparison.

Another non-classical scale abstracted from the Western chromatic is the Fregish scale used in Jewish liturgical and klezmer music, which is given its exotic charge by the inclusion of a minor second and by

* The same kind of academic condescension led some music theorists to attribute to 'lack of training' the distortions away from diatonic tunings used by many folk performers, such as those in the gypsy music of Eastern Europe. We can now see that in many cases, at least, these performers control intonation with immense skill and consistency, and that they do so for very good reasons which I explore in Chapter 10.

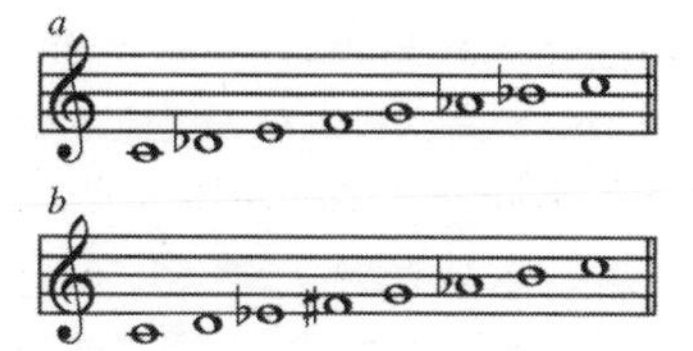

Figure 3.27 Fregish (*a*) and gypsy (*b*) scales.

the unusually large (three semitones) step between the second and third degrees of the scale (Figure 3.27*a*). Such a step also appears in the so-called gypsy scale (Figure 3.27*b*). Béla Bartók's compositions based on Hungarian and Romanian folk tunes made much use of non-diatonic scales like this one, as well as a 'major key' version that included a sharpened fourth. This unusual interval is also found in some traditional Swiss songs, where it may have an origin in the higher harmonics of the Swiss alphorn.

Several twentieth-century composers invented or adopted non-standard scales that created a highly personalized sound. Much of the floating, ethereal quality of Debussy's music can be attributed to his use of a so-called whole-tone scale, with six notes per octave and spaces of a whole tone separating them all (Figure 3.28*a*). Messiaen's aforementioned dabbling with high harmonics led him to experiment with eight-note (octatonic) scales (Figure 3.28*b*). And the Russian composer Alexander Scriabin made music that was *sui generis* using 'mystic' scales based on a largely inscrutable philosophy (Figure 3.28*c*).

A quick look at the landscape

I began this chapter by comparing listening to music to a journey through a landscape, and suggesting that the experience depends on what we can see and how we make sense of it. So what does this

Figure 3.28 Some of the alternative scales used by modern composers: (*a*) the whole-tone scale favoured by Debussy; (*b*) an octatonic scale of Olivier Messiaen; (*c*) Alexander Scriabin's 'mystic' scale.

landscape look like? It is too complex a question to answer at once, and in any case there is no unique answer. But we have now the tools to sketch out a preliminary map of the territory. I have been concerned here with *pitch space*: with the relationships between musical notes. To the acoustic physicist, this space might look like a smooth, bland uphill slope: pitch simply increases smoothly with increasing acoustic frequency. But our auditory system does something strange to this slope, for it seems to return us repeatedly to where we started from, or some place very much like it: every time the pitch rises an octave, we come back to a sound with the same 'flavour'. One way of depicting this is to twist the upward-sloping line into a spiral, in which pitches that sit vertically above one another correspond to pitches an octave apart. In this representation, pitch can be considered to have two dimensions: 'height', which is an objective quantity determined by the frequency of vibration, and what music theorists call *chroma*, meaning the pitch class: a circular property that arises purely in perception (Figure 3.29).

This, however, only begins to capture the subtle character of pitch. As we have seen, there are good fundamental reasons to posit a special relationship between one note and another a perfect fifth above it, whether this be through the Pythagorean cycle of fifths or the harmonic series. In terms of perception, it seems quite reasonable to suggest that C is somehow 'closer' to G than it is to C♯, even if the opposite is true purely in terms of frequency (or equivalently, proximity on a keyboard). So it doesn't seem terribly satisfactory that C and G sit on almost directly opposed sides of the chroma circle. Can we incorporate this cycle of fifths into a map of pitch space?

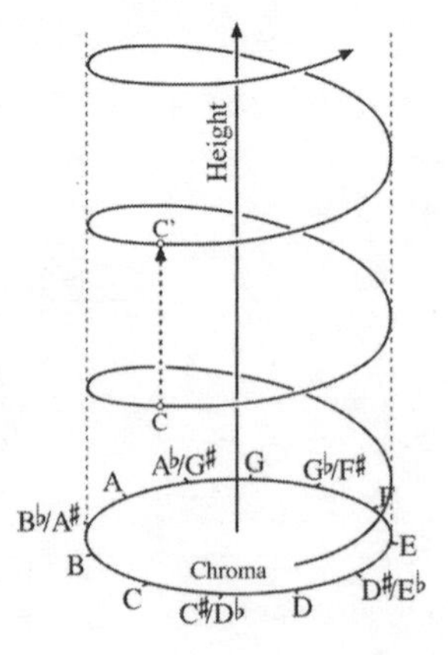

Figure 3.29 Each pitch can be assigned two qualities: register or 'height' (which octave it is in) and chroma (which pitch class it is – C, D, E and so on). These can be represented on a spiral.

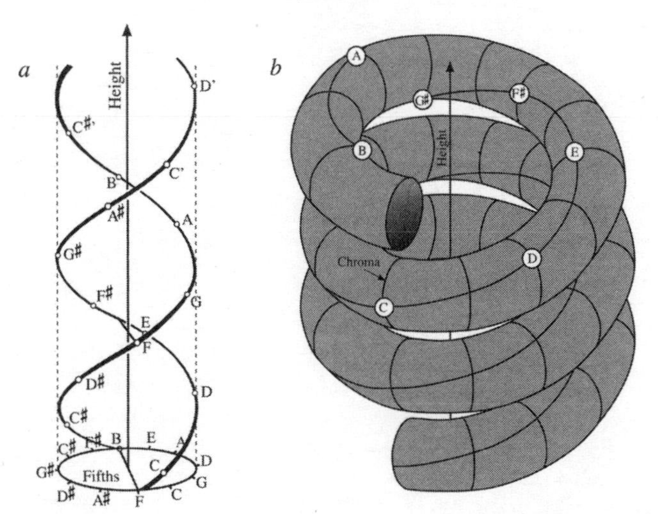

Figure 3.30 Other modes of pitch representation: double helix (*a*) and coiled tube (*b*).

This requires that we add another dimension, which makes our map something of a complex object. One proposal is to add a second helix to the spiral – a kind of musical DNA, if you like (Figure 3.30*a*). Or the cycle of fifths can be added 'locally' to each note on the spiral, turning the line of notes into a kind of spiralling tube, rather like the outer insulation of an old-fashioned telephone cable (Figure 3.30*b*). These are just two of the many possibilities, for there is no single best way of doing the cartography. For one thing, it isn't obvious whether the maps should be based on formal musical relationships, such as the cycle of fifths, or on perceptual qualities among pitches, reflecting the subjective degrees of similarity or association obtained from listening tests.

One of the earliest attempts at mapping was made in 1739 by the Swiss mathematician Leonhard Euler, who was concerned with finding a way to depict the pitch relationships of just intonation. This representation (Figure 3.31*a*) does away with the notion of pitch 'height' – rising frequency – altogether, and collapses all pitches into one octave. It is a flat map with two coordinate directions, here up-down and left-right. The former progresses in steps of a major third, the latter in fifths. So as we move along any row from left to right, we step through the cycle of fifths. But we saw earlier that for just intonation this cycle is not closed – we never quite come back to where we started. And there

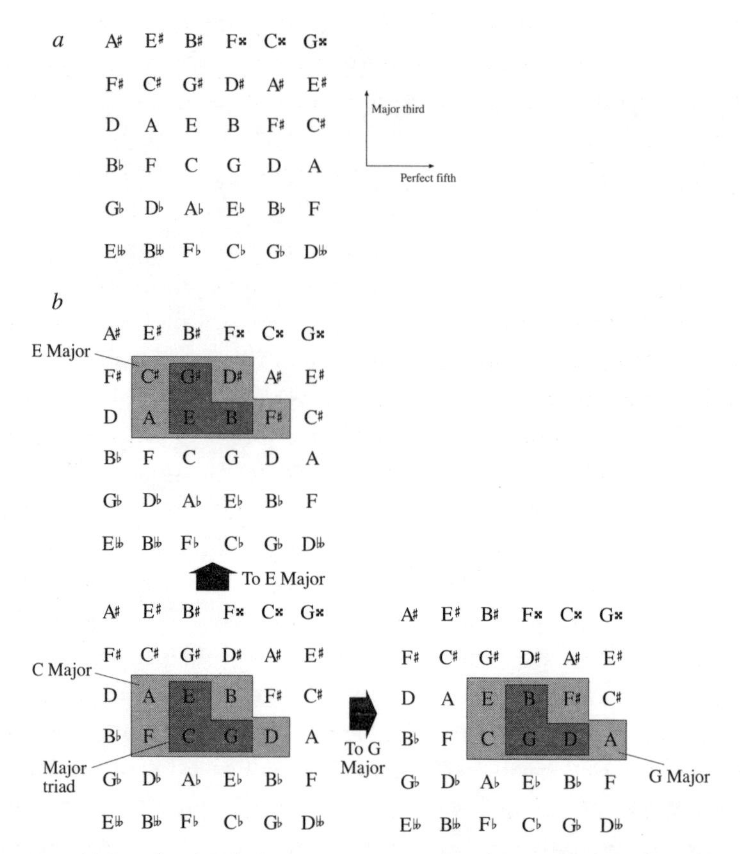

Figure 3.31 (*a*) Leonhard Euler's representation of pitch space, in which pitch changes in steps of a perfect fifth from left to right, and in steps of a major third from bottom to top. The 𝄪 sign is a double sharp. (*b*) Christopher Longuet-Higgins pointed out that the major triads and scales form L-shaped clusters in this space.

is another such near-cyclicity for the series of up-down steps of a major third. In just intonation, the major-third interval involves an increase in frequency by a factor of 5/4. So three such steps give an increase of $(5/4)^3 = 125/64$, which is very close to 2, or an octave. On today's equal-tempered piano, this near miss is corrected: three major-third steps from C take us to E, then to G♯/A♭, and finally to C'.

In other words, when applied to just intonation this map is a fragment of an infinite plane. That's why it here contains new symbols such as E♭♭ and F𝄪: double flats and double sharps. On a piano, E♭♭ is two semitones down from E, which brings us to D. But in just intonation, E♭♭ is not quite the same as D, differing by the factor of the syntonic comma, 1.0125 (see p. 59). Further out, there are triple and quadruple flats and sharps and so forth, each of them a distinct note. Equal temperament closes this infinite universe so that each edge eventually curls back to meet the opposite one.

Euler's map was used in the nineteenth century by Hermann von Helmholtz, and its properties were explored in the 1960s and 70s by the British mathematician Christopher Longuet-Higgins. He pointed out that it embodies the note relationships of both major scales and major triads. Every triad occurs in the map as a little L-shaped cluster (Figure 3.31*b*), while the map also groups all the notes of the scale together in a window with a kind of fattened L shape. A modulation of key then corresponds to a movement of this window: sliding it one space to the right, for example, moves the key by a perfect fifth (C to G, say), while sliding it back one space modulates by a perfect fourth (C to F). Sliding it up one space, meanwhile, modulates by a major third (C to E). Longuet-Higgins pointed out that within each 'scale box' the tonic note has the shortest average distance from all the others, which he suggested as a possible reason for its perceptual centrality.

You can see that each note appears more than once in this map. This reflects the way that each note can have more than one musical function. The A north-west of C falls into the cluster for the key of F, and so when it sounds in a C major tune, it might be accompanied by an F chord, as on the syllables of 'little' in 'Twinkle Twinkle Little Star'. But the A three steps to the right of C has a different provenance: it is reached by modulating the key to G, or to D (where A is part of the major triad). This distinction can be seen in the tune 'Do Re Mi', where the two different As are directly juxtaposed (Figure 3.32). These As are in a real sense different notes, even though on a piano they correspond to the same key – and in 'Do Re Mi', I'd maintain, you can genuinely feel that difference. They are, said Longuet-Higgins, musical homonyms, like words with the same spelling but different meanings (such as bear, or bank). Note that in just intonation these homonyms literally are different notes, because of the two different whole-tone step sizes of that system (see p. 59). The D north-west-west of C, for example, is attained via the 'small' major-second interval ratio of 10/9, while that two steps eastward is attained by the 'long' major second, 9/8.

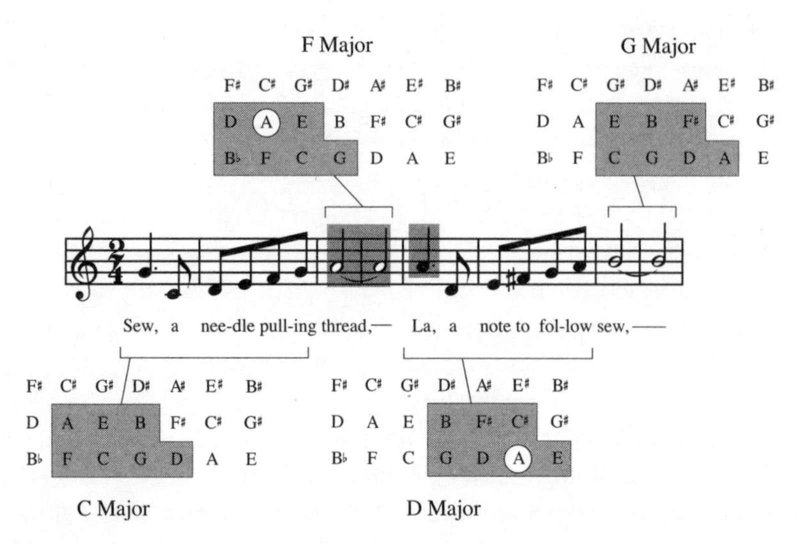

Figure 3.32 The same pitch can serve different functions in a melody. Here the two highlighted A's in 'Do Re Mi' have different locations in Euler's pitch space.

Here we are already starting to get ahead of ourselves by venturing into the territory of harmony. Let's just say that all these representations of pitch space offer a glimpse of the reason why music indeed takes us on a journey in which, hopping from note to note, we are aware of a host of other notes more or less closely related. It is this sensation of being somehow located in a musical environment, pressed about by notes and chords that we sense but don't hear, and uncertain of which we will encounter next – it is this that allows music, moment to moment, to excite our senses and our passions. In what follows, we will see how, even though most of us have never seen maps like these, this is possible.

4
Andante
What's In a Tune?

Do melodies follow rules, and if so, which?

In the BBC series *Face the Music*, the panellists were occasionally set the task of identifying a piece of music from a very short extract. Comedienne and singer Joyce Grenfell once succeeded given just a single note played on the piano. This is absurd, of course, and it got a laugh when actress Maureen Lipman reprised the feat in her Grenfell tribute performances. But it is funny for a complicated reason that the amused audience would probably find hard to explain. For the lone note is the sonorous D♭ that opens Debussy's 'The Girl With the Flaxen Hair'. If you know the melody, then you find yourself thinking not 'Oh come on, it could have been *any* tune', but rather, 'Ah yes, of course, she's right' – all the while knowing that it makes no sense to think so. Surely no melody is really betrayed by its first note! And yet the Grenfell sketch is somehow funny in a way that it wouldn't be if the D♭ had turned out to be the first note of 'Baa Baa Black Sheep'.

The real delight of this episode is that it reveals to us a guilty secret we never knew we had. No sooner have we heard a single note of a tune than our minds set to work, making deductions, assumptions, predictions. Where is the music going to go? Do we recognize it? *What is coming next?* We'll see in this chapter why even a *single note* is indeed enough to give us a fair idea of that – and why such anticipations are central to our ability to turn a string of notes into a melody, which is to say, into music.

What is a melody? It is not too gross a simplification to say that the melody of a piece of music is, broadly speaking, its tune – so long as we recognize that, on the one hand, there need be no connotation that tunes are trivial things without intellectual gravitas, and, on the other hand, not all music has or needs to have a 'tune' in the sense that 'Singin' in the Rain' does. While some tunes are harder to follow or less catchy than others, plenty of music has no single, extended tune that provides the basic melodic material. Take Bach's fugues: the melody here is not woven into a single skein that threads through the piece, but occurs in short, overlapping fragments, often unremarkable in themselves. And it's not clear that there is any tune as such in most of Gustav Holst's 'Neptune' from *The Planets* (while no one would say that of his 'Jupiter'), or in just about any piece by Steve Reich. That, and not musical snobbery, is why 'melody' is a better, more versatile term than 'tune'.

Yet some people who listen with bliss to Bach and Holst complain that much popular music, such as hip hop and rave, 'hasn't got any tunes'. Maybe they level the same accusation at 'modern' classical music by the likes of Stockhausen and Ligeti. Clearly, a lack of tune isn't the true cause of complaint (we'll see later what it might actually be). On the other hand, an elitist aversion to the notion of tunes probably stems from an overreaction to the same misconception that the 'tune' is where all the musical value of a composition resides. There's nothing wrong in liking nice tunes, but if that's your *sine qua non* of musical appreciation, large tracts of music will be closed to you – and your critical instincts may be dulled. The melody to the 'Ode to Joy' in Beethoven's Ninth Symphony would not be out of place in a nursery rhyme, but this doesn't make one equivalent to the other.

A melody, then, is essentially a connected series of notes of various pitch, duration and rhythm. Some melodies might sound simple to the point of monotony. Many Native American songs are extremely short: just a single musical phrase, perhaps encompassing no more than a semitone of pitch variation, or even none at all. At the other extreme, it may seem a stretch to regard the rapid-fire maelstrom of notes improvised by Charlie Parker or Ornette Coleman as 'melodies'. But they can be understood using the same set of conceptual tools that we bring to bear on 'Hickory Dickory Dock'.

Why do we find more to enjoy in some sequences of notes than others? Some people would pay a lot of money for the answer to that, since it would seem to offer a prescription for writing hits to order. Thankfully no one has found the formula, and it is doubtless naïve to believe that one exists, just as it would be foolish to think that a great tune guarantees a hit record, or that all popular songs have great tunes. One person's catchy hookline drives another wild with irritation.

Nevertheless, it seems clear that there *are* factors that many melodies share in common. Most songwriters and tunesmiths assimilate these 'rules' without knowing it, and could probably not tell you what they are if you asked them. Some principles of melodic composition are nonetheless more or less formally codified in the practices of particular musical traditions or genres – in the laws and rules of thumb for making 'good' music. While some composers, such as the medieval German abbess Hildegard of Bingen, have regarded their melodies as the products of inspiration channelled from mystical and divine sources, they are almost inevitably observing all sorts of unspoken injunctions and formulas. In contrast to that kind of numinous view, Paul Hindemith felt that the composer should craft sound structures by careful planning and adherence to rules, almost like an engineer. 'Melodies can be constructed rationally,' he said. 'We do not need to believe in benign fairies, bestowing angelic tunes upon their favourites.' One has to admit that Hildegard's working method, ostensibly guided by inspiration rather than technique, sounds a whole lot more attractive than Hindemith's; but the fact is that melody and music, however ecstatic their genesis, draw on principles rooted in our cognitive processes. In this chapter and the next, we'll begin to explore some of them.

Suppose we are that aspiring songsmith on Tin Pan Alley, looking for the 'perfect tune'. How do we begin the search? I suspect most composers like to imagine they find their melodies from a blend of intuition and inspiration, whereas in reality the tunes that emerge are amalgams and mutations of others they have heard. But let's say that in our crafting of melody we are determined to be genuinely original, and intend to work from the most basic of musical principles. Now, we know very well that you don't create a melody by just selecting notes at random.* As we saw in the previous chapter, most melodies in any musical tradition use pitches drawn from a scale, or a series of scales.

* Some experimental music *does* select pitches at random. We'll consider later whether the result can be considered melodic.

Figure 4.1 Two examples of random 'melodies' in the key of C.

So perhaps we need only order the notes of a scale in a sequence that no one else has tried, and there we have it: an original tune. Of course, there's the question of rhythm too, but never mind that: we'll worry about it in Chapter 6. For now, let's focus just on the notes. It rapidly dawns on us that there is an immense number of ways that we can arrange the notes of a scale. But never fear: we can enlist a computer to figure out all (or at least, a lot of) the permutations. And so we program our laptop to conduct this random shuffling of scales.

The results are terrible. Just look at them (Figure 4.1)! OK, so this is actually an absurd premise: even the most untutored songwriter surely doesn't imagine that tunes can be churned out by a random-number generator. But the dumb exercise illustrates two things. First, there seems to be only a subset of the possible permutations of notes that qualify as real melodies. That subset has no absolute boundaries – there's a continuum from great tunes through to mundane ones, odd ones and finally to ones that are too incoherent to seem like tunes at all. But what is changing in that transition?

Second, many if not most of these random sequences don't seem to have any centre to them. In musical terms, we get no real sense of there being a tonic, or equivalently, of the tune being in a particular key. This is odd, because the notes here are drawn strictly from the scale of C major. So why don't they sound like that? Are we sure that, by rigidly adhering to the diatonic scale, we have captured the right ingredients for a good melody? Are we really using the correct notes?

Maybe that's the wrong question.

A sense of right and wrong

There are not many laughs in music. True, there is joy and glee and gusto to be found in almost any genre. Performers from Tom Lehrer to Tom Waits and even Tom Jones know how to raise a chuckle. Some composers, such as Flanders and Swann, specialize in comic material, and Noel Coward's songs are shot through with wry humour. American composer Peter Schickele, aka. P. D. Q. Bach, has made a career out of musical comedy. And didn't I just begin this very chapter with a kind of musical joke? But there's a difference between funny lyrics, or cracking jokes mid-song, or inserting kazoos into well-known classical pieces, and humour that stems from the music itself.

A rare example of the latter was written by the famously mischievous Mozart in 1787 (curiously, the month after his father died). As he called it *Ein Musikalischer Spass* ('A Musical Joke'), his audience was well primed to realize that its eccentricities were supposed to make them laugh. You need to be pretty familiar with compositional practice in the Classical era to get some of the gags, for example in the way Mozart breaks well-established rules, although the general tone of buffoonery is clear enough. But one of the most obvious jokes comes right at the end, where the six different instruments come to a grinding halt in five different keys (Figure 4.2). The notes clang in a horrible din.

The joke is not the discordance itself – we have now grown accustomed to far more outrageous sounds from modernist composers. It becomes funny (though don't expect to split your sides) because it happens within an ostensibly tonal context. The notes are absurdly wrong because the music itself, and all our expectations about what

Figure 4.2 The final riotous cadence of Mozart's *A Musical Joke*, K522.

Figure 4.3 None of these 'white-note' melodies is in C major – all are well-known, and are in G major (*a*), F major (*b*) and D minor (*c*). I have purposely omitted key signatures.

a Classical composition should sound like, establishes a strong notion of what is 'right'.

Yet it is only rather recently that we have developed a good understanding of what underpins this sense of the rightness and wrongness of notes. At face value, it seems an obvious matter to even the greenest of music students. Tonal music – which meant just about all of Western music from the early Renaissance to the late nineteenth century – is music that has a key, and therefore an associated scale and tonic. This then tells you which are the 'right' notes: in the key of C major, say, they are all the white notes.

But a moment's consideration shows that this means nothing in itself. There are *no* intrinsically 'wrong' notes in the key of C (even if we set aside the possibility of modulating, within a C major composition, to other keys). In Bach's C major Fugue from *The Well-Tempered Clavier* Book I, no single pitch class is excluded: all the notes in the chromatic scale are used. Conversely, all of the familiar melodies in Figure 4.3 use only the white notes, but none of them is in the key of C major, and none sounds as though it is. The notes here are also drawn, respectively, from the scales of G major, F major and D minor*, and indeed *these* are the keys they sound as though they are in. But why those?

So sticking to the notes of the scale won't necessarily land us in the right key, and using out-of-scale notes won't necessarily take us away from it. Apparently, the key, or equivalently the tonic root, of a melody is not such an obvious property if we try to abstract it from the notes

* There is a modal flattened seventh (C natural) in the last of them.

of a piece of music itself, rather than from the key signature written at the start of the score. In fact the rules we use for establishing the tonality when we listen to a piece of music aren't ones that derive from any musical theory; we don't need to know any theory at all. They are purely statistical. And they are rules that we begin learning at birth, or possibly before, and have mostly mastered by the age of four.

What the key of a piece of tonal music determines is not 'which notes may be used', but the *probabilities* of the various notes it contains: the chance that any note in the music, picked at random, will belong to a specific pitch class. A composition in the key of C major is more likely to contain a G, say, than an F♯ or a C♯. The *probability distributions* of notes encode how many times each note occurs in a piece, or equivalently, the relative probability that a note chosen at random will belong to a particular pitch class. These distributions are easy to deduce simply by counting up notes. For Western classical music, they turn out to be remarkably stable across many periods and styles (Figure 4.4).

These statistics tell us what we'd intuitively expect. The common notes – the peaks in the distribution – are all in the diatonic (here the major) scale, and the troughs are all chromatic notes outside the scale. The latter are all used more or less equally rarely. The notes of the major triad (**1-3-5**, here C-E-G) are the most frequently used, although it is curious to see that the second note of the scale (**2**, here D) is

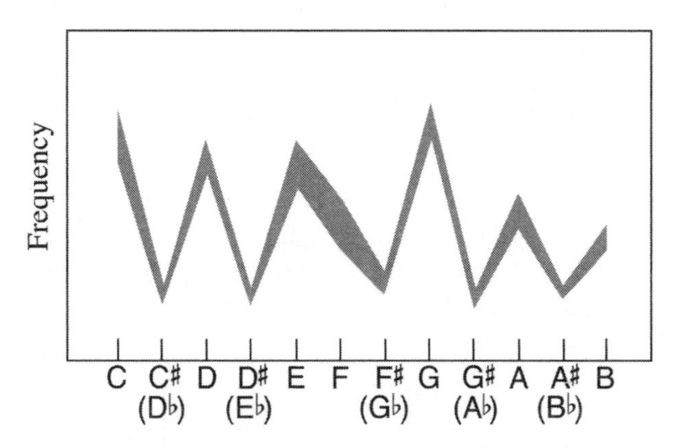

Figure 4.4 The frequency of occurrence of pitch classes for major-key Western tonal music of the eighteenth to early twentieth centuries. The sample (all transposed to C major) consists of: songs by Schubert and Schumann, arias by Mozart and Mendelssohn, lieder by Richard Strauss and cantatas by Johann Adolf Hasse. The width of the band spans the range of values.

equally prominent (we'll see shortly why that is). In this distribution we can identify a *hierarchy* of note status: first, the group C, D, E, G; then this group plus the other notes of the major scale, F, A and B; and finally the entire chromatic contingent.*

Although it is normally applied only to Western music, the word 'tonal' is appropriate for any music that recognizes a hierarchy that privileges notes to different degrees. That's true of the music of most cultures. In Indian music, the *Sa* note of a *that* scale functions as a tonic. It's not really known whether the modes of ancient Greece were really scales with a tonic centre, but it seems likely that each mode had at least one 'special' note, the *mese*, that, by occurring most often in melodies, functioned *perceptually* as a tonic.

This differentiation of notes is a cognitive crutch: it helps us interpret and remember a tune. The notes higher in a hierarchy offer landmarks that anchor the melody, so that we don't just hear it as a string of so many equivalent notes.

Music theorists say that notes higher in this hierarchy are more *stable*, by which they mean that they seem less likely to move off somewhere else. Because it is the most stable of all, the tonic is where many melodies come to rest. This is true of just about any children's song, and of most popular tunes or hymns, from 'Happy Birthday' to 'We Wish You a Merry Christmas' to 'I Wanna Hold Your Hand'. If a tune doesn't end on the tonic, the next most likely final note is the fifth (as in, for instance, Burt Bacharach's and Hal David's 'Alfie') or the third, as in the final 'amen' of hymns. You will struggle to find any popular song that doesn't end up on one of these three notes of the scale.

The notion that some notes are more stable than others can be turned on its head to say that some are more *active*, tending to push the melody off elsewhere. We can think of the pitch space as a sort of topographic landscape in which more stable notes correspond to the valleys (Figure 4.5). A melody is then like a stream of water that seeks the low ground.

From any point, it will tend to run towards the nearest depression: the nearest note of greater stability. (It's a limited metaphor, because the existence of pitch steps between non-adjacent notes implies that the

*We might add another level of the hierarchy for 'grace notes' which, in some instrumental and singing traditions, are microtonal, pitched outside the chromatic scale.

melodic stream can jump across peaks and valleys.) More stable notes exert a pull on nearby less stable ones. In C, an F is pulled down towards E, but also up towards G. An A gravitates down towards G, but a B tends to rise up to C. Chromatic notes are particularly unstable and are likely to move quickly on to more stable ones: an F♯ to a G, an E*b* to a D or an E. Such notes are generally just 'passing tones', gone in a flash, like the sharpened fifths and sixths in 'I Do Like To Be Beside the Seaside' (Figure 4.6). These attractions filter down through the hierarchy: ultimately, everything is drawn to the tonic.

This notion of active and stable notes is made explicit in Chinese music, where the most stable notes are those of the Chinese pentatonic scale. Notes outside this scale are denoted as *pièn* tones, meaning 'changing' or 'becoming' (*biàn* in the modern Pinyin system), and are named for the notes they are on the way to becoming: for example, an F in the pentatonic scale on C would be called '*pièn* G'. Western terminology is analogous: the **2** of a scale is called the 'supertonic', while **7** is called the 'leading tone' because it is deemed to lead up to the tonic.

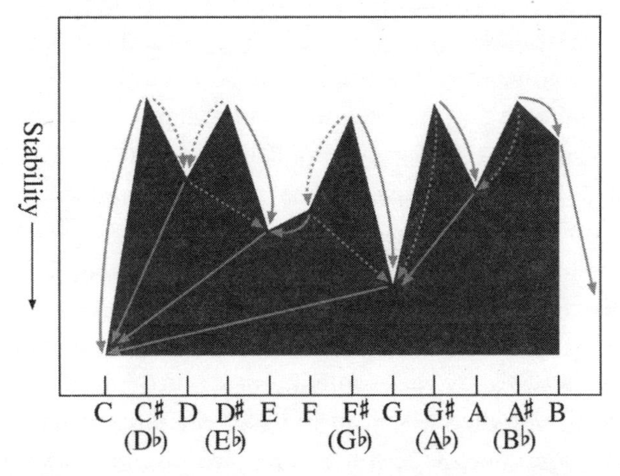

Figure 4.5 The hierarchy of notes can be inverted to produce a kind of 'stability landscape' in which the valleys correspond to the more stable notes. Notes at the peaks can then be considered to be 'pulled down' towards the nearest valleys. Strong attractions are shown as solid lines, weaker ones as dashed lines. Ultimately all notes gravitate towards the tonic. Astute readers may notice that the topography here is not exactly the inverted form of the pitch distribution in Figure 4.4 – in particular, the tonic is 'deeper' and the major second 'shallower'. That's because the perceived stability of a note is better gauged from the *perceptual* pitch hierarchy discussed on p. 102 and shown in Figure 4.8. The differences between the perceived hierarchy and that measured from music itself are generally small, however, and explained later.

Figure 4.6 All the chromatic ('out-of-scale') notes in 'I Do Like To Be Beside the Seaside' (shown with arrows) are transitory passing notes leading to a more stable tone.

Some early music theorists likened this attraction explicitly to gravity. In 1806 the Belgian-French composer Jérôme-Joseph de Momigny even went so far as to suggest an 'inverse square' law comparable to that of Newtonian gravitation, describing how the 'force' weakens with increasing distance between the notes. Of course the matter is nothing like so simple and reductive – for one thing, the 'force' isn't symmetrical, since E 'attracts' F but not vice versa. The music theorist Heinrich Schenker, given to an overly Hegelian turn of mind, interpreted the attraction in 1935 as a 'will of the tones'.

If music was simply a matter of following gravity-like attractions from note to note, there would be nothing for the composer to do: a melody would be as inevitable as the path of water rushing down a mountainside. The key to music is that these pulls can be resisted. *It is the job of the musician to know when and how to do so.* If there were no underlying tendencies, no implications within each note of which one will follow, we would be indifferent to the outcome, and all melodies would sound like the same random meandering. The effect of a tune is determined by whether it follows the attractions or resists them. This is one of the fundamental principles of how music exerts its *affective* power, how it stimulates or soothes us: it is a question of whether or not the music meets our expectations of what it will do next. I will explore this issue in more depth when I consider emotion in music. For now it is enough to say that the hierarchy and different stabilities of musical notes create a context of expectation and anticipation, which the composer or performer manipulates to make music come alive and convey something akin to meaning. If the melody moves from a note of lesser to greater stability, we sense a reduction of tension, as though some constraint has been relaxed.

If the tension inherent in less stable notes is quickly resolved by

Figure 4.7 'Maria' from *West Side Story*. The A natural here is a ♯4 in E♭ major.

letting them fall onto a nearby, more stable note, we may barely notice their presence. We saw above that 'I Do Like To Be Beside the Seaside' contains chromatic notes, outside the major scale. So do many familiar tunes, such as 'Head, Shoulders, Knees and Toes'. But we don't think of these tunes as being odd in any way, because the out-of-scale notes are gone in an instant, capitulating to more stable ones. If such outré notes are held for longer or given more emphasis, their effect is more striking – not necessarily discordant or unpleasant, but piquant and memorable. That's the case, for example, in the opening phrase of Leonard Bernstein's 'Maria' from *West Side Story*, where the first and second syllables of 'Ma-ri-a' rise from the tonic to a sharpened **4** (**♯4**), a very unusual interval in tonal music. And crucially, the **♯4** falls on a strong beat and is held long enough for it to register as such before finally resolving on to the **5** (Figure 4.7).* The uncommon note takes us by surprise, and our interest is aroused in a way that it isn't by the ephemeral chromatics of 'I Do Like To Be . . .'. This is why jazz musicians sometimes cover up wrong notes – ones that fall outside the correct scale or harmonization and are thus unstable – by quickly 'resolving' them on to nearby notes that are more stable. Then the wrong note sounds like a kind of ornamentation of the right one, and is quickly forgotten or perhaps not even registered at all by the listener. Psychoacoustics expert Albert Bregman has gone so far as to call jazz improvisation an 'ongoing accommodation of errors'.

It's one thing to enumerate the relative importance of notes in the chromatic scale by counting them up in musical scores. But it's not obvious that we will *judge* them this way – that musical practice conforms to subjective impression. Music psychologist Carol Krumhansl of Cornell University has conducted exhaustive listening tests to find out if this is the case. In a typical study, she and her co-workers established a tonal context – a sense of key – by playing a chord, a scale, or a short sequence of chords of the sort that might end a song (called a cadence). For

*Notice too that here repetition of the 'odd' note emphasizes the intentionality – see p. 296.

example, to create a C major context they might play the chords of C, F, G, C. They would then immediately play their subjects a note in the chromatic scale and ask them how well it seemed to 'fit' the context. The results were remarkably consistent, regardless of the extent of the listeners' musical training (Figure 4.8). Krumhansl calls this subjective evaluation of the 'rightness' of notes the *tonal hierarchy*.

This is very similar to the actual distribution of notes we saw earlier (Figure 4.4). The only significant differences are that the tonic note **1** registers a little higher, and the major second (**2**, here D) is rated lower. The latter is explained by the fact that in real melodies, **1** is very often followed by **2** because of a preference for small pitch steps (see p. 109). The perceptual tonal hierarchy has a five-tiered structure: the tonic, tonic plus fifth, the major triad (**1**, **3**, **5**), the diatonic scale, and the chromatic scale (Figure 4.9).

But what is cause here, and what is effect? What, ultimately, makes us decide that a G is a better fit in the key of C than is an F♯? Do we make that judgement based on what we've learnt from exposure to music, or is it determined by innate mental factors which composers have then simply put into practice?

It's often been implicitly assumed by music theorists that there is something 'natural' to these judgements, stemming from questions of consonance. In this view, the mathematical neatness of the relationship between the frequencies of different tones (whether based on Pythagorean ratios or the harmonic series) governs our perception of how well they fit. This issue of consonance and dissonance is complex and controversial, and I will postpone it until Chapter 6. For now, it's enough to say that it is hard to disentangle anything innate about the way we experience note relationships from the effects of mere exposure to the way they are applied. We can't just say (as some do) 'But C and G just *sound right* together, while C and F♯ don't' – because this may be simply what we have become accustomed to.

It *is* possible, however, to calculate an objective measure of the 'consonance' of two tones. Although there's no unique, generally accepted way of doing that, the various methods that have been proposed tend to give broadly similar results – and these match the tonal hierarchy quite well. But there are also some significant discrepancies. For example, **3** is higher in the tonal hierarchy than **4**, whereas their calculated degrees of consonance are the other way round. Likewise, the minor third is

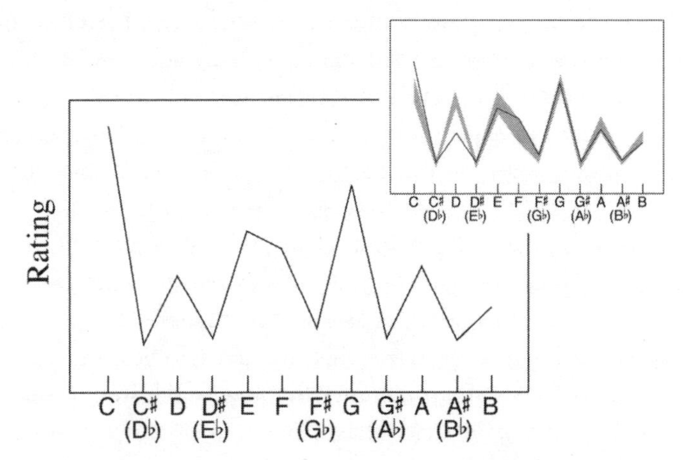

Figure 4.8 The 'tonal hierarchy': how people rate the 'fittingness' of notes within the context of C major. Inset: comparison with the actual note frequency distribution for Western tonal music, shown on p. 97.

only a moderately consonant interval, but features prominently in the tonal hierarchy of minor keys (which differs from the major) because we are used to hearing it in that context: convention has overridden acoustic 'fact'. After looking carefully at these issues, Krumhansl and her colleagues concluded that in fact learning of statistical probabilities is far more important than intrinsic consonance in deciding the preferences

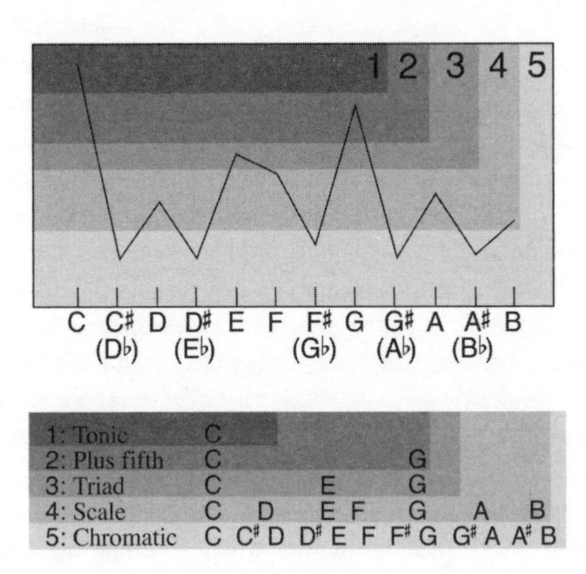

Figure 4.9 The tiers of the tonal hierarchy.

expressed in the tonal hierarchy. This implies that we should be able to learn new notions of 'rightness' if we hear them often enough.

We seem to learn the tonal hierarchy of our culture's music quickly and unconsciously during childhood. Psychologists Laurel Trainor and Sandra Trehub have found that children are generally able to identify notes that don't 'fit' in a tonal melody by the age of five. In contrast, while babies as young as eight months can spot alterations to a simple melody, they respond equally when the changed note is within the scale or outside it. They haven't yet developed preconceptions about tunes. (Other work by Trainor and Trehub challenges that general claim, but we'll come to it later.) Nonetheless, developmental psychologist Jenny Saffran and her co-workers have shown that learning of statistical regularities in sequences of notes is already underway in eight-month-olds. They found that when such infants were played a series of notes consisting of six different three-note 'words' assembled in random order, the infants subsequently showed greater interest in the 'words' when played back individually than in random three-note 'non-word' sequences: they had abstracted and remembered the 'words' from the initial sequences. This may be a key aspect of how young children identify real words within streams of syllables.

Not all music theorists accept that Krumhansl's tonal hierarchy tells us something deep about the way we process tones in real music. Composer and musicologist Fred Lerdahl points out that it provides a rather static and superficial picture, taking no account, for example, of the harmonic context underlying a melody. A C♯ in the key of C might sound peculiar in a nursery rhyme (I can't think of any that has such a thing), but classical composers will readily find ways to accommodate it as they modulate within a composition nominally in C major. So merely playing a C chord followed by a C♯ note doesn't constitute a very 'musical' test.

Another criticism is that Krumhansl's experiments are structured so as to encourage the listener to judge how well each tone *rounds off* the context, bringing it to a close. This is not the same as a judgement made about an ongoing musical extract. Bret Aarden of Ohio State University has found that people have different judgements and expectations about what fits or what comes next depending on whether the music is just beginning, in mid-flow, or ending. Using a cadence to set the context, as Krumhansl sometimes did, sends out a subliminal message of finality. In that case, Aarden found, people tend to give greater weight to the tonic note, which implies closure, and less weight to the **2**, which implies continuation.

All the same, Krumhansl's point that statistical learning guides or even governs our perception and expectation of the notes that make up a melody is widely accepted. The implication is that, whether we know it or not (and I suspect you did not, until now), we have a mental image of the tonal hierarchy in our head, and that we constantly refer to it to develop anticipations and judgements about a tune, whether for nursery rhymes or for Bach. This isn't at all implausible, for the human mind is spectacularly good at spotting patterns. It's one of our most highly evolved capabilities.

For example, when we hear a piece of music, we begin immediately trying to match it to a particular tonal hierarchy – in other words, to locate the key and tonic. It makes sense to assume that the first notes we hear are high in the tonal hierarchy, since these are the most likely. That's usually a good guess, because many tunes start on such notes, particularly the tonic ('Frère Jacques', say, or the first movement of Beethoven's Third), the third ('Three Blind Mice' and the first movement of Beethoven's Sixth) and the fifth ('London Bridge is Falling Down' and Beethoven's Piano Sonata No. 13 in E♭). 'The Girl With the Flaxen Hair' begins on the fifth, so Joyce Grenfell's deduction was anything but a shot in the dark.

We're really rather good at this game. Even people who have no musical training or specialized knowledge, and who may not even know what a key is, will typically deduce a tonal centre from just a few seconds of music. A general sense of tonality, which entails an ability to sing a song without constantly wandering out of key, develops in most people by the age of five or six without any formal training. By seven, many children can detect a key change – a switch to a different tonal hierarchy – in the middle of a familiar song. And you thought you weren't musical?

New rules

If the tonal hierarchy gets fixed in our minds at so young an age, does this mean we are incapable of appreciating music of other cultures that uses a different hierarchy? You might expect that, if we've never heard an Indian or a Balinese scale before, we will be thrown into confusion. Some anecdotal reports appear to confirm this. In his 1914 book *Music of Hindostan*, one of the first serious studies of Indian music by a Western musicologist, Arthur Henry Fox Strangeways claimed that a performance he heard seemed to Western ears to have a quite different tonic from the one Indian listeners perceived, and this 'wrong' frame of reference gave the Westerners a different sense of what the melody was doing.

But that may not be the general experience. Carol Krumhansl has compared the tonal hierarchies of Indian music constructed by people familiar with this tradition to those of Western listeners to whom it is alien. Although the Indian scales (*thats*) have different tunings from diatonic scales, they are likewise seven-note scales that give prominence to a kind of tonic (*Sa*) and a fifth (*Pa*). There is more to this hierarchy too. The music, while largely improvised, is organized into themes called ragas, which define some key melodic features on which the improvisation is based. Aside from the *Sa* and *Pa*, each raga has two key tones called the *vadi* and the *samvadi*, separated by an interval of a fourth or fifth.

Krumhansl asked listeners to identify the goodness of fit of the various notes of the relevant *that* after the context had been set by playing a thematic fragment of the raga. Both the Indian and the Western groups came up with very similar ratings, which generally identified the key features of the hierarchy: first the *Sa*, then the *Pa*, then the *vadi*, followed by other tones in the *that* (including the *samvadi*), and finally tones that were not in the *that*, equivalent to the chromatic tones of the diatonic scales.

Perhaps the Western listeners were merely guessing this hierarchy by analogy to the diatonic scales they knew, preferring tones that sounded closest to the important ones in Western scales? But careful testing showed that they seemed instead to be making their choices mostly on the basis of how often the various notes appeared in the extracts of the ragas that they heard initially. In other words, these inexperienced listeners very quickly deduced *from scratch* the way the notes were organized hierarchically. Of course, they didn't do that as expertly as the listeners familiar with Indian music, who were, for example, more adept at distinguishing *that* tones from non-*that* tones.

But it seems that not only do we acquire our basic tonal vocabulary by unconsciously assimilating the statistical expectations created by music itself, but we can set aside our preconceptions and learn a new vocabulary with rather little effort. Music from other cultures need not sound so weird if you give it a chance.

A similar ability was found by ethnomusicologist Christa Hansen and her colleagues among people in a remote village in Bali who had never seen a Westerner, let along heard Western music: tests showed that villagers, after being played three sixteen-note Western melodies, were able to figure out the tonal hierarchy. And Finnish music psychologist Tuomas Eerola found that statistical learning explained the tonal hierarchies offered by South African tribespeople when played songs called *yoiks* unique to northern Scandinavia, which make many unusually large melodic leaps.

Of course, there is much, much more to 'understanding' another culture's music than figuring out which notes matter most, as we'll see later. But acquiring the right tonal bearing is a crucial first step. And the statistical salience of different tones doesn't just offer a sound empirical approach, but also tends to reveal the 'theoretical' principles of the music: notes that are used more often tend to be the ones around which the scales, harmonies and musical structures are based. Naturally, the success of this strategy depends on whether the music supplies a sufficiently representative sample of the normal tonal hierarchy. When Westerners in the experiments of Christa Hansen were asked to deduce tonal hierarchies for the five-note Balinese *sléndro* scale based on just a short melody, they deduced a different hierarchy from the Balinese listeners because the principal tone or 'tonic' (called the *dong*) was not the most frequent in the melody chosen. A longer exposure to *sléndro* scales was needed to override the skewed information provided by this snatch of melody.

Perhaps the most important point about the tonal hierarchy is that it doesn't just help us understand and organize music: by doing so it helps us to *perceive it as music*. The random 'melodies' I created at the start of the chapter by plucking notes from the C major scale don't sound very musical at all, for they ignore the usual note probabilities of tonal music. Music that rejects such hierarchies is harder to process, as we'll see, and can seem merely bewildering.

The shape of melody

The distribution and hierarchy of note probabilities in Western music implies an unspoken consensus that a 'good tune' uses notes in these proportions. But there's got to be more to it than that. Our naïve songwriter searching for the hit formula might now adjust his melody-generating program so that it uses notes in these magical relative amounts, only to find that the results still sound tuneless and banal.*

What he is neglecting is that the pitch composition is only a *surface* property of melody, in the same way that the relative proportions of colours represent merely a surface property of a painting by van Gogh or Rembrandt. These features by themselves have no meaning, no significance, no artistic *information*. They are just a part of the rules of the game. When two teams play a soccer match, we expect that there will be eleven in each side, that one will be the goalkeeper and the others will be evenly distributed around the pitch, that they will abide by certain regulations. Otherwise, they simply aren't playing soccer at all. But none of that guarantees a good game.

Everything we value in a melody (and in a soccer match) comes from the relationships between the elements that constitute it, and the context we create for them from the knowledge and expectation we bring to the experience of perceiving it.

We've already seen some of the tricks that make a good tune more than just the right blend of notes. Some notes seem to lead naturally to others, according to their proximity and stability. A **7** tends to lead on to the tonic above; a **6** is drawn to a **5**. But not all the notes move so predictably. And a melody can't simply grind to a halt as soon as it reaches the maximal stability of the tonic.

What we're really saying here is that each note creates its own implications of what is to follow. In part, this is a matter of what we learn to expect about the size of *pitch steps* or so-called melodic intervals. And we can again answer that empirically by merely counting up the statistics (Figure 4.10*a*). The message here is clear: some step

* Dirk-Jan Povel of the University of Nijmegen has devised a melody-generating computer program that is considerably more sophisticated than this. All the same, I suspect you'd agree that the results, which can be heard at http://www.socsci.kun.nl/~povel/Melody/, do not exactly send your spirits soaring.

sizes between successive notes are more common than others, and broadly speaking an interval is used more rarely the bigger it is. In other words, melodies tend to progress smoothly up and down musical scales. In C major, there is a stronger probability that, say, a C will be followed by the D above it than by an F. There is a correspondingly stronger probability that a G will be followed by an A or F than by a C. As with predicting the weather day to day, we are most likely to forecast correctly the next note in a melody by assuming that it will be similar to the last one. The problem with our randomly generated melodies is that, even if they adhere to the conventional note probability distribution, they sound disjointed: there doesn't seem to be any logic to them. That's because the pitch steps are as likely to be big as they are small, which contravenes what we've learnt to expect from a melody.

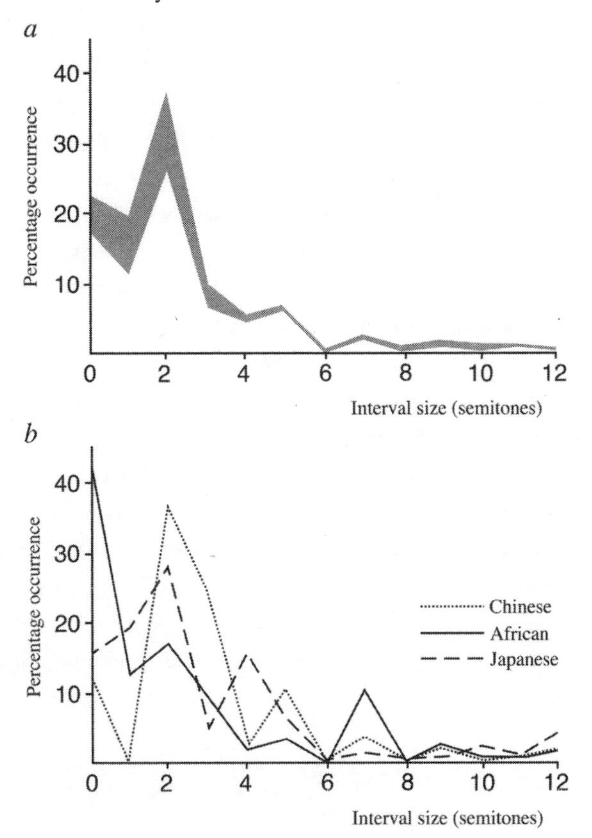

Figure 4.10 The distribution of pitch steps for Western (*a*) and non-Western (*b*) music.

These probability distributions of melodic pitch steps turn out to be pretty universal not only for Western tonal music but for that of many other traditions too (Figure 4.10*b*). This universality can help us make sense of unfamiliar music. Music psychologist David Huron and his co-workers conducted an experiment in which North American and Balinese participants were asked to bet, using poker chips, on what they considered to be the most likely next note in a Balinese melody as it unfolded one note at a time. While the size of the bets revealed the Americans to be understandably far less confident in making this prediction at the outset, by ten notes into the melody they were betting as confidently as the Balinese. But was the confidence of the Americans justified? Not particularly – as one might expect, the Balinese participants made considerably more accurate guesses. However, the Americans did significantly better than chance. If we look at the test melody (Figure 4.11), it's not hard to see why. Most of the pitch steps are relatively small, and the contour is rather smooth. And although the scale is not a common one in Western music, it bears some clear similarities to the Western minor scale. So by generalizing these aspects of Western music to the Balinese case, the Americans had some fairly sound guidelines.

The interval statistics encode what looks like a very cautious, conservative way of getting around musical space – as though composers have elected to limit themselves mostly to timid little steps, only very occasionally risking something as outré as a fifth. Why aren't they more adventurous?

One reason is probably mechanical. Most of the world's music is vocal, and it is much easier to sing small pitch steps than large ones: it requires less rearrangement of our vocal apparatus, so less muscle effort and control. It is also generally true for instrumental music that small pitch steps require less effort to execute: we're not forever leaping across the keyboard or the frets. But there is a less obvious reason too, which is cognitive: such leaps tend to break up melodies and hinder

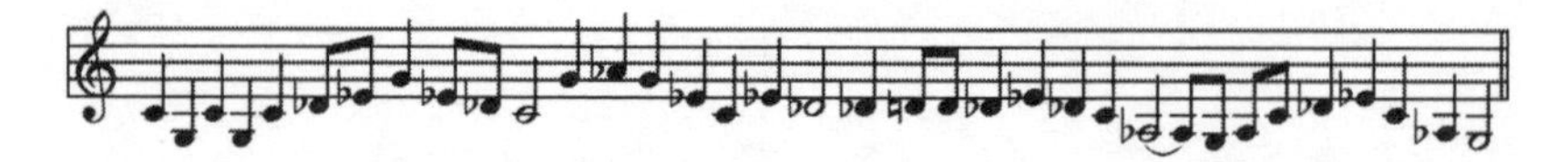

Figure 4.11 The Balinese melody used to test the expectations of listeners for how it will continue.

a

b

Figure 4.12 Pitch jumps in 'Somewhere Over the Rainbow' (*a*) and 'Alfie' (*b*).

us from hearing them as a coherent whole. Like sudden steps in an otherwise gradual incline, they are liable to trip us up.

Yet if that's the case, shouldn't we *always* avoid big intervals? Even if they are less common, they are by no means 'rare' – about one in fourteen pitch steps in Western music jumps by a perfect fourth, for example. Well, perhaps it stands to reason that if we were to use only small pitch steps then music would be rather dull, with melodies that were mere fragments of scales joined together. So the question is, when is a big pitch jump 'good', and when not?

Truly big intervals aren't hard to find: think of the first line of 'Over the Rainbow' (Figure 4.12*a*). 'Some-' comes on an ordinary enough note around the middle of the average vocal range (E♭ above middle C) – but straight after, '-where' soars way up high. Actually it soars a full octave, a jump that constitutes only one to two per cent of all pitch steps in folk and popular music. There's another such leap at the start of 'Singin' in the Rain'.*

These big jumps have to be kept under control, as it were, so that the song doesn't sound like wild yodelling. One of the ways that is done is by following them soon after with other large jumps. Composers and songwriters have come to recognize that large changes in pitch shouldn't be placed in isolation among little steps. The repetition is, in effect, a way for the composer to 'rewrite the rules': the message it conveys is something like 'yes, I know that these big jumps are unusual in general, but not here'. The octave

* In fact, either of these songs provides a handy mnemonic for the task of singing an octave. Ask many amateur musicians to sing an octave interval and they'll need a moment to figure it out. But ask any moderately competent singer to sing 'Over the Rainbow' and they'll get the jump spot on without thinking.

of 'Somewhere', for example, is echoed (though not exactly) on 'way up', and then on 'there's a [land]' (Figure 4.12*a*). Similarly, in 'Alfie' the leap on 'Al-fie' itself is followed by several others (Figure 4.12*b*).

We may thus be called upon to revise our normative expectations from past experience in the light of what a composition tells us about its own 'internal' norms. If the first jump on 'Alfie' were not echoed subsequently, we wouldn't necessarily find the tune bizarre or incomprehensible, but it would seem somehow less coherent. In effect, *every piece of music* creates its own mini-idiom.

There's more. In 'Over the Rainbow' and 'Singin' in the Rain', both octave jumps come on notes that are longer than most of the others. We leap and then stay suspended on '-where', before the melody sweeps on more smoothly in graceful little steps with 'over the rain-bow . . .'. This is another trick for binding the notes together so that the big jump doesn't snap the melodic thread. The melody waits at the top for the brain to catch up.

Underlying this perspective on the processing melodies is the brain's search for coherence in the stimuli it receives. We have already seen how we mentally and unconsciously 'bind' the harmonics of a complex tone into a single perceived pitch. In the perception of melody, the task is to bind together a string of notes into a unified acoustic entity, otherwise known as a tune. The cognitive principles that enable this organization of complex perceptual stimuli were discovered in the early twentieth century in the field of gestalt psychology, and I say more about them in the next chapter. A large interval creates a discontinuity of contour, like a cliff face. It also juxtaposes two notes that are quite dissimilar in pitch, implying that they don't 'belong' to the same melodic thread. Both challenge the brain's ability to perceive the melody as a single 'gestalt' (the German word *Gestalt* means 'pattern').

We'll see later some of the other impressive feats of organization that the gestalt principles achieve. There is good reason to believe that musicians and composers have encoded these principles unconsciously in their habits, rules and conventions, thereby structuring their music to take advantage of them. The result is music designed to be comprehensible.

Skips and arches

In both 'Over the Rainbow' and 'Singin' in the Rain', the octave jumps are immediately followed by reversal of melodic direction towards lower notes. It is as though, having taken a giant step in pitch space, the tune then sets about filling in the intervening gap. This is so common a feature – you can see it in 'Alfie' too – that some musicologists have given it the status of a universal trait: the influential American music theorist Leonard Meyer called melodies like this 'gap-filling'. The sixteenth-century Italian composer Giovanni Pierluigi da Palestrina advised that the pitch direction of a melody be reversed after big pitch steps (or 'skips'), and this was often reiterated in later times as a tenet of good compositional practice. Certainly it *appears* to have been widely observed: so-called 'post-skip reversal' happens about seventy per cent of the time for skips of three semitones or more, in many musical cultures.

But this doesn't mean that these 'post-skip reversals' are an *intentional* feature of music. That sounds contradictory, but it isn't. If you make a big pitch jump, the chances are that you end up on a note closer to the extreme of the pitch range than you started from. In 'Over the Rainbow' and 'Singin' in the Rain', the octave leap takes us up high. Given that on average most notes in a melody tend to fall around the middle of the typical gamut (few songs remain resolutely at the edge of a singer's range), chance alone dictates that the next note after such a high jump will be lower: there are simply more lower notes to choose from. 'Post-skip reversal' would only be revealed as an intentional device if skips that *start* at a more extreme position and end closer to the mid-range reverse as often as ones that take the melody from the centre to the edge of the range. But they don't, for music ranging from Chinese folk songs to the songs of sub-Saharan Africa. So it seems that chance alone is enough to make a soaring melody go back and fill in the gaps.

All the same, post-skip reversal *is* an empirical fact – and this means that we come to expect it as a likely feature of melodies. When musicians are asked how they think a melody will continue after a skip, they do anticipate reversals. And significantly, they do so regardless of where the jumps start and finish, even though the statistics show that rever-

Figure 4.13 In the theme of Mozart's Rondo in D major, thirty-eight of the fifty pitch steps of a major third or smaller are followed by others in the same direction. The reversals are indicated with bars.

sals are only favoured when the jump goes from a mid-range pitch towards an extreme. In other words, statistical learning of the 'shapes' of melodies seems to have generated an incomplete rule of thumb: 'anticipate post-skip reversal' rather than 'anticipate post-skip reversal *only if* the end note is more extreme than the start note'. That's understandable, however, since the first of these rules is much easier to retain and process, and is confirmed more often than it is contradicted. Musical expectations like this typically represent a compromise between accuracy and simplicity – they just have to be 'good enough'.

One consequence is that we tend to overgeneralize. This can be seen for another general organizational principle of melody proposed by Meyer: that small pitch steps tend to be followed by others in the same direction. Anecdotally this too seems plausible. In the main theme of Mozart's Rondo in D major, K485, for example, thirty-eight of the fifty steps of a major third or less are followed by others in the same direction (Figure 4.13). But again, the statistics don't bear it out: it's true only for descending lines, not ascending ones. And yet again, tests show that musicians expect the rule to apply in both directions.*

Entire melodic phrases have characteristic shapes. One of the most common is the arch, in which the pitch rises and then falls again – not necessarily with perfect smoothness, but perceptibly. This shape is very obvious in 'Twinkle Twinkle Little Star' (Figure 4.14*a*), and

* In both these cases, musicians were used to test expectations. Non-musicians don't show *any* systematic expectations of what a melody will do next: they haven't abstracted any 'rules' at all from their listening experience. This is significant when we come later to consider how expectation leads to musical affect: if we have no expectation, we won't be affected.

Figure 4.14 The melodic arch in 'Twinkle Twinkle Little Star' (*a*), and its repetition in 'We Wish You a Merry Christmas' (*b*). The latter also has an overall arch in which the others are nested.

also in the phrases of Beethoven's 'Ode to Joy'. (Didn't I say it is a nursery-rhyme tune?) Much Gregorian chant and many Western folk tunes follow a melodic arch too. Other melodies display a hierarchy of arches, small ones nesting inside a large one, as in 'We Wish You a Merry Christmas' (Figure 4.14*b*). Interestingly, although the arch seems to be exceedingly widespread in song, it doesn't seem to be 'internalized' in the expectations people develop for how a tune will go. Rather, tests show that we simply anticipate the second half of the arch: a descending line towards the end of a phrase or line. Again, this seems to be a matter of economy: it is simpler to mentally encode the expectation that melodies fall towards the end than that they first rise and then fall. This simpler expectation is proved right often enough that there is no motivation to refine it.

Why is the melodic arch used so much? We don't know, but can guess. For one thing, more melodies start on the tonic note than on any other, probably because this maximizes our chances of guessing the tonic very quickly – and most music, at least of the traditional and folk varieties, *wants* to be understood easily. And most melodies end on the tonic too, since this is the most stable resting place in the tonal hierarchy – it gives the greatest sense of closure. So if we start and finish a melody at the same place, and if for reasons of ease of production we tend to move in small pitch steps, the easiest solution is either an upward or a downward arch. Why the upward arch is generally preferred has been much debated. Some music theorists argue that a rising scale signifies a rising spirit, conveying an optimistic, seeking or yearning intent that is in keeping with the meaning music often serves to express. Others hint at an almost gravitational theory of music, whereby the rising of pitch mimics the familiar process in which objects rise and then fall again to the ground. I find these spatial metaphors dubious. It is true that there seems to be a mental link between pitch and spatial perception, in the sense that cognitive deficits in one of them are often accompanied by deficits in the other. But this says nothing about how 'high' notes might relate to spatial height. The connection seems so obvious that it is often taken for granted, and certainly some composers have used it symbolically (see p. 399). Yet it is mere convention that makes us call notes 'high' when their frequency is greater. To the ancient Greeks the spatial analogy was reversed: 'low' notes (*nete*, the root of 'nether') were those *higher* in pitch, because they were sounded by strings lower down on the kithara.*

It is conceivable that, because sung music is so prevalent, the preference for raised arches stems from similarities with the patterns of speech: a falling tone at the end of sentences is common in many languages. And lullabies are full of gently descending lines, mimicking the speech patterns of mothers soothing their babies.

How melodies take breaths

One of the common delusions of aspiring jazz and rock musicians is to imagine that, if they learn to play fast enough, they will sound like Charlie Parker or Jimmy Page. They won't, because what both those virtuosi possessed beyond awesome fluency was a sense of how to make a solo breathe: how to break up their flurries of notes into phrases that sound like coherent entities – musical thoughts, if you will.

In language, the phrase structure of sentences and clauses supplies an important mechanism for decoding syntactic meaning. In the sentence 'To fasten one's seat belt is wise because driving can be dangerous', we place mental brackets around 'to fasten one's seat belt' and 'driving can be dangerous': these are 'thoughts' whose relationship we then need to establish from the connecting words.

* In fact the Greeks used a completely different set of metaphors for our 'high' and 'low' notes. The former were *oxys* (sharp – the origin of the word for the musical accidental ♯). Interestingly, *oxys* also became, via its association with acids (sharp-tasting liquids), the root of 'oxygen', which comprises one-fifth of air. Our 'low' notes, meanwhile, were called *barys* (heavy) by the Greeks.

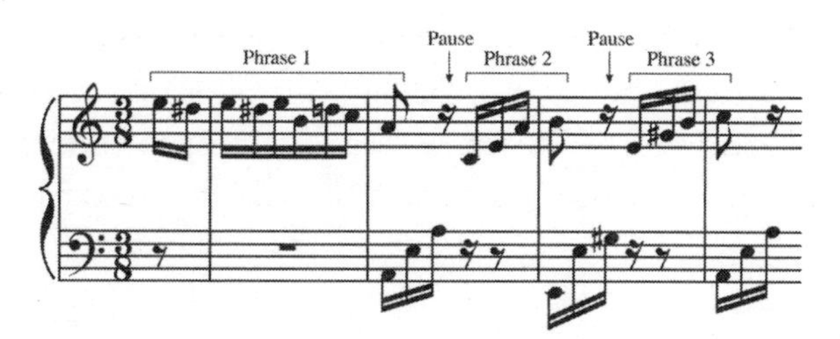

Figure 4.15 The articulation of melodic phrasing in Beethoven's 'Für Elise'.

A similar phrase structure is found in music. Composers aid our perception by using short, clearly articulated and continuous phrases divided by pauses, as in Beethoven's 'Für Elise' (Figure 4.15). Notice that this isn't just a matter of dividing up a melody into bite-sized chunks; the phrases must relate to one another. I think it is easy here to hear that the 'thought' in the second phrase is a response to the first, but that the statement isn't fully resolved until the third phrase, which clarifies the second.

If recordings of sentences with complex syntactic structure, like the one above, are played to people with clicks superimposed during particular words, listeners tend to shift the clicks perceptually to the boundaries between phrases, where they are less disruptive to the syntax. In other words, the clicks migrate to the weaker links between words, because our tendency to impose order on the stream of words is so strong. Click migration to phrase boundaries happens in analogous musical tests too – but because the boundaries of musical phrases aren't always as clearly defined as they are in language, click migration can be used to figure out *where* we perceive these boundaries to be. For example, look at the melody shown in Figure 4.16*a*. It's not obvious at a glance how to split this up into phrases. But tests showed that a click placed on note five in the second bar had a high propensity to migrate back to note four, implying that we discern a break between these two notes. (This is especially striking given that notes four and five have the same pitch, which might be expected, other things being equal, to bind them together perceptually.) What we seem to hear in this phrase is a division more like that made explicit in Figure 4.16*b* by the use of a longer note at the end of the first phrase.

Phrasing is closely linked to the rhythmic patterns of music, which

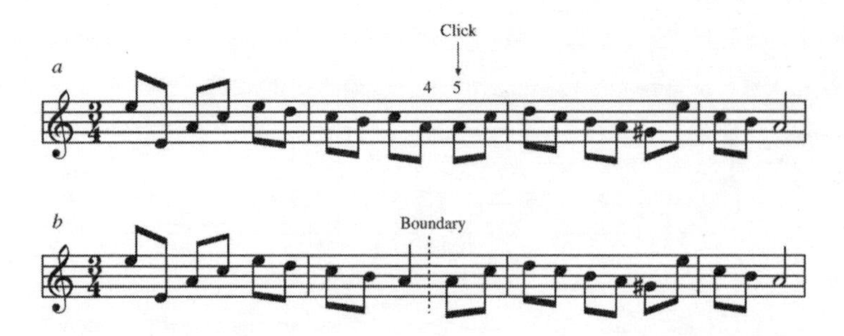

Figure 4.16 'Click migration' – the perceived relocation of a click from where it is actually sounded during a melody – can show us how we subconsciously divide melodies into phrases separated by boundaries. The clicks tend to migrate to the boundary at which one phrase ends, where they are less disruptive. In the example in (*a*), a click placed on note 5 in the second bar (arrow) is typically reported by listeners as having occurred on note 4, implying that we discern a break between 4 and 5. The line is mentally divided up as though it was more like the line shown in (*b*), where the longer note in the second bar signifies the end of a phrase.

are the subject of Chapter 7. By dividing up music into bars, which generally specify patterns of more and less strongly accented notes, we are already providing segmentation into bite-sized chunks. A phrase can occupy several bars, or less than one, but the bar divisions in themselves give music a kind of natural breathing rhythm. And 'breathing' seems to be much the right analogy, especially in vocal music: the average length of a musical bar in hymns is 3.4 seconds, close to the average line length in poetry of 2.7 seconds. If a bar is much shorter than this, there is insufficient time to develop any sense of an articulated phrase with a start and a finish. If it is much longer, we have trouble remembering, by the time we get to the end, what happened at the beginning, much as we have difficulties with long sentences. This bar length thus serves as a kind of frame around the 'musical present'. Of course, we can certainly identify structures on much longer timescales, but not in the sense of being able to 'see them all at once'. This series of 'windows on the present' gives music a kind of pulse even if there is no regular rhythm or pronounced metre. The easier it is for us to hear subdivisions on this sort of timescale, the easier we can organize what we hear. Music that undermines such division into temporal chunks can be hard to follow, which is one reason why some atonal compositions can sound like a disjointed series of unrelated events. But composers and musicians may intentionally play with that confusion: fragmentation might be precisely what the modernist composer is after. Conversely, a seamless, unarticulated drone-like flow of notes might induce trance-like, meditative listening. Or it might just be boring.

The song remains the same

To a considerable extent, what we get from music depends on our ability to extract structures from sequences of notes: to spot the patterns and links, the allusions and embellishments. That's why listening to music relies on memory. We make sense of what we hear by framing it in the context of what we have already heard; as Aaron Copland put it, 'It is insufficient merely to hear music in terms of the separate moments at which it exists. You must be able to relate what you hear at any given moment to what has just happened before and what is about to come afterward.' We've seen that this is how we figure out the tonality of a piece of music: by comparing it with the tonal hierarchy that we have built up through experience and stored in our memory. This sort of constant comparison and revision of our interpretation is also a key feature of how we process melody.

We don't simply have a 'memory box' into which we dump an entire melody, note for note. Rather, we remember structures and patterns over various timescales, with varying degrees of fidelity and which fade from memory at various rates. Whereas a computer 'remembers' a graphic image as a series of pixels of different colour, brightness and so on, we recall instead the unified items: a red rose, a mountainous landscape. We recall general shapes and colours, the identities of objects, but not precise details. Where we forget those details, we unconsciously fill in the gaps. We have sketchy, 'big picture' memories, which are spectacularly good at spotting patterns and similarities that computers struggle to identify, which is why it is so hard to make computers good at visual and speech recognition even though they can collect and securely retain data much more efficiently than our brains can.

Similarly, we group the 'pixels' of music into lumps with recognizable outlines, a process called chunking. To do this we use subconscious rules that help us decide if the notes 'belong' together. Unifying features such as small interval steps and shared tonality provide the glue that binds them.

Chunking is vital for cognition of music. If we had to encode it in our brains note by note, we'd struggle to make sense of anything more complex than the simplest children's songs. Of course, most accomplished musicians can play compositions containing many thousands of notes entirely from memory, without a note out of place. But this seemingly awesome feat of recall is made possible by remembering the musical *process*, not the individual notes as such. If you ask a pianist to start a Mozart sonata from bar forty-one, she'll probably have to mentally replay the music from the start until reaching that bar – the score is not simply laid out in her mind, to be read from any arbitrary point. It's rather like describing how you drive to work: you don't reel off the names of roads as an abstract list, but have to construct your route by mentally retreading it. When musicians make a mistake during rehearsal, they wind back to the start of a musical phrase ('let's take it from the second verse') before restarting.

So when we listen to a melody unfold, we hear each note in the light of many remembered things: what the previous note was, whether the melodic contour is going up or down, whether we've heard this phrase (or one like it) before in the piece, whether it seems like a response to the previous phrase or a completely new idea. We remember the key, and are thus alert for signs that it is changing. We remember that this is, say, the symphony's second movement. We may even be recalling some of the composer's other works, or those of another composer or performer: 'Oh, the Sinatra version was much better.' We remember so much. And yet, how well?

If people are played single notes and asked to say whether a second note, sounded later, is the same or different, they usually remember accurately when the delay is as much as fifteen seconds (a long time in music!). This short-term encoding and memorization of pitch has neural machinery dedicated to the task, situated partly in a brain region called the inferior frontal gyrus and encoded by specific genes. This seems to be a 'module' provided by evolution for non-musical purposes: there's a survival value to distinguishing different pitches, whether we use it for understanding speech and emotional intonation or for recognizing animal calls and other environmental sounds.

But not much music, even of the most minimalist kind, offers us single notes separated by long pauses. If other notes are interpolated between the two we are supposed to remember, our pitch memory is disrupted. We also remember pitch steps between two notes better when they are smaller: we're pretty good at recognizing that two successive whole-tone steps (C→D, say) are the same, but find it harder to tell apart a minor-sixth interval (C→A♭) from a major sixth (C→A). This suggests another reason for the distribution of interval sizes we saw earlier: we have a more robust melodic memory for, and thus easier comprehension of, music that uses predominantly small interval steps.

The contour of a melody – how it rises and falls in pitch – is one of the most important clues for memory and recognition. Babies as young as five months will show the 'startle response' of an altered heartbeat when a melody changes its contour. And most of the spontaneous, charmingly wayward songs that children begin to sing from around eighteen months contain brief phrases with an identical repeated contour, albeit without any tonal centre (Figure 4.17).

Both children and untrained adults often think that two melodies with the same contour but with some slightly altered intervals are identical: they don't notice the small differences. Musically untrained adults asked to sing back an unfamiliar melody might not get a single note right, yet will capture the basic contour. And familiar tunes remain recognizable when the melodic contour is 'compressed', as if reducing the vertical scale on a mountain range. This is essentially what young children do when they learn to sing a song: they make rather arbitrary (usually too small) guesses at the right pitch steps, but you can still tell if they're singing 'Old MacDonald' or 'Three Blind Mice'. In fact, even if all the up and down steps in a melody

Figure 4.17 A song by a 32-month-old child, approximately transcribed into conventional notation.

Figure 4.18 When familiar tunes are played with the notes assigned to random octaves, as with 'Mary Had a Little Lamb' here, they become impossible for most people to identify. This is because the melodic contour is severely disrupted by this octave-shifting.

are reduced to a *single semitone*, making the tune a rather tedious drone, people can still sometimes guess the song.* And conversely, if the notes of a well-known melody are played in randomly selected octaves, so that the pitch classes are right but the melodic contour is completely altered, the tune generally becomes very hard to identify (Figure 4.18).

Short fragments of melody with characteristic contours are often used as the architectural units of complex music. Music psychologist Diana Deutsch has found that the phrase in Figure 4.19*a* is more easily remembered when broken up by pauses as in 4.19*b*, even though they contain exactly the same sequences of notes, because the pauses divide the sequence into groups of notes with identical contours (sometimes called parallelism, because the successive melodic contours stay parallel to one another). In a very real sense, there is *less information* to recall in melody *b*, because the repeating pattern makes it possible to condense the information of the whole sequence into a more concise form. Instead of having to remember that 'C♯ follows the first D, and then D follows that . . .', we need simply recall that there are four repeating motifs, each consisting of a semitone step down and then back up, on each note of the G major chord.† In the same way, we can immediately see the concise 'formula' that captures the sequence 123123123123. . . in a way that we can't for 121322311322 . . . In technical terms, the sequence is algorithmically compressible. In contrast, if pauses divide up the melody in a way that cannot be transparently encoded, they can actually hinder memory: the phrase in Figure 4.19*c* is recalled significantly less efficiently than that in *a*.

* Of course, pitch contour isn't by any means the only clue to the tune's identity here – there are also strong hints in the rhythmic pattern.

† The fact that the starting notes of each three-note cluster in *b* are the notes of the major triad also aids recall, because this triad is another structure we have already encoded in our brains in the tonal hierarchy.

Figure 4.19 People can recall sequences of tones more accurately if they are grouped in ways that impose easily heard regularities, for example repetition of a pitch contour. The sequence in (*a*) is recalled more accurately if pauses are inserted between groups of three notes (*b*), emphasizing their identical contour. But if the pauses disrupt this repetitive structure, as in (*c*), recall is considerably worse: the sequence 'makes less sense'.

A contour need only be preserved roughly for us to be able to recognize it as a repeated building block, and thus to intuit this ordering principle in the music. Take, for example, the first three verse lines of 'Jingle Bells' (Figure 4.20*a*). None of these is an exact repetition of any other, but all have the same sawtooth shape. An even simpler, and devastatingly effective, repeated contour is the four-note cliff-edge motif of Beethoven's Fifth Symphony, the simple little hook that keeps cropping up all over the first movement.

Bach uses the same principle with a rather more sophisticated contour in the fifth fugue of *The Well-Tempered Clavier*, Book I (Figure

Figure 4.20 Repeating contour patterns help to bind a melody together even if they are not exact repeats, as in the verse of 'Jingle Bells' (*a*). The contour may need to be adapted as it is 'draped over' the pitch staircase, as in Bach's fugue in D major from Book I of *The Well-Tempered Clavier* (*b*).

Figure 4.21 Arpeggio 'alphabets' in Beethoven's 'Moonlight' Sonata (*a*) and Bach's C major Prelude (*b*).

4.20*b*). This is the same basic contour but stretched or compressed to fit the constraints of the D major scale, so that whole-tone steps appear in one phrase where semitones appear in another. Diana Deutsch calls these building blocks 'pitch alphabets'. She points out that they are compiled from pieces of other, more general alphabets such as major and minor scales and arpeggios (the triad notes played in succession). The first movement of Beethoven's 'Moonlight' Sonata and Bach's C major Prelude from Book I of *The Well-Tempered Clavier* are two of the most famous examples of melodies made from arpeggio alphabets (Figure 4.21).*

Musicians and composers help us remember their music by including such similarities and connections between different segments of melodic material: gentle reminders, even if not exact repetitions of themes and ideas. In fact, music is *extraordinarily* repetitive. In folk and popular songs, learning is made easy by constant repetition: verse, chorus, verse, chorus. The next time someone bemoans the 'thump-thump-thump' of modern popular dance music, let them consider this: around ninety-four per cent of any material lasting longer than a few seconds that appears in musical pieces of cultures ranging from Inuit throat-singing to Norwegian polkas to Navajo war dances recurs more than once – and that is only taking account of verbatim repeats.

* The arpeggios of the 'Moonlight' Sonata are arguably the accompaniment, not the melody. But the melody is such a sparse one that the arpeggios effectively form a kind of tune of their own.

Sounds boring, perhaps? But Leonard Meyer argues that repetition in music 'never exists psychologically' – that we never quite hear the same thing twice. It's clearly a different experience, for example, to hear a theme for the first time and then to find it returning some time later. Only on the latter occasion do we think 'ah, this again'. The repeat can restore a mood. To build to a climax, come to a shuddering halt, and then resume with a calm passage heard at the outset, is a trick employed to the point of cliché in rock music, but is so devastatingly effective that it's easily forgiven. Think, for example, of the reprised verses in Pink Floyd's 'Shine On You Crazy Diamond' after Dave Gilmour's ecstatic slide-guitar solo, or in Jimi Hendrix's 'Voodoo Chile' after a tumultuous instrumental workout between Jimi and Steve Winwood's unbridled Hammond organ. In songs, constant repetition of the melody is palliated by lyrics that carry the narrative moving ever forward: that's what keeps us listening to the otherwise formulaic structures of Bob Dylan and Leonard Cohen. On the other hand, in rave music, Sufi qawwali, and the minimalism of Philip Glass and Terry Riley, insistent repetition creates a progressively deepening trance-like experience.

Besides, predictability can be pleasurable rather than boring. We will belt out the chorus of 'The Wild Rover' with gusto, not with ennui. And when a theme reappears unexpectedly, it is like bumping into an old friend. Knowing when to bring back a reprise is part of the art of good composition: once the recognition dawns, we know what is going to come next, and that can be delightful. Music theorist David Huron argues that the evolutionary benefit of accurate prediction makes us predisposed to be pleased when our expectations are fulfilled. Repetition creates expectations specific to the composition or context, and when these expectations are met it binds the music into a coherent and satisfying whole.

In much of classical music, repetition is a formalized affair. In sonata form, for instance, the thematic material is laid out in the exposition, mutated in the development, and then repeated with only minor alteration in the recapitulation. You don't need to know anything about sonata form to enjoy this sequence: you need only be able to memorize a tune for a short time, and to recognize it in altered forms. Repetition is even more explicit in the technique of variation, where a single theme is subjected to a series of reinventions. In principle this is a much simpler form than the sonata: there's no grand overarching structure, but just a series of vignettes based on the same idea. But the variation was developed with exquisite sophistication in the Baroque period, most notably by J. S. Bach, whose tricks of symmetry and pattern in the 'Goldberg' Variations continue to enchant musicologists and mathematicians today.

The conventions of most Western composition actually impose rather more regularity than is required for us to notice the repetition. For example, a sonata recapitulates the initial melody in the same key, but listeners apparently aren't too bothered about that kind of consistency (unless they have perfect pitch). In a test where classical works were played to music students in an altered form, so as to end in a different key from the initial one, most of them didn't notice.

This illustrates a seeming paradox in musical memory. We generally forget things quickly, even if we have a great deal of sophisticated musical training. And yet we can perform the most extraordinary feats of memory without it. A single hearing of a piece of music can lodge in our minds, out of sight as it were, for virtually a lifetime. I once heard Ian Dury and the Blockheads play a new song at a concert, and recognized it instantly the next time I heard it fifteen years later. That, of course, is trifling compared with the jaw-dropping feats of memory that some great musicians have displayed. At the age of fourteen, Mozart is said to have written down from memory the entire score of the choral *Miserere* of Gregorio Allegri after hearing it once in the Sistine Chapel (he made some minor corrections after a second hearing two days later). Although the papacy had forbidden any transcription of the *Miserere*, Pope Clement XIV was so impressed by the young lad's genius that he gave him a knighthood. Felix Mendelssohn, who made another transcription of the *Miserere* in 1831, was also said to possess the gift of total musical recall.

This kind of feat is celebrated in the movie *The Legend of 1900*, in which Tim Roth plays a piano virtuoso who never leaves the ocean liner on which he was born. In a piano duel with the conceited jazz maestro Jelly Roll Morton, Roth casually plays back note for note the piece Jelly Roll has just improvised. It sounds exaggerated, but autistic musical savants have shown comparable capabilities. Understanding the interplay between musical structure and memory makes such displays less astonishing: what is being remembered here is not an arbitrary series of notes but a logical, hierarchical system of patterns. Not all of these pertain to the melody, of course, but that can serve as the framework on which the other elements are hung. Baroque music such as Allegri's *Miserere* tended to be highly formulaic, so that once you have the tune it is not so hard to deduce what the harmonies should be. That's not to downplay the prowess of the fourteen-year-old Mozart, but rather to emphasize that music is *meant* to be memorable, and is generally written that way.

Breaking the hierarchy

In the finale of his Second String Quartet, written in 1907, Arnold Schoenberg declined to indicate a key signature. The lack of any sharps or flats at the beginning of the score didn't mean, as it usually does, that the piece was in C major (or A minor). It meant that it was in no key at all. Schoenberg was admitting that the notion of a key had no meaning for this music, because the notes weren't structured around any scale or tonic. The piece was *atonal*.

In view of what we've learnt in this chapter, we can see more clearly what this notion implies. It is not the abandonment of a key signature per se that matters. Other composers, such as Erik Satie, had previously omitted any initial indication of key, finding it more convenient simply to annotate the various sharps and flats as they occurred. And conversely, Schoenberg could have written his finale with a key signature but merely added accidentals where needed. The reason this music can be considered atonal is not because it dispenses with a tonic in a formal sense, but because it does so in *perceptual* terms: we can't make out where the piece is rooted. In other words, the tonal hierarchy doesn't apply to this music – it isn't a good guide to what notes we should expect.

But in 1907, everyone in Western culture had grown up hearing and learning this tonal hierarchy. They still do today. And we will instinctively try to apply it to atonal music, which is why many people find such music baffling: they have no map for navigating it.

Didn't I say earlier that new tonal hierarchies can be learnt rather quickly, for example when listening to the music of other cultures? Yes – but the point about Schoenberg's atonality is not that it has a different tonal hierarchy. This music has *none*.

And that was quite deliberate. Schoenberg designed his compositional method to explicitly suppress such a hierarchy. He recognized how avidly we look for a tonic, and saw too that we do this on the basis of note statistics: we assign the most common note as the tonic. 'The emphasis given to a tone by a premature repetition is capable of heightening it to the rank of a tonic,' Schoenberg wrote in 1948. In order to remove all traces of tonality, it is not enough simply to use lots of notes that lie outside the diatonic scale. We have to make sure that no note is played more often than any other.

This is the objective of Schoenberg's *serial* or *twelve-tone* scheme. Every note in the chromatic scale is arranged in a particular sequence, and this series of twelve notes has to be sounded in its entirety before it may repeat. In this way, no tone can acquire any more significance than any other, and so there is no possibility of a sense of a tonic note emerging even by chance. All notes are equally salient: the hierarchy is flattened by fiat.

This might sound like an absurdly constrained way to compose, but Schoenberg provided rules for creating variety by manipulating the 'tone row' on which any specific piece is based. The order of notes may be reversed, for example, and each note can be sounded in any octave (Figure 4.22). Individual notes can also be repeated before the next one is sounded. And in Schoenberg's original scheme, the composer was free to choose rhythm, dynamics and so on (later serialists, notably Pierre Boulez, placed these parameters under strict, rule-bound constraint too).

Serialist atonalism purposely sets us adrift from any locus to which expectations of 'the next note' can be pinned. To many listeners, this simply provokes irritation, exasperation or boredom – the music seems incomprehensible, and they feel the composer is just being obtuse. To others, the effect is pleasantly arousing: like any other confusion of our expectations, it stimulates careful listening and a sense of tension. That tension can never be resolved in the way it is for tonal music – there is 'no way home' to a stable tonic centre – but atonalism can perform a delicious juggling act, offering little hints of structure and logic that keep our attention engaged.

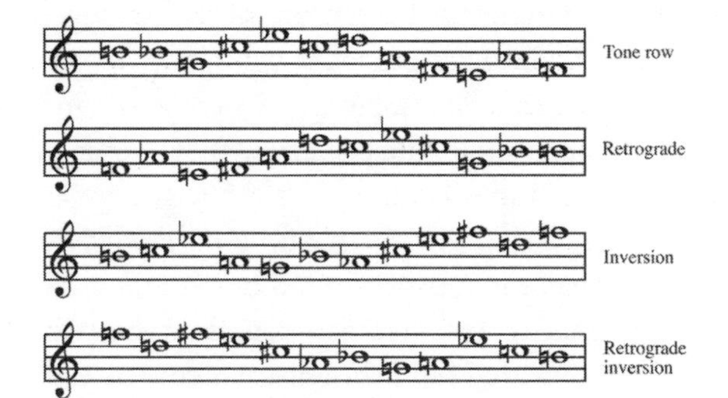

Figure 4.22 Permitted transmutations of the tone row in Schoenberg's serial method.

These confusions become clear when subjects are tested for their responses to serialist compositions using Carol Krumhansl's 'probe tone' method. Here, tone rows or atonal musical excerpts are used to establish the context, and then listeners are asked to judge the fittingness of each tone in the chromatic scale. But unlike the case for tonal music, their responses vary hugely and it is hard to find any general principles governing how they try to organize what they hear. Many listeners show signs of trying to apply the rules they've internalized from tonal music, only to find them frustrated. Subjects with more training in atonalism show the opposite behaviour: having learnt to expect no tonal root, their responses seem governed by an expectation that the music will shift away from any hint of tonality. At any rate, this music seems to be pretty effective at achieving its objective of banishing tonal organization.

The obvious question is: why do that? What did Schoenberg find so repugnant about tonality that he needed to stamp out all traces of it? Some musicologists portray this as an act not of prohibition but of emancipation: a 'liberation of the dissonance', a demand that we cease to consider some combinations of notes to be cacophonous and forbidden. To this extent, Schoenberg's method seems to be the logical end point of a trend that had been growing in music for almost a century: a gradual experimentation with chromaticism and unconventional harmony, taking a route from Beethoven and Chopin to Debussy, Wagner and Richard Strauss. When you listen to Schoenberg's composition *Verklärte Nacht*, written in 1899 before he devised the twelve-tone technique, you hear the sound of conventional Western tonality in its anguished death throes: music on the brink of falling apart. One critic said of it (accurately, though with insulting intent) that 'it sounds as if someone has smeared the score of *Tristan* while it was still wet'.

But by the start of the twentieth century, composers could use just about any amount of dissonance they fancied. True, they wouldn't always be thanked for it – the audience rioted at the premiere of Schoenberg's Second String Quartet in Vienna in 1908, and even *Verklärte Nacht* was controversial when it premiered in 1902. But the musical public was often far more receptive to new sounds at this time than is implied by the much repeated (and misrepresented) story of the *Rite of Spring* riot in 1913.

Yet Schoenberg created serialism not so much in order to do something new, as to avoid doing something old. This is the fundamental problem with it: for all the talk of liberation, it was in fact a system designed to *exclude*. What it wanted to exclude was any vestige of tonality. And the reason for that was not musical, but philosophical – some would even say political. To Schoenberg, tonality needed to be banished because it had become a tired, clichéd reflex, the musical language of a complacent and decadent social stratum.

I look more closely at what he meant by this in Chapter 11, but let me allow here that he was probably right in many ways – the Beethoven-like formulas of Sibelius can sound anachronistic, like a musical dead end. Yet there was no good *musical* argument for why twelve-tone composition was the inevitable answer. Indeed, there is good reason to consider it an *anti-musical* device. I realize that this sounds terribly reactionary, but I say it as someone who enjoys at least some of this music. It's not the threadbare complaint that Vivaldi had more tunes; what I mean is that serialism actively undermines some of the basic cognitive principles by which notes become music in our brains. Schoenberg was correct to say that his method provides a rough and ready means to eliminate tonality; but he seems to have had no real notion of what to put in its place.

For we've seen now that the concept of tonality is not some arbitrary convention; it plays a *cognitive* role in music. Tonal organization creates a framework for making sense of notes, arranging them in a hierarchy that gives us both a sense of orientation and a series of landmarks for navigating a melodic line. It helps us at the very level of perception: at age six or seven, children are found to be better at distinguishing when a tone has been altered between two melodies if they are tonal rather than atonal – the tonal hierarchy provides structural information to aid the comparison. This conceptual toolkit is supplemented by our implicit knowledge of the statistics of interval sizes and by the way we encode melodies according to their contours. Twelve-tone composition undermines these aids to cognition too. It contradicts our intuition that small intervals are more common, and thus more expected, than large ones. Indeed, it insists that there is no distinction between the two, since each note in the row can be sounded in any octave. And as a result, the serial 'melody' typically has jagged, irregular contours, constantly undermining any sense of continuity. It is for this reason (and *not* because it eschews the diatonic scales) that detractors of Schoenberg's music are right in a sense to say that it has no melody, no 'tune'.

They are *not* right to say that, in consequence, it is not music. As we'll discover, there are many other ways than melody to create coherence in a musical form. Yet did Schoenberg and his followers provide any?

I would say that sometimes they did. But this was not because the twelve-tone method supplies it, but rather because it neglects to eliminate it – or perhaps, more generously, because of the irrepressible musicality of some of those who used the technique. Personally, I experience no cognitive difficulty in listening to Alban Berg's *Lyric Suite* (1925–6), which is full of intelligible order in its forms and dynamics, and is indeed rather thrilling. Somehow Berg marshals the tone row in a way that creates a genuine sense of melody, despite the lack of any centre to which it might be pinned.

And twelve-tone music may occasionally fail to stamp out tonality after all, despite the fact that strict adherence to a tone row must by definition flatten out the probability distribution of pitch classes. Consider the tone row in Figure 4.23, which obeys Schoenberg's rules. It starts with the ascending major scale of C, and ends with the descending pentatonic scale in F♯. So it will create two *local* sensations of tonality, in C and F♯ – not because of note statistics, but because of our learnt associations of groups of notes with diatonic scales and pitch steps, in the same way that the sequence E, F, G implies a key of C.

If one were to choose tone rows at random, one would quite often find little groupings like this (if rather less extreme) that create a momentary sense of tonality. Some twelve-tone composers, including

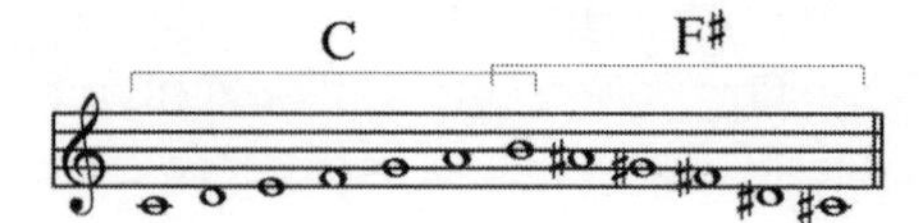

Figure 4.23 A tone row that splits into two distinct tonalities, in C major and F♯ pentatonic.

Stravinsky and indeed even Schoenberg himself in some late works, used rows that create momentary tonal effects in this way.* But David Huron has looked at the statistics of Schoenberg's tone rows, and finds that on average they have fewer local groups of notes that give a hint of tonality than a random selection would offer. In other words, it seems that Schoenberg tended to select the rows that were most effective at suppressing tonality. For this reason, Huron argues that serial composition should not be regarded as 'atonal' but as deliberately *contratonal*: not casually ignoring tonality, but taking pains to eliminate all vestiges of it. It seems that Schoenberg did this unconsciously – there is no sign he was aware that the tone rows needed pruning to achieve his contratonal objective.

Does twelve-tone 'melody' really lack schemes to make it intelligible in itself? One obvious candidate for an organizational principle is the tone row itself. Won't it, simply by dint of repetition, supply a new coherent structure?

But the tone row is simply an arrangement of notes, whereas a genuine melody displays some logic in the way one note follows another – a relationship between what has come before and what comes next. It is, in this sense, an *elaboration*, just as language and movies are not mere permutations of words or images. I don't mean that, by possessing these properties, a melody will have a 'nice tune' (nor need it); but it will be cognitively coherent.

All the same, can't we get used to the tone row if we hear it again

* Both Schoenberg and Alban Berg in fact manipulated tone rows ingeniously to reconstruct 'tonal' effects such as cadences (see p. 181). Berg even managed to recreate the opening of the Prelude to *Tristan und Isolde* in his *Lyric Suite*, and part of a Bach chorale in his Violin Concerto. This seems bizarre for composers who are allegedly trying to escape tonality, especially as it is so hard to do. But Schoenberg's school was keen to show that there was a link between twelve-tone music and the earlier (especially the Germanic) tradition, so that the new technique was somehow validated as an extension and generalization of the old ways. In many ways Schoenberg was no revolutionary but a staunch traditionalist.

and again? Apparently not. For one thing, twelve items seems to be rather too many for the human brain easily to recall in sequence – just try memorizing a random series of a dozen numbers. And once it is rearranged by one of the manipulations permitted by Schoenberg, the tone row is transformed out of recognition (now try reciting those dozen numbers backwards). Writing in 1951, Schoenberg appeared to be optimistically imagining otherwise by analogy with vision: 'Just as our mind always recognizes, for instance, a knife, a bottle, or a watch, regardless of its position,' he wrote, 'even so a musical creator's mind can operate in the imagination in every possible position, regardless of their direction, regardless of the way in which a mirror might show the mutual relations, which remain a given quantity.' According to Diana Deutsch, Schoenberg's assumptions of 'perceptual equivalence under transposition, retrogression, inversion, and octave displacement are fundamental to the theory of twelve-tone composition'.

But tests show that tone rows altered in this way are rarely recognized as equivalent even by experts in twelve-tone music. We don't encode melodies in ways that allow it. The various transformations are apt to alter the melodic contour, which seems to supply our first and crudest mnemonic device. It's one thing to *know* that one sequence of notes is an inversion of another, but quite another to *hear* it.

Actually, it's not clear that Schoenberg ever really intended the permutations of a tone row to be heard by the listener. They were simply a *compositional principle*, a way of creating building blocks for assembling into music.* In the quote above, he speaks only of the 'musical creator's mind', not the listener's. The tone row was not a musical idea so much as a set of musical atoms.

The music theorist Allen Forte has claimed that serial music is organized according to so-called pitch-class sets, which are small groups of notes (more properly, of pitch classes, taking no account of octave) that recur in clusters either simultaneously (in chords) or sequentially

* Schoenberg seems somewhat ambivalent about this. He claimed that 'consciously used, the motif [tone row] should produce unity, relationship, coherence, logic, comprehensibility and fluency'. But even this doesn't explicitly suggest that these things are to be experienced by the audience, as opposed to the composer.

(in melody). Rather like the tone row itself, these sets are transformed in the composition according to various symmetry operations, such as inversions or cyclic permutations. The problem with this rather mathematical analysis is that it focuses only on the musical score and again takes no account of whether the sets are actually perceived. There is no indication that they are, and when you look at how some of them are plucked out of a deep embedding in the musical structure, it's not surprising (Figure 4.24). Whether or not pitch-class set theory elucidates any formal structure in atonal music, it apparently says nothing about how that music is heard – nothing, indeed, about it *as music*.

Perhaps that needn't surprise us very much, because it was in the very nature of serialism that it concerned itself ever less with what a listener heard and became ever more a glass-bead game for arranging notes. There's no better demonstration of this than Pierre Boulez's *Le Marteau sans Maître* (1954), a composition that sets the surrealist poems of René Char for an alto voice and six instruments, including guitar and vibraphone. Widely acclaimed when it was first performed, it nevertheless posed a puzzle: Boulez indicated that it was a serial piece, but no one could work out how. It wasn't until 1977 that the theorist Lev Koblyakov figured out the unconventional serial process Boulez had used. In other words, for over twenty years no one could deduce, let alone hear, the organizational 'structure' of this masterpiece. This doesn't mean that *Le Marteau* is wholly unlistenable: the unusual sonorities are, for a

Figure 4.24 Some of the (three-tone) pitch class sets in Schoenberg's *Three Piano Pieces*, Op. 11 (circled). According to the theory of Allen Forte (which derives from an earlier suggestion by Milton Babbitt), atonal music is structured around such sets, which act as motifs. But as you can see from this example – which is atonal but not serial – the pitch class sets do not really occur in a way that is audible; they can be perceived only by analysis of the score.

time, diverting in their own way. But it shows that there is no intelligible organization of pitch (and, one might add, of rhythm either). One can hardly blame audiences for suspecting that what is left is musically rather sparse.

Worth a try

Paul Hindemith was hardly timid in experimenting with chromaticism, but he had no truck with atonalism. Trying to avoid tonality, he sniffed, was 'as promising as attempts at avoiding the effects of gravitation'. The result was like 'those sickeningly wonderful merry-go-rounds on fairgrounds and in amusement parks, in which the pleasure-seeking visitor is tossed around simultaneously in circles, and up and down, and sideways'. Future generations, he concluded, 'will probably never understand why music ever went into competition with so powerful an adversary'.

I don't share Hindemith's acerbic view of Schoenberg's experiment. One can hardly blame Schoenberg for not heeding principles of music cognition that had yet to be discovered. And experimentation in music should be welcomed, however extreme. But it must be regarded as precisely that: an experiment, which by definition may or may not work. Schoenberg's experiment did work to the extent that it led to new sonorities, new possibilities for finding form in music. It *did* liberate composers in some ways, and in the hands of artists like Stravinsky, Messiaen and Penderecki atonalism could become a vibrant force. The real problem was, as so often it is in art, that innovations driven primarily by philosophical or ideological motivations lack a tradition from which to draw. Good art works not because some theory says it should but because it is embedded in a web of reference and allusion, as well as convention – it takes what we know, and changes it. But it has to start from something we know, however iconoclastic. You can't just make it up – or if you do, you can't expect your new rules to be cognitively coherent. You can't simply break down centuries of experience in hearing the organization of notes by removing a tonic or a tonal hierarchy. With practice, we can change the way we listen. But some extreme forms of serialism may become tolerable only because we become inured to it, and not because there is really anything to hear – nothing except notes and silence, a meandering uniformity with no natural mechanisms for creating tension and release, or for beginning and ending. As Roger Scruton puts it, 'When the music goes everywhere, it also goes nowhere.'

If this sounds reactionary, let me add that the same considerations strike a blow at a more conventional tradition. Operatic recitative has a respectable *raison d'être*, being an attempt by Italian composers in the Renaissance to imitate what they thought to be the speech-like singing style of ancient Greek melodrama. But it is plainly not 'musical' in any real sense: its indefinite contours, irregular rhythms and almost absent melodies don't do anything to aid cognition. It is, in short, an anachronistic affection, and it is hard to see how its inclusion in opera, as well as oratorios and cantatas, can be considered to enhance their musical value. No doubt one can become inured to it, and clearly many people are. But it is a curious compositional method that relies on the listener switching off to unmelodic material in what is supposed to be a tonal composition. As the nineteenth-century music theorist Eduard Hanslick said, 'in the recitative music degenerates into a mere shadow and relinquishes its individual sphere of action altogether.'

As we come to understand more about the ways we process melody, we gain the advantage of being able to discuss these things objectively: to explain *why* we think a piece of music sounds odd or tuneless or difficult. But it would be sad if it did no more than whet our critical scalpel. I hope you can see that it may also encourage us towards new ways of listening and composing (and away from unproductive ones). It may help us to *hear* more, and hear better. We've already got the tools.

5
Legato
Keeping It Together

How do we decode the sound?

It seems quiet in the attic where I work – until I listen. Then I realize just how much is going on around me. There is the sound of children's voices from neighbouring gardens. I can hear that my own daughter is watching *The Sound of Music* (for the fiftieth time) downstairs. A helicopter is buzzing somewhere above, and there is a steady grumble of traffic from the city, punctuated by the whooshes of nearby vehicles and the wails of ambulance sirens. The trees are whispering, and so, lower pitched, is my computer. There goes an aeroplane, descending towards Heathrow. And the birds! I'd hardly noticed them, but they are in full voice at the end of the afternoon.

I can see almost none of these things – I detect their presence by noise alone. But I don't exactly *hear* a helicopter or a laugh: I pluck these things from the steady, unbroken stream of sound that washes around me.

Music is like that. The music of Charles Ives is sometimes very much like that. Whereas the massed instruments of an orchestra generally produce sounds that relate to one another, Ives wrote for what seemed to be several simultaneous voices that had nothing in common. In some compositions he used two separate orchestras playing different tunes in conflicting keys or rhythms, almost daring us to see if we can weave them into a coherent entity. The most renowned example is *Central Park in the Dark*, composed contemporaneously with Schoenberg's break with tonality and arguably a more radical gesture than anything in atonalism. Ives said of this that he wanted to recreate the ambient barrage of conflicting sounds we often hear in the world, calling his composition

> a picture-in-sounds of the sounds of nature and of happenings that man would hear some thirty or so years ago (before the combustion engine and radio monopolized the earth and air), when sitting on a bench in Central Park on a hot summer night. The strings represent the night sounds and silent darkness – interrupted by sounds [the rest of the orchestra] from the casino over the pond – of street singers coming up from the [Columbus] Circle singing, in spots, the tunes of those days – of some 'night owls' from Healy's whistling the latest [hit] or the Freshman March – of the 'occasional elevated', a street parade, or a 'break-down' in the distance – of newsboys crying 'uxtries' ['Extra!'] – of pianolas having a ragtime war in the apartment house 'over the garden wall', a street car and a street band join in the chorus – a fire engine, a cab horse runs away, lands 'over the fence and out', the wayfarers shout – again the darkness is heard – an echo over the pond – and we walk home.

It seems like a recipe for chaos. But what it sounds like is life.

Aaron Copland attempted to give modern listeners a graded list of contemporary music, so that they had some idea of what they might be letting themselves in for. Shostakovich is 'very easy', he said, and Britten 'quite approachable'. Bartók was 'fairly difficult' – but Ives was 'very tough'. I'm not so sure. What we have to do in listening to *Central Park in the Dark* is what we do every day. It was what I was doing just now, with so little effort that I was barely aware of it. What might be tough is accepting that music is allowed to do this sort of thing. But once we've managed that, we have all the cognitive apparatus we need to appreciate it.

Copland said that what composers would like to know is this: are listeners hearing everything that is going on? Listen to Ives, to Bartók's string quartets, to John Coltrane's *Blue Train*, to Javanese gamelan, and you can reasonably say one thing for sure: there is an awful lot going on. The raw sound you're receiving is almost absurdly complex, a mishmash of many pitches and timbres and rhythms. The acoustic events pile up one after the other, overlapping and merging. The mystery is not why some types of music are 'hard' to understand, but why we can make sense of any music at all.

Any? Surely a child singing a nursery rhyme in a pure, high voice is simplicity itself. But there is great acoustic richness even in that ingenuous signal. As we've seen already, to pick out a simple tune like this, our brains must work some clever magic: deciding how to group harmonics into single notes, how to determine a tonal centre, how to break up the melody into phrases.

I hinted in the previous chapter at the role that the so-called gestalt principles of perception play in this disentangling. They are the means by which we create pattern and order from complex stimuli, both in vision and in audition. In this chapter I will look at these principles in more detail. We'll see that, not only are they innate to the way we experience the world but they have also been assimilated and codified in musical practice for a long time, accounting for many of its rules and conventions. It is our reliance on these principles that makes us inherently musical beings.

Simplifying the world

When Albert Einstein said that things should be made as simple as possible but not simpler, he was talking about scientific theory. But he might as well have been referring to life. The art of appropriate simplification is part of the art of survival. We once needed to learn what particular species of dangerous animal sounded like, or how to distinguish juicy berries from poisonous ones. The trouble is that the distinctions aren't always clear. It's best to err on the side of caution, but not too much – or we'll be cowering from harmless creatures and passing up nutritious fare. We don't always need to be right; we just need senses that are good enough.

What we're really seeking here are meaningful associations between groups of stimuli: does *this* belong with *that*? And our default position is that it probably does, unless there's good reason to think otherwise: we are pattern-seekers. 'The mind', says Leonard Meyer, 'will tend to apprehend a group of stimuli as a pattern or shape if there is any possible way of relating the stimuli to one another.' We'll even apply a little distortion to what is actually seen or heard, if that makes the pattern simpler. The gestalt psychologists called this the law of *Prägnanz*, meaning succinctness or conciseness.

This group of German-based psychologists was centred around Max Wertheimer, Christian von Ehrenfels, Wolfgang Köhler and Kurt Koffka, who were active in the late nineteenth and early twentieth centuries.

They argued that the mind possesses holistic organizing tendencies that make perceived experience more than the sum of its parts. A visual scene is in one sense just a jumble of coloured patches.* By grouping and separating the patches into discrete objects, we make the world an intelligible place. And we learn to generalize and classify these objects. We do not need to have seen a tree before to recognize it as such, even though it differs from every other tree we have seen: we know what 'treeness' looks like. A cartoon cat, or a child's drawing of one, are recognizable even though they are clearly far from photorealist portraits. We form continuous objects from fragments, for example if more distant objects are broken up by intervening ones. And we learn to assume continuity of movement. An object that passes out of sight of a small child has ceased to exist for them, but we soon come to expect that when an aeroplane passes behind a cloud, it will appear on the other side: we are not so easily fooled by the 'surface' of the perceptual field.

The gestalt principles are given rather grand-sounding names, but they are easy to understand with visual analogies. The principle of similarity is obvious enough: we might associate together all the white objects we can see, say, or all round ones (Figure 5.1*a*). Or we group objects that are close together, owing to the principle of proximity (Figure 5.1*b*). The principle of good continuation makes us predisposed to splice objects together if they have smooth contours, like a telegraph line passing behind a pole (Figure 5.1*c*). This is the reason why we see an X as two diagonal lines crossing, and not as a V on top of an inverted V: the former interpretation 'enables' both lines to keep going in the same direction. It is easy to see why an assumption of continuity was helpful to hunters following their prey as it darts between rocks and trees. An ability to weave together pieces into a coherent but partially hidden whole may depend, however, on being able to see what is obscuring it: compare the effect in Figure 5.2*a* with that when the obscuring template is made explicit in Figure 5.2*b*. This is sometimes called the picket-fence effect.

Finally, we group moving objects that are travelling in the same direction: the principle of common fate. If a lone bird in a flock begins

*There is distance information in that signal, at least if we have binocular vision, because of the parallax we experience from two eyes. But that just means we see three-dimensional coloured patches.

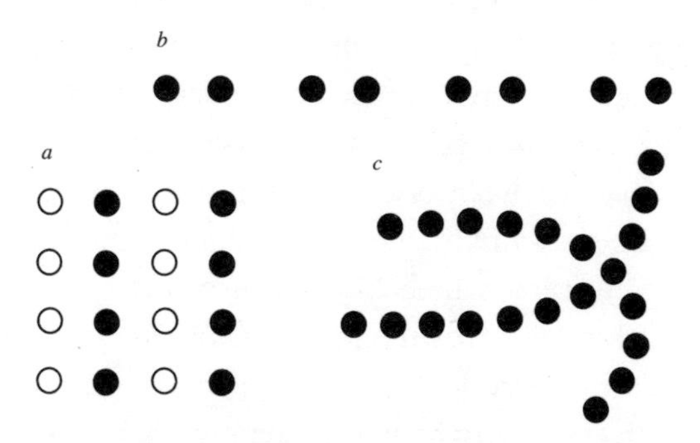

Figure 5.1 The gestalt principles of similarity (*a*), proximity (*b*), good continuation (*c*).

to follow a different trajectory from the others, we can identify it immediately. This is harder to convey in a static image, but it can be revealed very simply by drawing a cloud of dots on a transparent sheet and placing this over another cloud drawn on paper. The two sets of dots seem to form a single mass while static, but they become two easily distinguishable masses when the top sheet is moved slowly. Because music invokes a sense of motion in the trajectories of voices and melodies, there is a psychoacoustic analogue of this.

You'll notice that our comprehension of the visual scenes here requires no conscious effort: we see the patterns at a glance. In cognitive terms, we can *parse* the stimuli very readily. This word derives from linguistics, where again we find that unconscious assimilation of rules enables us to extract meaning without effort: you are not having laboriously to identify the verbs, nouns, prepositions and so

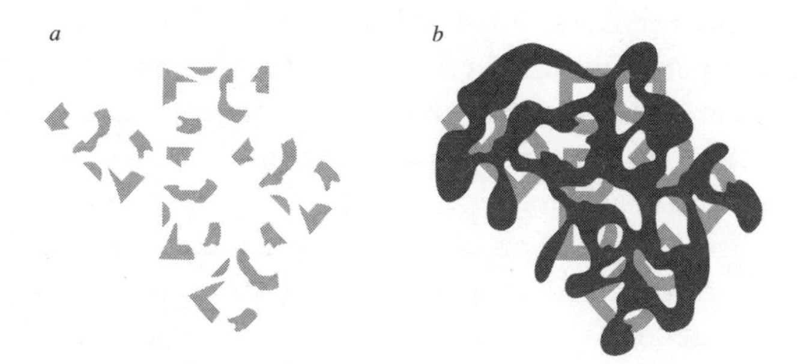

Figure 5.2 The picket-fence effect. The fragments in (*a*) seem to have no obvious structure, but when the occluding surface is made explicit in (*b*), they become ordered into groups so as to imply familiar, continuous objects behind it.

forth in these sentences before you can understand what I am saying. The gestalt principles operate 'out of sight', as it were.

All of these visual principles have sonic analogues, and this means that music is only indirectly related to the acoustic signals generated by the performers. What we 'hear' is an *interpretation*, a best guess in which our brains seek to simplify the complicated soundscape by applying the gestalt principles, which have been found from experience to do a fairly reliable job of turning the sound into a hypothesis about the processes that created it.

Grouping by similarity is evident in how we separate notes according to timbre, hearing the overlapping voices of violin and piano in a duet as two separate lines rather than a soup of notes. Grouping by proximity happens, for example, in the way we assign notes to the same melody if they are close in pitch, but may interpret big pitch jumps as a new start. We might also group according to the spatial proximity of sound sources. Our auditory system is rather good at locating the source of a sound by analysing the differences in the sound arriving at the left and the right ear, comparable to the parallax of binocular vision. The adaptive advantage of being able to tell where a sound is coming from is pretty obvious, whether one is in the role of predator or prey. We assume that sounds that appear to come from the same point in space originate from the same physical source: we bind those sounds mentally together in a 'stream'.

We encountered the principle of good continuation in the last chapter in the perception of melody. There is an auditory picket-fence effect too, and for much the same pragmatic reasons: in a noisy environment, it is advantageous to be able to keep track of a single sound even if it is temporarily obscured. When segments of rising and falling pitch are heard in isolation, separated by silence, they are perceived as discrete stimuli (Figure 5.3). But if the same segments are interrupted by bursts of noise, they seem to constitute a continuous tone that is rising and falling but is periodically drowned in noise.

Even babies can split complex sounds into distinct streams in this way, disentangling their mother's voice from background noises (probably on the basis of timbre) so that, when they begin to mimic the sounds of their mother's voice, they don't try to reproduce all the ambient sounds that might accompany it. We've already seen the stubborn tenacity with which these grouping principles can be applied: if

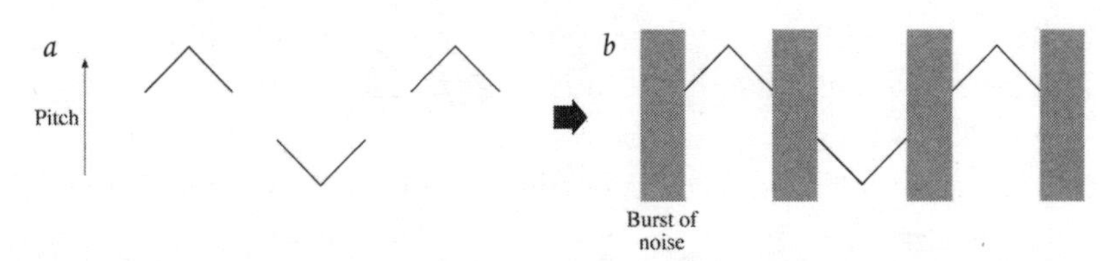

Figure 5.3 The auditory picket-fence effect: fragments of rising and falling pitch are heard as isolated when separated by silence (*a*), but when they are separated by an 'occluding' burst of noise, they are heard as a continuous tone of smoothly varying pitch (*b*).

an overtone of a complex tone is detuned, making it anomalous in the harmonic series, the brain is reluctant to let it go free (p. 67). Instead, what we hear is a pitch shift of the fundamental frequency, as the brain struggles to do the best it can in simplifying conflicting information.

In listening to music, streaming of sound is conducted in a very selective and sophisticated way. In general, we can distinguish a singer's voice from the sound of the backing band, and will usually be able to tell apart the guitar, piano and so forth. But the separation must not be too extreme. We want to retain a perception that the performers are all playing together rather than doodling individually. We want to be able to hear harmonies. And as we saw earlier, the notes themselves are each a tightly bound package of harmonics, so that the overtones of one note don't get jumbled up with those of another. It seems clear that this delicate balance of streaming (separating) and binding of the sound is something the brain must do quite deliberately. Oliver Sacks tells of a highly musical patient who lost the ability to hear harmony after a severe head injury, so that she could no longer integrate the four voices of a string quartet – she described them as 'four thin, sharp laser beams, beaming to four different directions'. With a full orchestra there were twenty such beams, which the agonized patient could no longer combine into a structure that made sense. This, one imagines, might be what music would sound like for all of us if it were not for our skill in juggling separation and integration. In fact music often teeters on the brink of this accommodation, and interesting or obstructive effects can follow from moving back and forth across the boundaries of coherence.

For instance, the various gestalt principles can compete with one

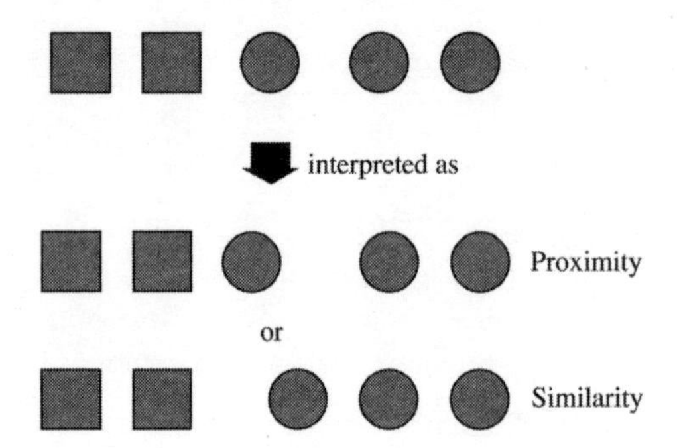

Figure 5.4 In the top arrangement, grouping of objects by similarity is in tension with grouping by proximity: there are two possible grouping schemes, and it is not clear which should predominate.

another, presenting us with alternative or ambiguous ways to conduct the grouping. In Figure 5.4 our tendency to group by similarity (squares versus circles) struggles against our tendency to group by proximity. This offers two different 'interpretations' of the image. Music theorist David Temperley believes that we may experience and interpret music by analogous 'preference rules' which bring order to the mass of auditory data, enabling us to figure out the beat and metre, the key, mode and so forth. These preferences are reinforced or contradicted by musical elements such as dynamics, timbre, length of notes, or their register (whether they have high or low pitch). Any interpretation is subject to constant updating as we receive new information, and so we may switch back and forth between different ways of hearing. Later we'll see several examples of such ambiguities in music.

Hearing voices

Many forms of music throughout the world are *monophonic*: they contain a single voice, whether that is literally a human singing voice or an instrument, or perhaps (as seems to have been the case in ancient Greece) the two sounding in unison. Although even here we must perform some grouping – to unify harmonic series, say, and to recognize melody – the sound is relatively simple to interpret.

But from around the ninth century, Western music began to use

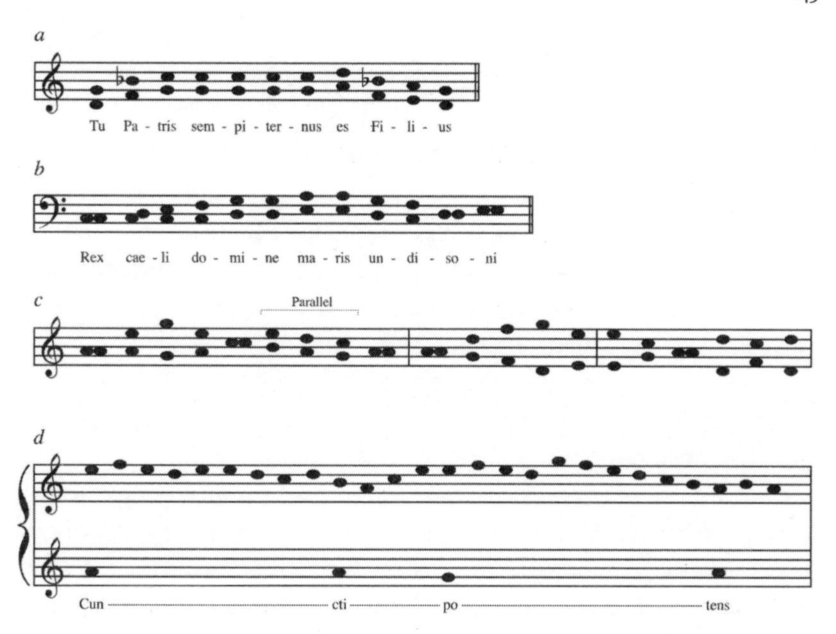

Figure 5.5 Organum in parallel fourths (*a*), with diverging voices (*b*), with 'mirror' lines (*c*), and with melisma, sometimes called florid organum (*d*).

several simultaneous voices: it was *polyphonic.** At first this was restricted to identical melodies moving in parallel a certain interval apart, generally in one of the classical consonances of a fourth, a fifth or an octave (Figure 5.5*a*). This form was called organum. A slightly more adventurous type of organum later developed in which voices started in unison but diverged to a fixed interval before eventually converging again (Figure 5.5*b*). By the end of the eleventh century, the variants were more sophisticated, combining these parallel and 'oblique' (divergent) motions with contrary motion, where one voice might move in mirror image of the other (Figure 5.5*c*). In the twelfth century, so-called florid organum introduced differences in rhythm too: the lower voice might take sustained notes in the original plain-chant melody while the upper voice took off on a freer, undulating

* Polyphony is a slightly ambiguous term with more than one connotation. Strictly speaking, it should be applied to any music in which more than just a single note is sounded at any instant. (Music in which several voices, or voices and instruments, play in unison is still monophonic.) But often the term is used to refer to music in which there is more than one simultaneous melody, as opposed to, say, merely a harmonization of a single melody. The latter is sometimes called instead *homophony* – see p. 159.

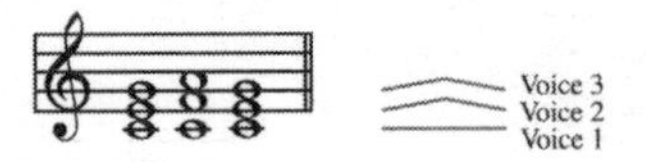

Figure 5.6 Voice-leading: each tier of the chords follows a melodic contour.

journey (Figure 5.5*d*). If both voices use the same words, then clearly the upper one will employ several notes for each syllable while the lower voice uses just one. This practice of prolonging syllables over several notes is called melisma, and is perhaps most familiar from the carol 'Ding Dong Merrily on High', where the word 'Gloria' is extended in a rather extreme melisma. Melismatic music is also the norm in many other cultures, particularly in the Middle East, where the words of songs are drawn out in long, wavering lines.

The art of Western polyphonic music became increasingly refined during the Middle Ages, exemplified by a famous four-part setting of the Mass, *Messe de Notre Dame*, by the French composer Guillaume de Machaut in 1364. Now listeners had to keep track of several voices at the same time. We know how hard that can be: if we hear several conversations at once, it is difficult to focus on one without being distracted by the others. (Psychoacoustics researchers know this by the graphic name of the cocktail-party effect.) But in polyphonic music, the whole point is that you *don't* just fix on one voice and ignore the others; the idea is that they are all blended into a harmonious, integrated whole. How is that possible?

In the Baroque era polyphonic music arguably reached its most refined state, and the governing rules became codified in the technique known as counterpoint. In these working practices we can see that composers had developed an empirical understanding of the factors that allow several voices to be combined in comprehensible fashion. For these contrapuntal composers were already making use of the gestalt principles.

The aim in this music is to create an uncluttered, unambiguous sonic 'scene' populated with clearly visible and distinguishable auditory 'objects' or streams. The composer needs to maintain the continuity and coherence of each voice while making sure that they don't merge. But the scene mustn't be so clear and obvious that it becomes boring.

This art comes from careful planning of the relationship between the movements of the different voices or melodic lines, which musicologists call (none too helpfully) 'voice-leading'. The very simple musical figure in Figure 5.6 can be heard as a series of three chords:

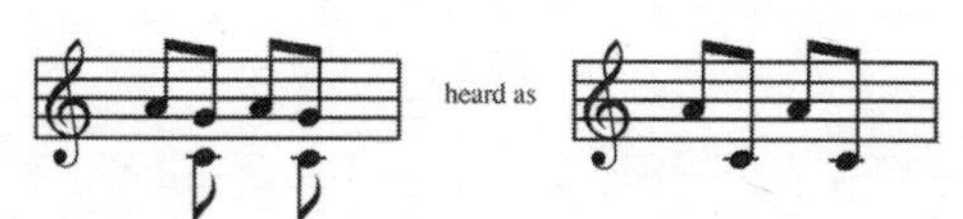

Figure 5.7 Close harmonies of an octave or a fifth can lead to the fusion of voices, whereby the fundamental captures higher notes that are heard as overtones.

C to F and back to C. But it can also be considered as a voice-leading of three voices: the top one moves from G to A and back again, the middle one from E to F and back, while the lowest voice remains on C. Good contrapuntal practice insists that these three voices remain distinct, so that we don't just hear the sequence as three 'block chords'.

This requires that the voices don't get perceptually entangled. Merging can happen because our auditory system combines overtones into single notes: it is a hazard of having *too much* consonance between the voices. So although Pythagorean theory, which provided the principal intellectual framework for understanding music during the early Middle Ages, seemed to recommend uniform consonances such as fourths, fifth and octaves, musicians discovered that voices spaced in these intervals and moving rigidly in parallel tend be heard as one. In fact, whenever two voices come together in a fifth or an octave – or worst of all, a unison – there is a danger that they will fuse like two converging rivers, so that we lose track of one or the other. In Figure 5.7, for example, the higher note (G) of the harmony can be 'captured' by the lower one (C), with the mind deciding that it is a mere overtone, so that what is heard is just an alternation of two single notes. It is probably for this reason that J. S. Bach's counterpoint uses these highly consonant intervals less frequently than we would expect at random: he seems to avoid them on purpose. Moreover, he does so increasingly as the tendency for fusion increases, so that there are fewer octaves than fifths, and fewer unisons than octaves. And when Bach does use these intervals that are liable to fuse, he often prevents merging by making sure that the two notes start at different moments – an auditory cue that tends to keep notes segregated into different streams.

For the same reason of preventing fusion, counterpoint usually avoids parallel (or even approximately parallel) motions of voices. This came to be seen as a guiding rule of composition that was to be observed even when composers were not writing counterpoint. And so we very rarely find chord sequences in the Classical and Romantic

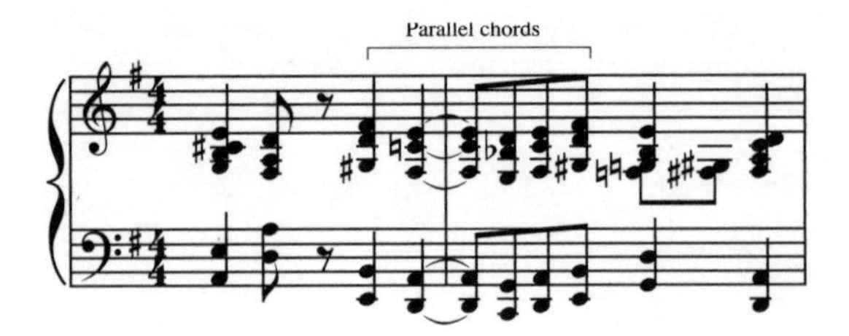

Figure 5.8 The parallel chords in Ravel's *Pavane pour une infante défunte* break the classical rules of voice leading.

eras that involve simultaneous, identical movements of all the notes.* There was actually no good reason for this prohibition, however, when the intention was not to maintain separate voices, although it wasn't until the late nineteenth century that composers such as Claude Debussy and Maurice Ravel dared to ignore it (Figure 5.8).

Another way to avoid fusion of contrapuntal voices is to start them at different times. That's why fugues, in which a single melodic subject is expounded by several simultaneous voices, employ staggered entries of the voices (Figure 5.9). And voices can become confused if they cross over in pitch, for example if one descends while the other ascends (Figure 5.10), just as it isn't clear, when roads meet at a crossroads, which is a 'continuation' of the other. The answer is to ensure that crossings never happen, and Bach makes sure that they don't.

The 'streaming' that remains feasible for voices marshalled within distinct pitch ranges was demonstrated in the 1970s by music psychologist Jay Dowling at the University of Texas at Dallas. He interleaved

Figure 5.9 The staggered entry of voices repeating the theme of fugues helps each to be identified and tracked independently. The example here is Bach's E Major Fugue from Book II of *The Well-Tempered Clavier*.

* It was considered particularly important to avoid parallel movement of fifth and octaves, since these create the greatest danger of fusion. Parallel thirds and sixths are much more common, because these are less consonant, and so we're less likely to interpret them as overtones.

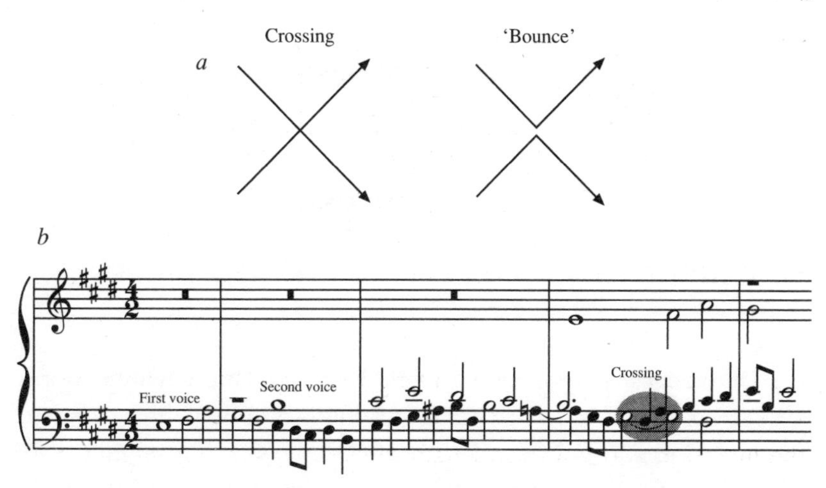

Figure 5.10 If two voices cross (*a*, left), they can become confused, because they then tend to be heard as a 'bounce' (*a*, right) in which the identities are exchanged at the point of 'contact'. Contrapuntal technique tends to ensure that such crossings are avoided. For example, in Bach's E Major Fugue shown earlier, the second voice could have been written as shown in (*b*), more precisely following the trajectories of the first and third voices. But this would have created a point of crossing, and so Bach avoided the F♯ that drops the second voice below the first.

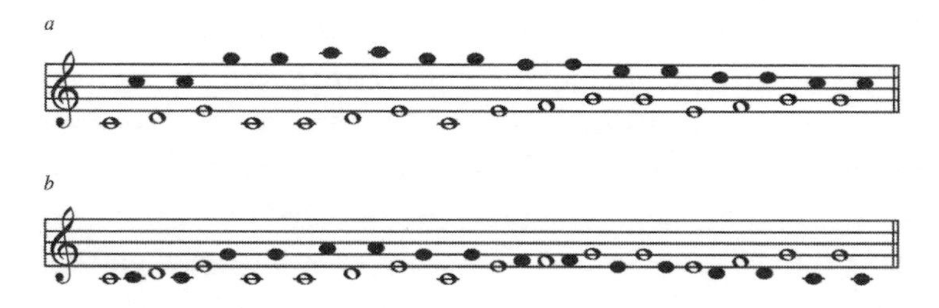

Figure 5.11 When familiar tunes are interleaved in registers that don't overlap – for example, an octave apart – they are both easy to identify aurally (*a*). But if the registers overlap, they become all but incomprehensible (*b*).

two familiar melodies, such as 'Frère Jacques' and 'Twinkle Twinkle Little Star', by playing alternating notes from each (Figure 5.11), and asked listeners to identify them. They found it almost impossible when the pitches of the two tunes overlapped significantly, but fairly easy if the tunes were shifted into different pitch ranges.*

* Dowling found, however, that overlapping melodies could be perceived, albeit with difficulty, when listeners were told to look out for them specifically.

Figure 5.12 Virtual polyphony: in this extract from the E♭ Prelude in Book II of *The Well-Tempered Clavier*, repeated large pitch jumps separate the higher notes, marked in grey, into a distinct stream relative to the lower notes.

It might seem surprising that the tunes in these experiments could be recognized at all. After all, consider this:

> To the be lord or is not my to shepherd be, I that shall is not the want question

OK, so you figured it out, but it wasn't easy. Yet in music, it seems we can parse such a mixed-up pair of well-known phrases without too much effort. This shows not just how strongly our mind *wants* to find some sense in auditory stimuli, but how much assistance it gains from clues such as pitch – which might be compared to this:

> To ^{the} be ^{lord} or ^{is} not ^{my} to ^{shepherd} be, ^{I} that ^{shall} is ^{not} the ^{want} question

We saw earlier that large pitch jumps in melodies are less common than small ones, and they tend to fragment melodic continuity. In Dowling's experiments this fragmentation created two entirely separate streams from a single sequence of notes: not only did the listeners separate out the peaks from the troughs, but they then strung together all the peaks, and all the troughs, to make two melodies.

Baroque composers exploited this streaming effect to produce the auditory equivalent of optical illusions, by alternating large jumps in a melodic line to separate it into two streams that are heard more or less simultaneously. This is called virtual polyphony,* and there's an example in the Prelude in E♭ from Book II of Bach's *The Well-Tempered Clavier* (Figure 5.12). This splitting-off of pitch jumps into separate streams depends both on the size of the steps and the rate of alternation: for the effect to work, the jumps have to be quite big and follow in rapid succes-

* Strictly speaking, virtual polyphony refers to the situation in which there is strict alternation between the two streams. When the alternation is less orderly and more clustered, it is known as implied polyphony.

sion. Only if the notes last less than about a tenth of a second, and the leaps are bigger than three semitones, can the splitting be guaranteed.

If, on the other hand, the jumps are *small* and rapid, the discrete notes become blurred. That's really what a trill is, and it typically sounds like a single wavering note rather than an alternating series of two. The effect is rather like two lights alternately blinking on and off in a dark space: we perceive this as a single light moving back and forth between the two positions. In the same way, the mind 'simplifies' very fast pitch steps in a single direction into an upward sweep (glissando) of continuously varying pitch. That's how we tend to hear the rapid four-note runs on the flute at the start of the 'infernal dance' movement of Stravinsky's *Firebird*. Here the visual analogy is that of the scrolling LED message displays that often now announce station stops on trains. However hard we try, we can't see the letters moving smoothly across the screen for what they really are: a series of static dots blinking on and off.

It is often important during the course of a piece of music for a single voice or instrument to emerge from the backing. We *could* simply achieve that by making the solo voice much louder, but that's not very satisfactory – it might ruin the dynamic balance, and can leave the soloist little room for expression.* The gestalt principles provide alternative, more subtle ways to split a solo into a separate stream. In *The Unanswered Question*, Charles Ives uses the principle of grouping by proximity (here spatial location) to attain distinctness in the plaintive, repeated 'question' asked by the trumpet: the instrumentalist is often placed in a different part of the performance space from the rest of the orchestra, such as a gallery. The Greek composer Iannis Xenakis has also experimented with placing musicians among the audience, although here the intention may be more political – to eliminate boundaries between performer and listener – than musical.

A rather neat trick that instrumentalists will use if they don't play fixed-pitch instruments (such as the piano) is to play their solo slightly sharp – not so much as to sound out of tune, but enough that the ear groups together the harmonics of the solo voice in a different stream from those of the accompaniment. Alternatively, jazz soloists might offset

*Rock musicians do it anyway, since nuance is seldom their forte. Who can forget how guitarist Nigel Tufnell, in the rock spoof movie *This is Spinal Tap*, used amplifiers with volume controls that 'go up to 11'?

the start of their runs and licks from the phrases played by the rest of the band, in the same way that Bach does for his fugal voices. Or the entire rhythm can be cunningly shifted out of step: delaying or advancing the onset and ending of notes by just thirty to fifty milliseconds relative to the backing will be enough to make them sound distinct (if the musician is skilled enough to do it). A soloist might also suppress merging by playing phrases that avoid any parallel motion with those of the accompaniment, for example ascending when the accompaniment is descending. Here the separation is encouraged by the principle of common fate: the solo becomes a bird that defies the movement of the flock.

Opera singers are particularly adept at carving out a sonic stream for themselves. Even with their formidable vocal power, they face a difficult job in making themselves heard without amplification against a full orchestra. They do this by altering the shape of their vocal tract to focus a lot of the energy in their voice into a particular frequency band. By opening their mouth wide and controlling their throat muscles, they enlarge the pharynx and lower the glottis (the part of the throat containing the vocal cords), boosting the energy in a frequency range (2,000–3,000 Hz for a female soprano) where the sound output of the orchestra is rather modest. This is basically a brute-force 'loudness' technique, but achieved in a remarkably skilful way.

The effectiveness of this technique depends on the nature of the vowel being enunciated (vowels being the principal carriers of the vocal energy). So singers have to balance intelligibility – the extent to which a vowel sound can be identified – and audibility, and this explains why, at the very highest frequencies of female soprano singing, all vowels tend to sound like an 'a' as in 'ah'. This also explains why opera can seem, for listeners unfamiliar with it, to deal in rather artificial, stylized forms of vocalized emotion: the singers' voices are quite literally different from everyday spoken or sung voices, and so don't obviously convey the same emotional nuances. There is reason to believe that Wagner made allowances for this trade-off between volume and textual intelligibility: he seems to have matched the pitch of vowels in his libretti (which he wrote himself) to the frequencies at which the vocal tract resonates when they are enunciated in normal speech. That's to say, vowels with lower resonant frequencies are sung by sopranos in Wagner's operas on lower notes more often than would be expected from pure chance, making them easier to identify. This doesn't seem

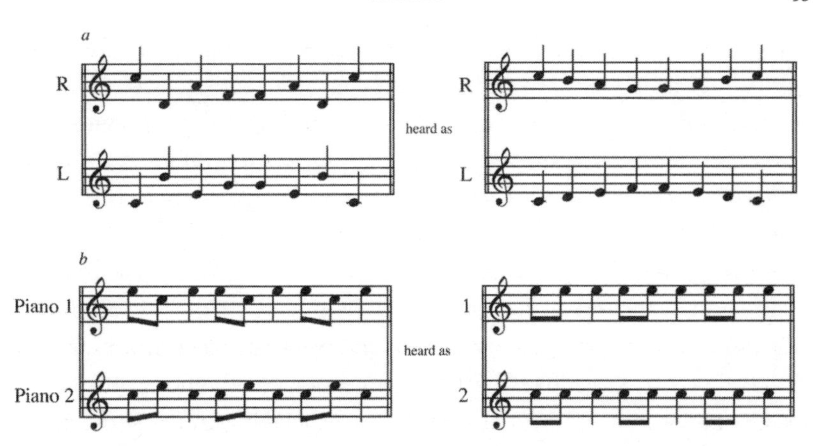

Figure 5.13 (*a*) The scale illusion: the brain sorts these two lines, played into the left and right ear, into two smooth scales. (*b*) An analogous illusion occurs in the second movement of Rachmaninov's *Second Suite for Two Pianos*, where the interweaving lines of the two pianos seem to separate into two repeating notes.

to happen in some operas by other composers (such as Rossini and Mozart), implying that there was something deliberate, and presumably intuitive, about Wagner's choices of melody.

As well as keeping voices distinct in polyphony, it's crucial that the notes in each individual voice be bound together into an integrated whole. We saw earlier that coherent melodies tend to use mostly small pitch steps and to have smooth contours. One corollary is that the mind will be predisposed to *create* melodies from notes that follow one another with these properties. Diana Deutsch has illustrated this with a dramatic auditory illusion in which she played two different sequences of notes to listeners through headphones, one in each ear. Notes heard in a single ear are interpreted as coming from the same physical location, and we've seen that, thanks to the principle of proximity, these tend to be grouped together.* So the notes in the right and left headphone would be expected to segregate into two separate streams.

But Deutsch chose notes that formed two separate but interdigitated scales, one going up and then down again and the other with the opposite contour (Figure 5.13*a*). This is the most 'logical' way of grouping

* There is some reason to think that this is a relatively weak grouping principle: after all, echoes and reverberations in natural environments might confuse interpretation if the mind insisted too strongly on using the apparent sound direction as a criterion of grouping.

the notes into coherent patterns – by themselves, the left and right signals create two rather odd, jumpy (and palindromic) melodies. And so that is what the listeners perceived: they reported a smooth down-and-up scale coming from one headphone and an up-and-down scale in the other. The strength of this grouping mechanism – in essence, an example of the gestalt principle of good continuation – is so strong that it persists even when the notes from each speaker have different timbres.

In some very rare instances, composers have made use of this so-called scale illusion. One such appears in Rachmaninov's *Second Suite for Two Pianos*, Op. 17, in which the two pianos play a sequence of up-down jumps that sound just like two series of repeated notes (Figure 5.13*b*). According to music psychologist John Sloboda, the effect of the illusion is disconcerting for the musicians themselves: he says it is hard to believe that you're not actually playing the repetitions yourself, despite seeing what your fingers are doing. 'It was as if someone had suddenly returned the lower note I was playing to the same pitch as the upper one,' Sloboda says. He suggests that Rachmaninov wrote the piece this way not to perpetrate an elaborate trick on the pianists but because the rapid repetition of quavers would be demanding for them to achieve.

Working in harmony

Writing polyphonic music with two voices that remain clearly separated and simultaneously audible is not so hard. With three such voices, our cognitive processes have a heavier load, but it's possible. How far can this be taken? Obviously, even the most dextrous keyboardist would struggle to play a separate voice with every finger, but in principle one could imagine polyphonic music in which each member of a thirty-strong choir or a thirty-piece string section plays a different melody. Would we have any hope of hearing each of these voices distinctly?

Paul Hindemith claimed that no one, however well trained, can follow more than three polyphonic voices. But Bach seemed to feel no such limitation, often writing for four voices and sometimes for as many as six. However, as he multiplies the voices, he appears to recognize that the listener needs more help, and he tends to keep only a subset of voices active at any one time.

All the same, it seems remarkable that we can apparently make sense of several simultaneous voices in music while we struggle to attend even to just two in speech. When we hear two verbal messages simultaneously, we can interpret and retain one of them only at the expense of ignoring the other almost entirely – it can even switch language without our noticing.

One possible reason for this greater receptiveness to simultaneous streams in music is that the streams are generally not independent. They are (in traditional counterpoint) in the same key, and are related to one another harmonically even if the melodies are quite different. John Sloboda and his student Judy Edworthy have shown how harmony supports the mental juggling act. They played subjects two melodies, first individually and then simultaneously (in non-overlapping pitch ranges). In the simultaneous tests, one of the melodies contained an error, and the subjects were asked to identify in which melody it was located – and ideally, where. They were able to do this most effectively when the melodies were both in the same key. The accuracy was slightly poorer when the melodies were a perfect fifth apart, and worst when they were a tritone (sharpened fourth) apart. So harmonic concordance seemed systematically to assist cognition. Sloboda and Edworthy figured that a common key enables the music to be more easily knitted together: even if only one of the voices is actively attended to, the other can be encoded in memory as a kind of harmonization of it, so that wrong notes stick out. But when there is no harmonic relationship between the voices, wrong notes don't sound any 'worse' than correct ones.

Maintaining conventional harmony between the voices in polyphony can be difficult, however. Each simultaneous melody has its own exigency, and occasional dissonant clashes may be inevitable. In early medieval polyphony it was considered better to compromise or simplify the melody than to incur dissonance. But composers became increasingly concerned that each voice should maintain a good melody, rather than simply fitting with the others in a harmonic jigsaw. So when melody conflicted with harmony, it was melody that tended to predominate. Yet if some dissonances were unavoidable, nonetheless composers felt a need to keep them under some kind of control.

This led to a series of more or less systematic rules governing the dissonances that were and were not permissible. In the counterpoint of Palestrina, a central melody (the *cantus firmus*) would be written first, and the other voices would have to be harmonically anchored

Figure 5.14 Some of the allowed dissonances in Palestrina's counterpoint. The dissonant intervals are shown with arrows – note that they all occur on 'passing tones' leading towards a consonance.

to it at certain key points. It was especially important that clear, strong consonances were achieved at the beginnings and ends of phrases.* One might say that the main objective in counterpoint is to maintain *horizontal* coherence – to make sure that the thread of each voice is intact, for example by the use of small pitch steps – while enforcing a judicious amount of *vertical* integration that persuades the listener that the separate voices are all part of the same composition.

Some of the dissonant figures that the Palestrina style permits are shown in Figure 5.14. The 'dissonant' intervals of sevenths and ninths don't seem too jarring, because streaming disguises the harmonic relationships between the voices at these points. That's to say, because the melodies are coherent in themselves, we don't really perceive them to be in harmonic relation at all. The dissonant notes are merely 'passing tones': transitory moments in a series of small steps between one relatively stable pitch and another. You might say that we accept the dissonances because they are clearly signalled as ephemeral. The same thing happens in the first line of 'Three Blind Mice', where the tune starts on the stable major third (E in the key of C) and moves to the even more stable tonic, via two whole-tone steps: E→D→C. D here is a passing note – it is a potentially dissonant major second when sounded against an accompanying chord of C major, but it

* Legend has it that Palestrina rescued polyphony from the threat of a total ban by the Council of Trent in the mid-sixteenth century, and it is sometimes implied that this ban was motivated by a proliferation of undignified dissonance. Others have said that the Council's objection was that a multiplicity of voices undermined the notion of God's unity. But it seems that their main concern was that the increasing complexity of counterpoint had made the words of the liturgy hard to hear. Palestrina's rules of counterpoint not only controlled the dissonances but also improved comprehensibility, allegedly by making the melodies more congruent with the rising and falling cadences of ordinary human speech.

Figure 5.15 The offset of voices in No. 18 of Bach's 'Goldberg' Variations helps to keep the two streams perceptually distinct, so that potentially dissonant intervals – here major and minor sevenths – are not registered as such.

sounds fine here because the D is strongly 'captured' in the melody stream.

In none of these cases does the potentially dissonant note in the upper voice start at the same time as a clashing note in the lower voice. Such timing offsets also enhance the streaming, and Bach uses the principle in parts of the 'Goldberg' Variations (Figure 5.15). Note that, although no two notes are sounded simultaneously, this is not the same as the splitting of a melody line in virtual polyphony, because the rate of events is too slow. Rather, we separate the two voices here quite consciously, and indeed use the first to draw inferences about the second, since they echo one another.

Another trick used to strengthen streaming for the purpose of disguising dissonance in polyphony was to use repetition of small phrases – so-called ostinati – to bind the notes into a simple perceptual unit, insulated from harmonic clashes around it. There is a nice visual analogy for how this works: in Figure 5.16, image *c* is perceived as *b* plus a series of vertical lines. This image in fact contains *a* embedded within it, but we can no longer see this closed square because the repetition of the vertical has 'pulled' the corresponding line out of that pattern and into a new one. The repetition suppresses any tendency to try to integrate this vertical with the lines that make up *b*.

Streaming offers a robust barrier to the perception of dissonance. Bach's fugues contain some strikingly dissonant collisions that go more

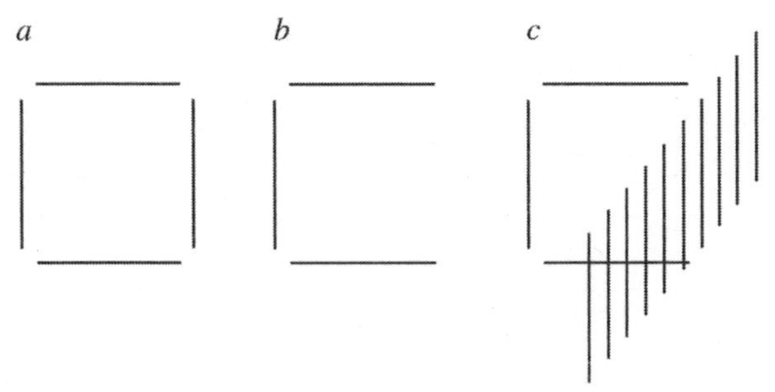

Figure 5.16 Repetition to bind figures together: a visual analogy. (*c*) contains (*a*) embedded within it, but we see it instead as (*b*) plus a series of vertical lines.

Figure 5.17 Transient dissonances may not be registered as such in polyphonic music in which the voices are each sufficiently distinct that they are heard independently. In the C major Fugue from Book I of Bach's *Well-Tempered Clavier*, the minor ninth interval marked in grey would normally be considered very jarring, but does not really seem so here.

or less unnoticed except to the most attendant listener – it can be rather shocking to discover them in the score. The C major Fugue in Book I of *The Well-Tempered Clavier*, for example, contains a G in the upper register against an F♯ in the bass, which is about as 'dissonant' as you can get (Figure 5.17).

This role of streaming in polyphony seems to challenge Sloboda's and Edworthy's view that harmony makes the edifice hang together, and perhaps in a sense it does. But it seems more likely that we don't simply listen to this music in the same manner throughout – sometimes we can juggle the voices independently, at other times they morph into a melody plus harmonic accompaniment. It might even be the case that these shifts in our modes of perception contribute to the pleasure of listening.

Indeed, the composer doesn't always *want* polyphonic voices to be clearly defined. In hymn singing, there is little concern to create the kind of elaborate interweaving of voices that Bach writes; one wants the harmonies to be heard, but also to create a sense of unity. It's rather like the difference between seeing a stream of people taking several different paths towards a common destination, and seeing them all move along much the same route. David Huron distinguishes between these types of music in textural terms: hymn harmonization, he says, is not so much polyphonic in texture as *homophonic*, in which several voices combine to carry a single melody. The extreme form of homophony is when all the voices sound in unison, so that there is no polyphony left at all. Huron shows that these different textures can be created by adjusting two things: the relation between the pitch motions of each voice, and the simultaneity of the notes. He finds that Bach's contra-

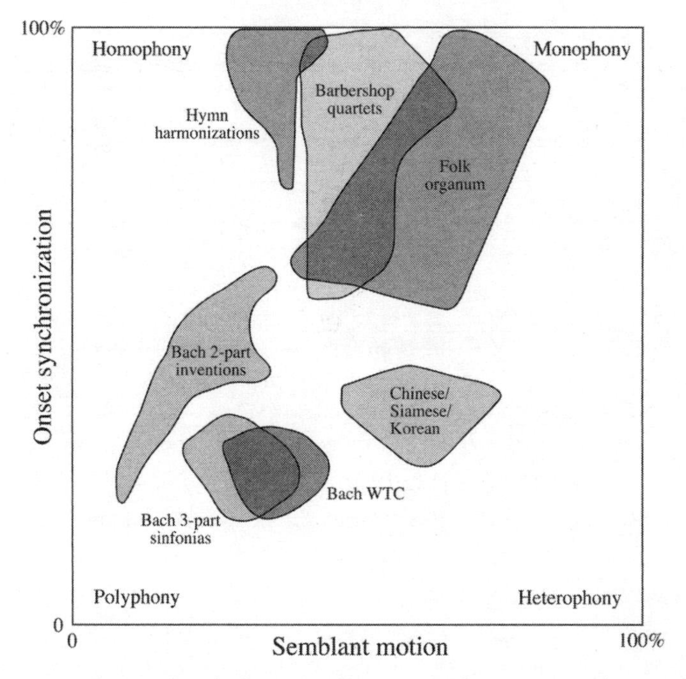

Figure 5.18 David Huron's polyphonic 'texture space'. There are two 'directions' in this space: the degree of onset synchronization of notes or events in different voices, and 'semblant motion', which measures the degree to which the pitch contours of concurrent voices resemble one another. 'WTC' is *The Well-Tempered Clavier.*

puntal music occupies different regions in this 'texture space' from hymn harmonization, and that the latter is distinct from the rather more monophonic harmonization of barbershop quartets (Figure 5.18). Curiously, the fourth corner of this space – so-called *heterophony*, where voices copy each other but never overlap in time – seems not to be explored by any current music, although some Far East Asian music comes close.

Experiments in confusion

During the nineteenth century, composers began to experiment with the new timbres that can be produced by merging the voices of different instruments. This fusion can sound like some fantastic new sort of instrument rather than a combination of familiar ones. Ravel achieves that in his *Bolero* by using parallel voice motions of the celesta, flutes and French horn to promote their fusion (Figure 5.19).

Figure 5.19 The parallel motions in Ravel's *Bolero* cause the celesta, French horns and flutes to blend into a single composite instrument.

Instead of making clear distinctions between voices, conflicting clues about how to separate them can produce a rich, shimmering musical surface. Some composers have used timbre this way, playing it off against other grouping cues. Anton Webern orchestrated a ricercar, a kind of precursor to the fugue form, from Bach's *Musical Offering* in which the voices were split between different timbres, giving mixed messages about whether grouping should be based on pitch and melodic line or on timbre (see p. 236).

The polyphony of *Central Park in the Dark* seems to threaten all sense of integration: the voices observe no obligation even to match their key or rhythm. This might appear to frustrate the gestalt principles altogether. But somehow it all hangs together rather than degenerating into confusion, at least until the final crescendo. That's a testament to Ives' skill, for writing a genuinely musical piece for truly independent voices is harder than it seems. As Albert Bregman has said:

> Even music of the type written by Charles Ives must have important musical relations between the parts; if this did not have to be true, we could create an endless fund of such music by simply combining the works of the existing repertoire two at a time, three at a time, and so on.

In fact, the gestalt principles are probably still very much at work here, enabling us to disentangle the conflicting melodic and harmonic streams of Ives' parts so that we hear them as distinct and not clashing in much the same way as we do for Bach fugues. As Roger Scruton puts it, what we hear when we listen to Ives' simultaneous bands is 'each enjoying its local harmony, so to speak, and quite satisfied with that'.

Although composers and musicians have manipulated some of the gestalt principles by trial and error, it seems likely that they have only scratched the surface of the musical bounty on offer. Some modern composers have dismantled the apparatus of polyphony to make music in which individual voices and melodies are submerged beneath a generalized sonic mass. In pieces such as György Ligeti's *Atmosphères* (1961), made famous by its use in Stanley Kubrick's movie *2001: A Space Odyssey*, the instruments play simultaneous notes across most of the chromatic range, creating sound structures that can't meaningfully be described as chords and which renounce all conventional principles of melody, harmony and rhythm. The opening sequence involves fifty-six string players all playing different notes.

Ligeti himself described the music in terms that seem still to insist on classical concepts of polyphony, harmony and so forth:

> The complex polyphony of the individual parts is embodied in a harmonic-musical flow, in which the harmonies do not change suddenly, but merge into one another; one clearly discernible interval combination is gradually blurred, and from this cloudiness it is possible to discern a new interval combination taking shape.

But it isn't very meaningful to call the result 'polyphony' at all, since no voices as such are audible. Because Ligeti wrote out the parts and could see them plainly in the score, it seems he continued to consider them as such, whereas to the listener these voices are not a perceptible reality. It is all too easy for composers to think that, because they have planned a certain musical structure, it will be heard; what determines that, however, is not the score but the principles of auditory cognition.

In Ligeti's case this doesn't obviously matter: it is not clear that he really expected anyone to hear distinct harmonic events in this music. It has been claimed, in fact, that the complex sonic texture of *Atmosphères* creates aural illusions, tricking the ear into hearing things that are not there. From the cognitive point of view, however, the main effect of this density of harmonic texture is to weave all the voices together into one block of shifting sound, a rumbling, reverberant, mesmerizing mass. Faced with such acoustic complexity, the best the mind can do is to lump it all together, creating a single 'object' that, in Bregman's well-chosen words, is 'perceptually precarious' – and therefore interesting. These dense sonic objects can be manipulated in dramatic ways, for example allowing a voice to segregate momentarily from the mass – to emerge and 'poke out', so to speak, before sinking back again. The French composer Jean-Claude Risset used electronic alteration of sound to create such effects in his *Inharmonique* (1977) for soprano and recorded tape.

The precarious and sometimes illusory quality of stream fusion, akin to the disconcerting optical effects of op art, is also evident in the music of Steve Reich. In contrast to Ligeti, Reich's music, exemplified by compositions such as *Desert Music* (1984) and *Music for 18 Musicians* (1974–6), is mostly rooted in conventional tonality and employs rhythmic repetition. But the rhythms are overlaid in complicated ways, creating a constantly changing soundscape within which the mind is constantly finding and then abandoning new patterns. This cognitive hunger for pattern is vital to Reich's music: it means that the listener is ever vigilant for 'better' ways of organizing the sound, but is forced into constant reappraisal as the repeated phrases enter new relationships. Some find the result infuriating, presumably because of this perpetual frustration of a 'right' way to hear the music; for others, the effect of forever discovering new structure is deeply satisfying. It would be fair to say that Reich's music doesn't merely accommodate the demands of cognition, as for example does Baroque counterpoint; rather, it commandeers cognitive mechanisms and sets them an active role in music-making.

6

Tutti

All Together Now

How do we use more than one note at a time?

The German mathematician and astronomer Johannes Kepler is remembered chiefly for adducing mathematical laws governing the motions of the planets in their orbits that helped Isaac Newton formulate his gravitational theory. That is enough to make him revered in modern astronomy: a space telescope launched by NASA in 2009 to search for planets around other stars was named after him.

But the book in which Kepler laid out his third law of planetary motion* in 1619 makes uncomfortable reading for today's astronomers. Here's an example:

> Now there is need, Urania, of a grander sound, while I ascend by the harmonic stair of the celestial motions to higher things, where the true archetype of the fabric of the world is laid up and preserved. Follow me, modern musicians, and attribute it to your arts, unknown to antiquity: in these last centuries, Nature, always prodigal of herself, has at last brought forth, after an incubation of twice a thousand years, you, the first true offprints of the universal whole. By your harmonizing of various voices, and through your ears, she has whispered of herself, as she is in her innermost bosom, to the human mind, most beloved daughter of God the Creator.

What is all this about music?

Kepler's book was called *Harmonia mundi*, 'Harmony of the World', and it put forward a peculiar idea. The planets, he said, are singing a

* The ratio of the squares of the times taken for two planets to complete an orbit is equal to the ratio of the cube of their mean distances from the Sun.

polyphonic motet. Their 'vocal' range increases with the distance from the Sun: Mercury takes the soprano, Earth and Venus the alto, Mars the tenor, Saturn and Jupiter the bass (no other planets were known at that time). The heavens were filled with celestial harmony.

This wasn't a new idea. A musical cosmology was sketched in the *Harmonia* of Claudius Ptolemy, the Egyptian-Roman astronomer and mathematician of the second century AD who gave the Middle Ages its Earth-centred universe. Ptolemy's *Harmonia* was a book about music, and it argued in favour of the mathematical basis of consonance developed by the school of Pythagoras. The Greek word *harmonia* is not an essentially musical term: it simply means 'fitting together', and its root is the same as that of both 'arithmetic' and 'rhythm'. It implies a kind of numerical order. And under the title 'What music can do' in his *Etymologies*, the seventh-century theologian Isidore of Seville claimed that 'The universe itself is said to have been put together with a certain harmony of sounds, and the sky itself revolves under harmonious music.'

Kepler's own musical universe was a product of its time: in the early seventeenth century Neoplatonism, which attributes a fundamentally geometric character to the universe, was enjoying a resurgence, and nowhere more so than in the court of the Holy Roman Emperor Rudolf II in Prague, where Kepler worked as an advisor from 1600 to 1612. There was, however, a decidedly new aspect to Kepler's heavenly music. Whereas Greek music was, as we have seen, monophonic, Kepler's musical world embraced the polyphonic opulence of Palestrina and Monteverdi. His planets were singing together, and singing in strict harmony.

He was convinced that the movements of the planets encoded simple ratios like those found in musical intervals. Kepler favoured Zarlino's system of just intonation (p. 58), which assigns small-number ratios to intervals such as the major and minor third (5:4 and 6:5 respectively) that were more complex in the Pythagorean system of tuning. Here Kepler relied on the judgements of his ear, saying that the main argument for these just intervals is not their neatness but the fact that they sound better.

Harmonia mundi shows just how deeply both the theory and the practice of polyphonic music had penetrated intellectual culture by the end of the Renaissance. It illustrates how one of the most important questions in music was now that of *how well notes sounded together*: in other words, what is harmonious and what is not? Kepler's treatise can be considered to raise the stakes: at root, the question of harmony is not simply a matter of how to make good and proper music, but of how the world is constructed. As the poet John Dryden put it in his 'Ode for Saint Cecilia's Day' in 1687 (the year in which Newton published his *Principia*):

> From Harmony, from heavenly Harmony
> This universal frame began . . .

Yet although we now know how and why planets move the way they do, we are not agreed about musical harmony. We don't know why some combinations of notes please us and some don't.

There is much more to harmony, however, than the contentious issue of consonance and dissonance. In polyphonic music, harmony is what fills out the musical landscape. If melody is the path, harmony is the terrain: the path exists only in that context. We will see here what the terrain looks like.

While the most common form of music in the world is monophonic song, harmony is used in many cultures – it attains great sophistication, for example, in the communal singing of sub-Saharan African music. But it has arguably been cultivated most highly in Western music, and it is on this tradition that I will largely draw here.

Dissent on dissonance

As the Greek implies, harmony is about fitting notes together. Most people would agree that some fit well, and others less well. Conventionally, we'd call the former combinations consonant and the latter dissonant. In the most reductive formulation, consonance is good and pleasing, dissonance bad and unsettling. It all sounds perfectly clear in theory.

Well, it isn't.

Many classical-music traditionalists would deny that they enjoy dissonance. The word conjures up the jarring sound-worlds of

Stockhausen and Boulez, who, to these listeners, seem to delight in finding ways to combine notes that set your teeth on edge. Oh no, they'll say, give me Chopin or Beethoven any day. This merely illustrates why consonance and dissonance are two of the most misunderstood and misrepresented concepts in music. To the opponents of modernism, who rebelled against Debussy and Stravinsky and aren't through with their rebelling even now, the use of dissonance was an affront to music itself, a violation of natural laws dictating the way music should be composed. Until cognitive testing and ethnomusicology began to challenge the assumption in the later twentieth century, it was common for music theorists and composers in the West to claim that consonance was merely a matter of acoustic physics and physiology, and so would be judged the same way by anyone. Meanwhile, champions of atonalism such as Schoenberg and Boulez dismissed this notion that we have an innate aversion to dissonance, claiming instead that it was culturally learnt – and could thus presumably be unlearnt.

But the question of consonance and dissonance is not a battleground between tonalism and atonalism. It's much more subtle than that. Take Chopin, that delicate soul sometimes caricatured as a composer for simpering ladies in cloistered parlours. His music is *riddled* with dissonance. At its most extreme, Chopin's dissonance probably deserves to be called ugly (but not therefore bad!). Even in some of his most popular pieces, horrors lurk for those who crave simple, clean consonance. Or take gamelan, which the uninformed listener might dismiss as a clangorous riot but which may leave Javanese people smiling joyously. Or consider the smoochiest, most clichéd of hotel-lobby pianists, whose saccharine versions of popular songs are undoubtedly laden with intervals that music theorists would have no hesitation in pronouncing dissonant.

Of course, there is not a great deal of point in telling people they should like combinations of notes that sound unpleasant to them, and that's not my intention. I simply want to explain that these likes and dislikes probably involve very little that is innate, but are rather mostly the products of learning. And it is in any case probably not the 'dissonances' themselves that are disliked, but what is being done with them: how they are being combined into music. Dissonance has become something of a fall guy, shouldering the blame for what many people perceive as the 'difficult' character of modern classical music. When listeners complain about contemporary compositions, they tend to suggest that the 'fault' lies with a perverse selection of note combinations, which they accuse the composers of having chosen purely to create a jarring effect. But it is not primarily the 'dissonant' intervals of this music that jangles nerves and frustrates the identification of musical themes; instead, a whole range of musical factors, including rhythm, melodic continuity and timbre, combine to create music that is not easy on the ears of a listener brought up on Mozart. We can afford to let go of the idea that certain juxtapositions of notes are intrinsically horrible, and indeed we'll find much more to enjoy in music if we do.

For a start, there is more than one kind of dissonance. When musicians talk about dissonance they are generally referring explicitly to a convention: to *musical consonance*, a culturally determined definition of which tones do and don't work together. But isn't this obvious, you might ask? Play a perfect fifth on the piano (C-G) and then a tritone (C-F♯). Surely the first is harmonious, the second a bit of a din? Well, we'll look shortly at the extent to which that is or isn't true, but just consider that in tenth-century Europe a perfect fifth was generally *not* deemed consonant – only the octave was.* And when harmonizing in fifths did become common, fourths were considered equally consonant, which is not how we see them today. The major third (C-E), meanwhile, part of the 'harmonious' major triad, was rarely used even by the early fourteenth century, and was not fully accepted as consonant until the High Renaissance. And the tritone interval that is supposedly wincingly dissonant becomes a most pleasing and harmonious pairing when played as part of a so-called dominant seventh chord (we'll see later what that is). Just add a D bass to that C-F♯, and you'll see what I mean: there's nothing discomforting about that. (The fact that the harmoniousness of a chord is not simply the sum of the consonance or dissonance of the intervals it contains is widely recognized but poorly understood.) Many people make a song and dance about the fact that the tritone is supposed to sound so awful that it was called *diabolus in musica*, the devil in music, in the Middle Ages,

* Fifths were used from at least the ninth century, but weren't formally codified as consonant until later.

and was banned from sacred music. But this prohibition was essentially for theoretical reasons, and not because the tritone was the 'worst sounding' of all intervals: it is at this interval that the Pythagorean system goes seriously awry because the circle of fifths doesn't close there, as I explained in Chapter 3. There's actually much more of a dissonant crunch in the minor-second interval than in the tritone, but no one made *that* diabolical.

Frankly, the whole polarizing terminology of consonance and dissonance is a rather unfortunate legacy of music theory, forcing us, for example, to describe as mildly to highly dissonant the major sixth, seventh and ninth intervals that in the hands of Debussy and Ravel create the most luxuriant sounds (and in the hands of lounge-bar pianists and drippy singer-songwriters, the sickliest slush).

This is not to say that whether we experience note combinations as smooth or grating is *only* a matter of convention. There is a genuinely physiological aspect of dissonance, distinguished from the dictates of musical convention by being called *sensory* or *tonal dissonance*. This refers to the rough, rattle-like auditory sensation produced by two tones closely spaced in pitch. If two pure tones (single frequencies) are played simultaneously with frequencies that are only very slightly different, the acoustic waves interfere. At one moment the acoustic signals reinforce each other, boosting the loudness; a moment later they partly cancel, reducing the loudness (Figure 6.1*a*). The result is a periodic rise and fall in loudness superimposed on the two notes, called beating. The beats get faster as the two tones get further apart in pitch, and when their frequency difference exceeds about 20 Hz the ear can no longer follow the rapid beating fluctuations, but hears them instead as 'roughness': as sensory dissonance.

So sensory dissonance occurs for two tones that fall within a certain critical range of frequency difference. Too close, and the tones sound identical but with beats superimposed. Too far, and they sound like clearly distinct tones. In between, they create roughness (Figure 6.1*b*).

One of the surprising, even disturbing characteristics of this effect is that the width of the dissonant region depends on the *absolute* frequencies of the two notes (Figure 6.1*c*). This means that an interval that might be consonant in a high register, high up on the piano keyboard, may become dissonant in a lower register. In other words, *there is no such thing as a tonally dissonant interval* – it all depends on where you

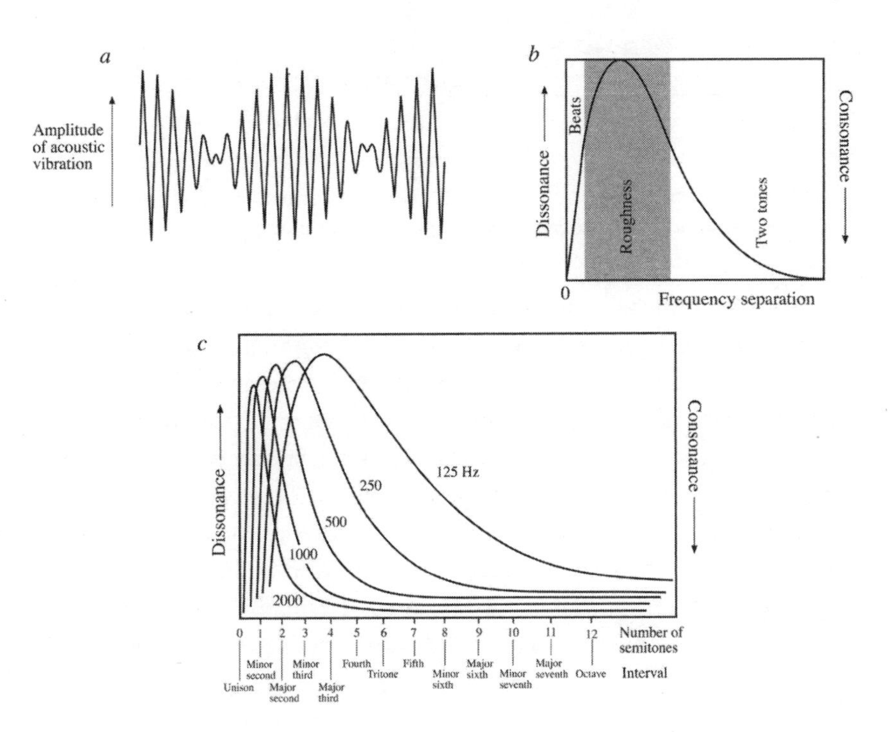

Figure 6.1 Sensory dissonance is caused by 'beats' – the rise and fall in loudness of two tones that differ only slightly in frequency (*a*). If the rate of these beats exceeds about 20 per second, they create an audible sensation of roughness. But once the frequency difference becomes large enough, this gives way to the perception of two separate tones (*b*). The width and the position of the peak in the band of sensory dissonance depends on the absolute frequencies of the tones, getting wider as the tones get lower (*c*). These curves were produced in 1965 by Reinier Plomp and W. J. M. Levelt in the Netherlands by asking subjects to rate the perceived dissonance levels of two pure, single-frequency (sine) tones with the mean frequencies indicated.

play it, using high notes or low. In the mid-range of the piano, intervals of a minor third (three semitones) generally lie beyond the band of roughness, evading sensory dissonance. For high notes, even a semitone (minor second) interval does not create roughness. But in the bass, rather wide intervals can become rough, and even an interval like a perfect fifth that is considered highly musically consonant becomes dissonant in sensory terms. This explains the 'gruffness' of chords sounded low down with the left hand, and helps us understand why Western music almost universally shows voicings (combinations of notes) that

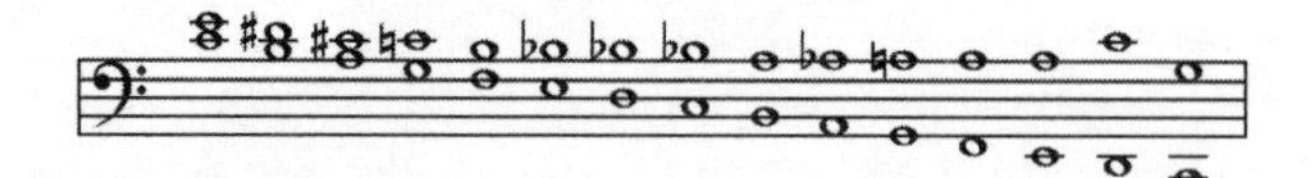

Figure 6.2 The average span of bass chords in the music of Haydn and Bach increases as the notes get lower, to avoid sensory dissonance. The intervals shown here are averages, and have no musical significance in themselves.

become more widely spaced the lower they are sounded. The left hand of a pianist will typically play voicings spanning almost an octave or more when the lowest note is around the octave below middle C, while it might include some fourths and fifths if this note is higher than the E below middle C. The right hand, meanwhile, merrily bangs out chords containing thirds and even seconds. David Huron has shown that the average distances between the two lower pitches for chords in Hadyn string quartets and Bach keyboard works get steadily wider as the lowest note gets lower (Figure 6.2). These composers had clearly figured out for themselves that this was necessary to avoid sensory dissonance in the bass.

Helmholtz's peaks and troughs

At face value, sensory dissonance seems to tell us nothing about whether particular combinations of notes will sound consonant or dissonant, beyond the roughness that sets in if they are too close together. *All* intervals beyond that critical pitch separation ought to sound equally fine. But they don't.

This is partly because the tones played by musical instruments are complex: combinations of several harmonics. So any two notes sounded simultaneously offer many different possibilities for two overtones to clash and produce sensory dissonance if they are close enough in frequency. The German physicist and physiologist Hermann von Helmholtz realized this in the nineteenth century, and he performed calculations that supplied one of the first persuasive, scientific accounts of how dissonance arises.

Helmholtz was a man of diverse interests. In part this was a result of circumstance: in Germany at that time, the state provided funding for medical students, which prompted the young Helmholtz, a man of modest means, to channel his passion for physics into medicine. This forced marriage yielded interesting results: Helmholtz worked on the electrical nature of muscle action and united optical physics with physiology in a theory of vision. Acoustics offered similar cross-disciplinary prospects, and Helmholtz combined his thorough knowledge of the workings of the ear with his mathematical understanding of vibration to figure out how we hear tones. His 1863 book *On the Sensations of Tone as a Physiological Basis for the Theory of Music* is one of the first truly scientific expositions on music cognition, and in no respect the work of a dilettante.

Helmholtz knew that the question of consonance was ancient and profound. Notwithstanding the vicissitudes of musical convention, it was widely accepted by the nineteenth century that the most 'harmonious' intervals of the diatonic scale were those favoured by the Pythagoreans, with simple frequency ratios of 1:2 (octave), 2:3 (fifth) and 3:4 (fourth). Countless eminent scientists, including Galileo, Francis Bacon and Marin Mersenne, had attempted to explain why these intervals sounded 'good', but none had succeeded in finding a satisfactory reason for why, as Galileo put it in 1638, 'some Pairs are heard with great Delight, others with less; and . . . others are very offensive to the ear'.

One of the main difficulties was that simplistic notions based on the ideal Pythagorean proportions were hard to reconcile with musical practice. Our hearing is rather tolerant to deviations from these ideals: we will accept equal-tempered intervals as consonant even though their frequency ratios differ, sometimes significantly, from the Pythagorean values. And Helmholtz himself noted that some intervals sound more or less consonant depending on which instruments they are played on: a major third D-F♯ played on clarinet and oboe, he said, sounds better when the clarinet plays the lower note than when the oboe does.

Helmholtz understood the phenomenon of roughness caused by the beating of two tones of similar frequency. He calculated the total roughness for all overtone combinations (up to the first five) as two complex tones become steadily separated in their fundamental frequency between unison (ratio 1:1) and an octave (1:2). This generated a curve of roughness, or sensory dissonance, with dips at various intervals, each of which can be assigned to one of the intervals of the chromatic scale (Figure 6.3*a*). The steeper a consonant 'valley', the less tolerant the interval is to mistuning. Helmholtz's dissonance curve has been updated using information collected in 1965 by Dutch psychoacoustics researchers Reinier Plomp and W. J. M. (Pim) Levelt, who asked listeners to rate the degree of dissonance between closely spaced pure tones over a wide range of average frequencies (Figure 6.3*b*).

It's impossible to look at these graphs and not suspect that Helmholtz was on to something. The fact that nearly every dip more or less matches a diatonic or chromatic interval, and that some of the biggest dips are those for the conventionally consonant intervals of an octave, fifth, fourth and third, can't be sheer coincidence. What's more, the graph offers some rationale for the pleasing sound of the minor-third interval, which doesn't pop up anywhere in the harmonic series. The dips are broad enough to accommodate some leeway in tuning. And Helmholtz's approach also explains why the specific combination of overtones – and thus the timbre of the instruments playing the notes – can be crucial to the sensation of consonance.

But evidently this isn't the whole story. Notice that the depths of several of the 'consonant' valleys don't differ by very much. The octave and fifth are particularly deep, and indeed we've seen that there is theoretical reason (from the harmonic series) and some empirical evidence in support of the idea that they are both physiologically 'preferred' intervals. But there is little to distinguish the major third, perfect fourth, and major sixth. Indeed, in the modern curve (Figure 6.3*b*) just about all the intervals between the major second and the major seventh lie within a narrow band of dissonance levels, except for the perfect fifth – and this includes the 'microtonal tones' in between the diatonic notes. And the perfect fourth has about the same dissonance rating as intervals midway between a minor and major sixth, or a major sixth and minor seventh. Even more strikingly, the supposedly awkward tritone interval here appears *less* dissonant than a major or minor third. In short, the margins that discriminate between these intervals are small, and easily modified or manipulated by experience and acculturation. All we could really expect on this basis is that fifths and octaves will sound good, minor seconds will sound pretty awful, and the rest is much of a muchness.

The greatest sensory dissonance is found close to the unison interval

– in particular, the minor second (C-C♯, say) is predicted to sound fairly nasty. Now, again this will vary depending on the register – Helmholtz's calculations are for middle C – but in any case we shouldn't imagine that such intervals are going to be musically useless, or even necessarily unpleasant. In particular, they can be used to create interesting timbral or colouristic effects. The 'crushing' of two notes separated by a semitone – the grace note or acciaccatura – is a common

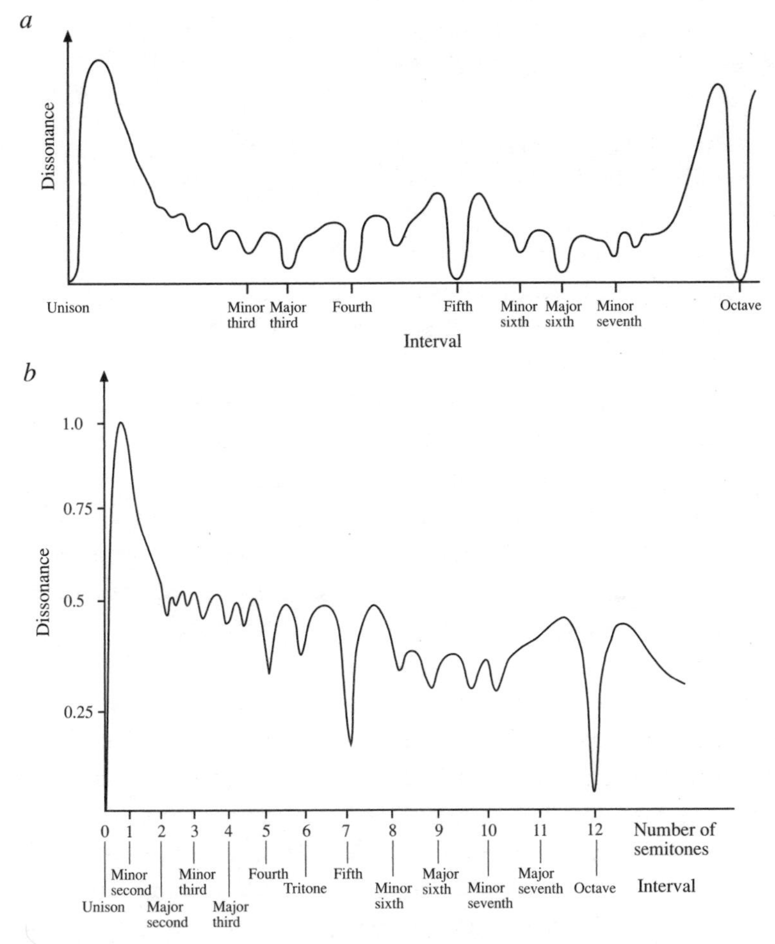

Figure 6.3 By adding up all the sensory dissonance from clashes of overtones, Hermann von Helmholtz calculated the roughness of all intervals for complex tones in the span of an octave. Dips occur on each note of the diatonic scale (*a*). More precise graphs of this sort have been calculated using the perceptual dissonance measurements for pure tones by Plomp and Levelt shown earlier (Figure 6.1*c*). Here I show the result for complex tones of nine harmonics (*b*).

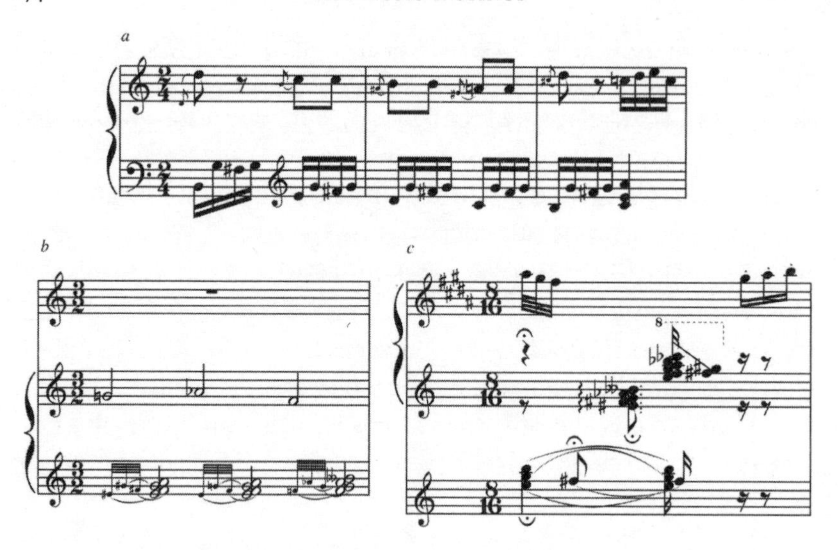

Figure 6.4 (*a*) The grace notes or acciaccaturas in Mozart's Sonata No. 1 in C major (K 279) create 'crushed' minor-second intervals. (*b, c*) 'Dissonant' tone clusters in 'Musiques Nocturnes' from Béla Bartók's *Out of Doors* suite.

feature of the music of Mozart and Haydn (Figure 6.4*a*). Although here it is nominally a sequential rather than simultaneous (that is, melodic rather than harmonic) interval, there will always be some sensory overlap of the two notes – especially when played on the piano, where it is usual to play both notes at the same time but to release the grace note first. The result is not horrible, but merely quirky, producing a pleasingly wonky sound that was often exploited also by Prokofiev and by jazz musicians.

And the minor-second interval achieves, by virtue of its sensory dissonance, a kind of percussive quality and a harmonic 'thickness' that sounds akin to many inharmonic natural sounds, such as wooden objects striking together. Béla Bartók piled them up in 'tone clusters' for creating his trademark 'night music', a sonority that features for example in his *Out of Doors* suite (Figure 6.4*b,c*) and in his string quartets. Here they create a sense of meditative strangeness, vaguely recalling the nocturnal animal sounds of cicadas, frogs and birds. They are eerie sounds, but not at all jarring. One might argue that these cluster chords aren't exactly dissonances at all, but a *spreading out of pitch* so that the note becomes ambiguous or indeterminate, much as it does for the inharmonic sounds of xylophones, marimbas and bells.

Indeed, there is at least one musical tradition that has assimilated major and minor seconds so thoroughly that they are considered pleasing, apparently because of the very acoustic interference that is supposed to render them dissonant. Writing of the *ganga* songs of the mountain people of Bosnia and Herzegovina, ethnomusicologist John Blacking says 'Chords of major and minor seconds that would be considered discordant in terms of acoustical theories of musical structure, were regarded as harmonious . . . This was enhanced by the way they were performed: singers stood very close to each other, and the vibrations from the loudly sung, close intervals induced pleasant bodily sensations.' In this and other ways, sensory dissonance does not have a strongly proscriptive function even in tonal music: acculturation can overcome it.

I should mention that there is a wholly different possible explanation for consonance that unites Pythagoreanism with neuroscience. It argues that the brain prefers combinations of frequencies in simple ratios because these generate stronger neural responses, thanks to their overlapping periodicities: a kind of constructive interference, although it should be said that the synchronization of interacting oscillators such as neural circuits is a complicated matter that is still being unravelled. It's too early to assess these ideas, although they do seem to leave a lot unexplained: why, for example, unison and octave intervals are rather sensitive to precise tuning (neural synchronization seems to predict the opposite), and why stronger neural firing should result in aesthetic preferences.

Do you know what you like?

It is one thing to try to quantify consonance and dissonance mathematically. But you might reasonably think that what matters more in music is how we actually experience them. Do we in fact find intervals classified (by whatever means) as consonant innately preferable to those that are dissonant?

This is highly contentious. The music critic Charles Rosen claims that dissonances are 'to most ears more attractive than consonances' – but this seems to be the anecdotal experience of a connoisseur of Western music, not a conclusion based on testing, say, people who sing tribal songs or who listen to FM radio. In the 1950s the French musicologist Robert Francès found evidence that musicians do favour dissonant chords over consonant ones, and that non-musicians might do so too. On the other hand, Helmholtz claimed that sensory dissonance 'exhausts the ear' and is for this reason deemed unpleasant. Some discussions on the topic have an element of tautology about them, equating consonance with what is experienced as pleasant, and dissonance with what is reportedly unpleasant.

Unfortunately, these questions can't be resolved by playing people a bunch of intervals and seeing which ones they like best. For one thing, there is no objective measure of dissonance: different criteria give different rankings for the intervals beyond the octave, fifth and fourth. And hearing an interval or chord in isolation is not necessarily a good indicator of how it will be perceived in a musical content: putatively dissonant chords can sound exquisitely 'right' in the right place. Besides, we've seen already that people tend to like best what is most familiar. Since most Western music is tonal, people within this tradition will be acclimatized to octaves, fifths, thirds and so on, and will hear less common intervals as more odd.

This is obvious and well understood by music psychologists, but they have not always acknowledged how deep these influences run. It seems a fair assumption that, while adult Westerners are conditioned by exposure to conventional views about consonance, children will be more like blank slates. One can also imagine conducting cognitive experiments on people who have not heard Western scales and do not have strong harmonic traditions of their own. But that is complicated not only by the difficulty of finding such people any more, but by the usual problem in ethnographic studies of being sure that the answers you get are genuinely addressed to the questions you want to probe.

In any event, the case commonly made is that many infants show a preference for consonant intervals over dissonant ones. But these children are by no means too young to have learnt such a preference from exposure to music. They generally hear lullabies and play songs from the moment of birth, and indeed even before: babies can hear and assimilate sound while in the womb. We know that musical preferences are strongly influenced by exposure, and also that babies are voraciously acquisitive of information from their environment. Neuroscientist Aniruddh Patel says that by thirty weeks of gestation, babies 'have already learnt a good deal about their auditory environment'.

It is certainly possible that harmonic preferences have an innate element, but the evidence remains ambiguous. For example psychologists Marcel Zentner and Jerome Kagan of Harvard University examined the responses of four-month-old infants to two harmonized melodies, one in 'consonant' parallel major thirds and the other in 'dissonant' parallel minor seconds – that is, with the simultaneous melody notes a semitone apart. The latter is decidedly odd to our ears: every note sounds wonky. The infants' attention was held much more firmly for the major-third tunes, and they showed more agitation, moving their arms and legs, for the minor-second ones. Any suspicion that learnt preferences are at work here was somewhat allayed (one can't quite say ruled out) by the fact that questions put to the children's carers revealed no obvious dependence of the infants' responses on their amount of prior musical exposure.

In one of the most stringent tests so far, Japanese researcher Nobuo Masataka played a thirty-second Mozart minuet, along with a modified version of it that contained many dissonant intervals, to two-day-old (hearing) babies of deaf parents, who had presumably heard no songs sung by their parents *in utero*. He found essentially the same slight preference for the consonant version as was exhibited by infants of hearing parents. But this preference was so small – the babies looked at the consonant source for twenty-seven per cent longer on average – that it's unclear what may be concluded from it, especially as Masataka admitted that even here one couldn't be sure that the infants heard no music at all while in the womb.

Another argument for an innate predisposition to favour consonance is that non-human primates, such as macaques, can distinguish it from dissonance. Consonant and dissonant intervals generate different patterns of neural activity in the auditory cortex of both monkeys and humans, and some researchers believe that there are specific populations of neurons that 'fire' in response to consonance or dissonance. That may help to explain why, for example, young children can more readily tell when a consonant interval changes to a dissonant one than when it changes to another consonant one. But it says nothing about *preferences*. In experiments in which monkeys were placed in a maze with V-shaped passages and offered the choice of sitting in one branch where a consonant interval was audible, or the other in which a dissonant interval could be heard, they showed no preference either way. In a separate study, however, two-month-old babies in much the same environment chose consonance more often than should happen by chance.

In short, the debate is far from settled. Unfortunately, it has sometimes been burdened with the temper of a battle for the soul of music: can we defend modern dissonance, or bury it as a contravention of human biology? But a resolution would have little to say about that. For one thing, it seems clear that if innate preferences exist at all, they are weak and quite easily rewritten by experience or convention (we need only recall that Mozart's contemporaries were baffled by some of the 'dissonances' he introduced). And if it were all just a matter of sensory roughness, as Helmholtz's explanation implied, we would expect register and timbre to feature more strongly in the way music theory handles dissonance, and for us to be rather more sensitive to different systems of tuning.

I believe there is good reason to think that in harmony, as in so much else in music, we mostly like what we learn to like. I look forward to a time when we stop fretting about whether or not music is dissonant, and focus more broadly on the characteristics that determine cognitive coherence. By the same token, complex new systems of tuning to 'improve' scales and harmonies seem to me to be rather fruitless in themselves; what matters is not the individual notes and chords, but the relationships between them. That's what I want to look at now.

The right formula

In ancient Greece the singer might also play a lyre or a kithara, tempting us to imagine the performer as a kind of ancient Bob Dylan strumming to his lyrics. But there was an important difference. The Greek musician simply duplicated the vocal melody on the instrument, which therefore was used not so much for accompaniment but as a parallel voice.

If, in contrast, you were to hear what Dylan plays on the guitar without his vocals, you might be hard pushed to guess the song at all. It's just a series of chords – mostly rather simple ones, strung together into patterns that are in truth often quite formulaic. Asking a musician 'Do you know "Tangled Up in Blue"?' is not like asking 'Do you know Beethoven's "Appassionata" Sonata'? In the former case, you're asking about a sequence of chords, a so-called harmonic or chord progression. In the latter, you're asking if they know where each note belongs.

In the classic singer-songwriter mode of composition, melodies are typically placed against a chord sequence that 'colours' the notes, giving them a particular local context. The process of fitting melodic lines to chords is called harmonization. It is often here, rather than in the nature of the melody itself, that music comes alive. A good jazz musician can take just about any tune, even the most anodyne of children's songs, and give it substance with an interesting choice of backing chords, where a nursery pianist may plod boringly (to adult ears) back and forth between two or three simple ones.

Harmonization is generally more sophisticated in classical than in popular music, in part because it tends to use voice-leading: the main melody is accompanied by other voices that harmonize while possessing their own impetus and logic, rather than being monolithic chords. Take the elegant opening of the Andante grazioso movement from Mozart's Piano Sonata No. 11 (Figure 6.5*a*). One *could* harmonize this with simple block chords: A in the first bar, E in the second and so on (Figure 6.5*b*). But how dull that would be in comparison. Mozart's accompaniment echoes the tune melodically and rhythmically while, by using chords in *inversion* (that is, so that the lowest note is not the tonic), maintaining a steady, repeating E at the top. I should add that Mozart wasn't simply guided here by 'what sounded good' – there are rather well-defined rules, such as the principles of voice-leading (see p. 146), governing (or perhaps one should say guiding)

Figure 6.5 Mozart's harmonization of the Andante grazioso movement of his Piano Sonata No. 11 (*a*) is far more elegant than a simple harmonization using block chords (*b*), although the latter doesn't sound in any way 'wrong'.

this sort of classical composition. These, however, may themselves draw unconsciously on perceptual considerations.

Despite this sophistication, the basic harmonic structure of a great deal of classical music, particularly from the Baroque and Classical eras, is as formulaic as pop, based on the cycle of fifths that we encountered in Chapter 3. A composition in the key of C may modulate up the cycle to G – so that a G major chord will accompany the melody – and down the cycle to F. An immense number of tunes, including most nursery rhymes and all conventional twelve-bar blues songs, can be played using just these three chords – or rather, this same pattern anywhere in the cycle of fifths. The key a perfect fifth above the tonic is called the dominant and its major chord is denoted V. (The corresponding minor chord is denoted with lower case, v.) A perfect fifth below the tonic, which is equivalent to a perfect fourth above, brings us to the subdominant, IV. The tonic chord is written I.

Chords formed from triads on the other notes of the scale are likewise denoted by the corresponding interval to their root: II/ii, III/iii and so forth (Figure 6.6). This means that the tonic chord I is surrounded by a cluster of chords that are in some sense closely related – 'within easy reach', you might say, being made up entirely of notes contained within the major scale of the tonic. Some of these are major chords (IV, V), and some are minor (ii, iii, vi – the latter is called the *relative minor* key of the tonic major). Only one (denoted vii^0) is neither, being instead a so-called diminished chord – in C, it is the triad B-D-F. But even that triad can function in a different role: set against a G root note, it becomes a dominant seventh V^7 (G-B-D-F), containing a minor seventh interval (G-F) with respect to the root.

Vast swathes of popular music involve a conservative, even hackneyed, excursion around the harmonic centre of the tonic whose main events are transitions to the IV and V chords. This is true of (to take a few random examples) 'I'm a Believer' by the Monkees, 'Da Doo Ron Ron' by the Crystals, 'I Saw Her Standing There' by the Beatles, Bob Dylan's 'Blowing in the Wind', and the Sex Pistols' 'Anarchy in the UK'. Like the composers of the Classical era, pop musicians turn this formulaic predictability into a virtue, using it as the reassuring framework on

Figure 6.6 The chords formed from triads on the notes of the diatonic major scale.

which to hang rhythmic, melodic, timbral and lyrical invention. Some of the other related chords, most notably ii and vi (D minor and A minor in the key of C major) are occasionally added in sequences that have been recycled endlessly, such as the 'turnaround' progressions I-vi-IV-V ('Wimoweh' by the Weavers,* Rodgers and Hart's 'Blue Moon', the Everly Brothers' 'All I Have To Do Is Dream') and I-vi-ii-V (chorus of the Beatles' 'You're Gonna Lose That Girl'). These structures are, as we'll see, soundly rooted in traditional harmonic theory, and are used in much the same form in classical music – a slightly more sophisticated variant of the I-ii-V turnaround appears, for example, at the start of the Rondo in Beethoven's Piano Sonata No.19, Op. 49 No. 1.

We saw earlier that most popular melodies find their way back to the tonic note at the end. By the same token, the chord progressions that accompany them end on the tonic chord. If they don't, we're left with a sense of incompleteness, of being up in the air. Imagine ending a tune like 'The Grand Old Duke of York' after the second line ('He had ten thousand men'). Almost painful, isn't it? You're stranded on a dominant chord, with the implication that we're only halfway through a journey.†

The sequence of chords that ends a musical phrase is called a *cadence*. The word comes from the Latin *cadentia*, 'falling', reflecting the fact that musical phrases commonly end with a descending melody line. A cadence is a kind of closure, a rounding off, which may be more or less complete depending on where in the music it arrives. Just about any children's song, and most folk songs and indeed most classical compositions until the mid-nineteenth century, end in the so-called authentic (or closed) cadence, which brings matters to a conclusion with the tonic chord. In the authentic cadence this is preceded by the dominant chord: V to I (Figure 6.7). This seems to Western ears to have such a sense of finality and closure about it that one can't help suspecting the operation of some 'natural law' : the dominant chord seems to beg for *resolution* by the

* This much-covered song, also known as 'The Lion Sleeps Tonight', is based on an African song, which the ethnomusicologist Alan Lomax introduced to the Weavers' leader Pete Seeger. A pioneer of American folk and protest music, Seeger is now often (and unjustly) remembered as the man who threatened to put an axe through the cables at Bob Dylan's famous electrified performance at the 1965 Newport Folk Festival.

† If we're simply *singing* the tune, then of course there are no chords at all. But nonetheless, the strong harmonic tradition of Western music means that we 'sense' them, a phenomenon called implied harmony. We'll see shortly why we do that.

Figure 6.7 The authentic cadence.

tonic. There is a popular story that a young composer was once compelled to rise from his bed in great agitation and rush to the piano to resolve a dominant chord that had been left hanging.

But it's not clear that someone who has never heard Western music would experience the same urge. The very ubiquity of the authentic cadence should be enough to create an overwhelming expectation of it for Western listeners, but that's no reason to consider it a fact of nature. Rather, an authentic cadence functions as a learnt clue about where the music is going, or in this case, where it is stopping: we recognize it as a stop signal. The authentic cadence became a near-universal formula during the Classical era: it was almost inconceivable that a composer such as Mozart or Haydn would end a sonata or symphony without one, often in the form of two big, isolated chords uncluttered by melody and arranged so that the final, crashing tonic chord comes on a strong beat (Figures 6.8*a*,*b*). By this time the tonic chord generally conformed to the major or minor mode of the whole piece, but in Baroque music the minor tonic chord was considered less stable than the major, and so the authentic cadence of a

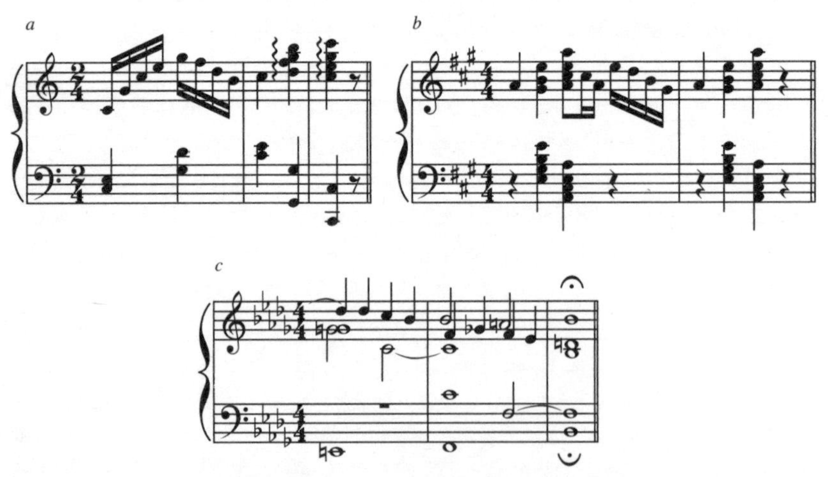

Figure 6.8 (*a, b*) The authentic cadence in Mozart's Sonata No. 1 in C major, and in Variation VI of Sonata No. 11 in A major. (*c*) A 'Picardy third' cadence in the Fugue in B♭ minor from Book I of Bach's *Well-Tempered Clavier*.

Figure 6.9
The plagal cadence.

minor-key piece would often switch to the major chord – a so-called Picardy cadence. Bach's music is full of them (Figure 6.8*c*).

Modernism often dispensed with these formulas, so that endings were not so clearly signposted. Compositions might come to a sudden halt almost unheralded, the carpet suddenly whisked from under the listeners' feet, as in Stravinsky's *Petrushka*; or the proceedings might end not with a bang but with a whimper, as in Samuel Barber's *Adagio for Strings*.

While the authentic cadence is the usual means of rounding off a piece of tonal music, the progression of subdominant to tonic (IV to I, or F to C in the key of C) also creates a gentler sense of closure and finality. This is called a plagal cadence (Figure 6.9), and is familiar from the 'amen' sung at the end of many Christian hymns, or the cadence that ends the verses of the Christmas carol 'Good King Wenceslas'.

Cadences don't have to end on the tonic: they can signify a pause, ending a phrase with a promise of more to come. An imperfect, open or half cadence ends on a V chord, typically reached from a IV or I chord. There is one at the end of the first phrase in Beethoven's Sixth Symphony (the 'Pastoral') (Figure 6.10) – a little opening gesture that, having clearly signalled that there's more to follow, pauses to take a breath. Meanwhile, a deceptive cadence moves from a V chord to any chord except the tonic. It's called deceptive because a V chord coming towards the end of a phrase is normally a preparation for the tonic in an authentic cadence – we're taken by surprise when the tonic doesn't arrive. The effect can be exquisitely piquant. One of my favourites happens in Bach's Prelude in E♭ minor from Book I of *The*

Figure 6.10 The imperfect or open cadence in Beethoven's Sixth Symphony.

Figure 6.11 The deceptive cadence in the Prelude in E♭ from Book I of Bach's *Well-Tempered Clavier.*

Well-Tempered Clavier, where the tonic chord is supplanted by an altered VI chord (Figure 6.11). It is a moment that captures all the magic of what music can do; I can't play this simple sequence without being moved by it. Precisely the same chord progression punctuates Chopin's 'Raindrop' Prelude No. 15 in D♭, where its unexpected outcome brings forth sunbeams from behind the storm clouds.

Debussy sometimes modified his cadential tonic chords to include other notes of the scale, such as a sixth (Figure 6.12), leaving a lingering hint of incompleteness. This idea of modifying the cadence with extra notes was taken up by early jazz musicians, and then in the pop music of the 1960s, where it appears for example in the final chords of the Beatles' 'She Loves You' and 'Help!'. In the latter, the unresolved quality perfectly matches the plaintive vocal 'Oooh' and preceding plea of 'Help me! Help me!' This imperfect closure chimes with the spirit of modern popular music, which typically embellishes its closing tonic chords with all kinds of harmonies to stave off the stodgy, emphatic dourness of Beethoven's stomping cadences. Rock musicians often mix a flattened seventh into the tonic chord, a trick learnt from the blues, as though to say, 'I'm not through yet!' And jazz musicians use closing chords of fantastic complexity, from the relatively harmonious sixths and ninths to stacks of chromatics that hang in the air like the cigarette smoke of an underground club (Figure 6.13).

In fact, rock has also found its own cadential formulas, which in classical terms are decidedly peculiar. One that is especially common

Figure 6.12 A cadence containing a sixth, from Debussy's *Valse Romantique*.

Figure 6.13 Some of the elaborate cadences typically used in jazz.

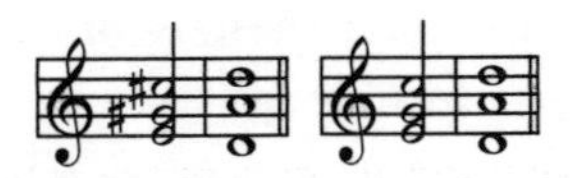

Figure 6.14 Some of the cadences common in late medieval music.

is the ♭VII-I cadence – or extended perhaps in a ♭VI-♭VII-I cadence, such as that which ends the verses of the Beatles' 'Lady Madonna'. Recorded music, meanwhile, made possible the ultimate in imperfect closure: the fade-out, where the music circles eternally without ever reaching its final resting point.

Anyone who feels that the authentic cadence is 'natural' should bear in mind that before Western scales and keys were formalized, several other cadences were in standard use that sound rather odd to our ears now (Figure 6.14). And although cadences are usually discussed in terms of chord movements, they don't strictly speaking have to involve any chords at all: they are simply standardized formulas for closing phrases. They feature as such in some non-Western music. Performances on the Chinese *bawu* bamboo flute, for example, typically finish with a long trill, while the songs of many traditions end, like Beethoven's symphonies, with a series of repeated notes (Figure 6.15).

Figure 6.15 A formulaic repeated-note cadence in a Native American Pawnee song (*a*), and the repeated-chord ending typical of Beethoven's orchestral music (here from the Second Symphony) (*b*).

What are you implying?

A musician's ability to improvise an accompaniment to a melody can seem miraculous to a non-musician, as though the music is being plucked out of thin air on the spot. But the performer is generally relying on some fairly simple rules of thumb, based around the idea that an appropriate chord contains the most salient melody notes.

Consider 'London Bridge is Falling Down'. In C major the melody starts with the notes G, A, G, F, E, F, G (Figure 6.16). As with nearly all children's songs, we are safe in assuming that this one starts on the tonic chord, C major. The phrase hovers around the fifth note of this scale, G, on which it begins and ends, and so it fits nicely with the tonic major triad, C-E-G. (The clashes on the notes F and A are negligible, because these are passing notes in the manner discussed on p. 100.) But then the first repetition of 'falling down' happens on the notes D, E, F. These are also in the scale of C, but they neither start nor finish on notes in the C major chord. So we need a chord with F in it, and perhaps D too. D minor fits those criteria, and it sounds OK. But so will the dominant seventh (G-B-D-F), which is better, because it involves a stronger movement from the tonic to the dominant. The next repetition happens on E, F, G – that's easy, because it brings us back to the C-E-G triad.

It might sound rather complicated, but it very quickly becomes second nature when you've done this sort of thing a few times. In a simple tune like this, we'd expect little more than the tonic, dominant and subdominant chords to appear, and can quickly train our ear to hear when a chord change is needed and which it is likely to be.

You may not play an instrument or have the slightest notion of chord structure, but you're probably fairly adept already at harmonization, simply because of the strong emphasis on harmony in Western music. Western listeners tend to hear the notes of a bare melody not as a succession of single tones but as a trajectory through the harmonic space that chords reveal explicitly. This is called *implied harmony*. Hindemith suggested

Figure 6.16 A simple harmonization of 'London Bridge is Falling Down'.

that Western listeners can't hear the 'pure' melodies of cultures with a monophonic musical tradition, because for them the music becomes irrevocably coloured by the learnt instinct to add implied harmony.

Music psychologists Laurel Trainor and Sandra Trehub have shown how our subconscious sense of implied harmony affects the way we judge melodic similarity. They tested the ability of adults and children (two groups aged five and seven) to detect the change of a single note in a simple ten-note melody in C major (Figure 6.17*a*). The altered note in each of three variants was always the sixth, a G. In one case it was raised a semitone to A♭ – a note that falls outside the C major scale. In another, it was raised a tone to A, which stays within the major scale. And in the third case, the G was raised two tones, to B, which is also still 'in scale' (Figure 6.17*b-d*). All these changes retain the same general melodic contour. Unsurprisingly, the adults spotted all the changes consistently better than five- and seven-year olds. But both the adults and seven-year-olds found it slightly harder to detect the third change, from G to B. This seems to be because the implied harmony – a G^7 chord underlying notes 4-6 (F-D-G) – in the original is not altered in that case, and so the melody is heard as being more similar to the original, even though the change in pitch of note 6 is actually the largest of all three alterations.

This shows that implied harmony is already established in Western listeners by the age of seven. But the five-year-olds in this experiment could only detect the change when it introduced an out-of-scale note (Figure 6.17*b*); they barely noticed the other two changes at all, and didn't significantly differentiate between cases where the implied harmony was retained or altered. So it seems that an ability to infer

Figure 6.17 In an experiment to investigate the developmental onset of implied harmony in Western listeners, the simple tune in C major in (*a*) was subjected to three variations of the sixth note (*b-d*). In (*b*) the new note falls outside the C major scale. In (*c*) and (*d*), both notes are in the correct scale, but only (*d*) preserves the implied harmony. Adults and seven-year-olds found it harder to detect the change when the implied harmony was retained, whereas five-year-olds could only detect the out-of-scale change (*b*).

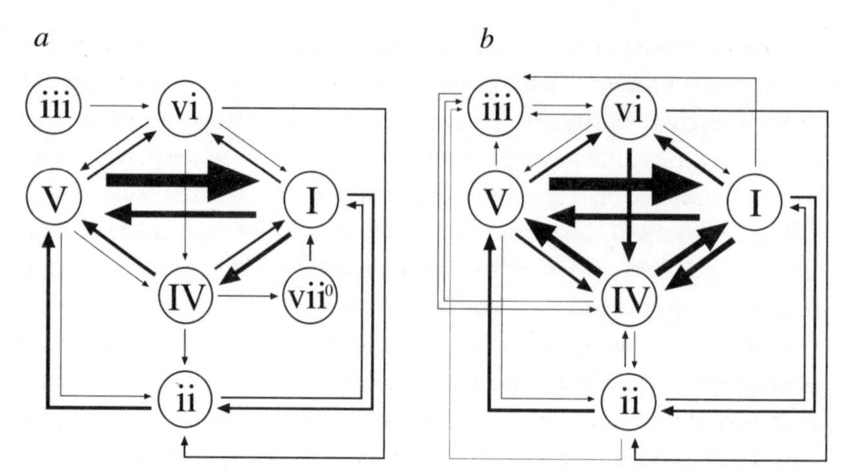

Figure 6.18 The probabilities of different chord transitions in harmonic progressions is similar for Baroque (*a*) and popular (*b*) music. Here the width of the arrows denote the size of the probabilities. Notice that the diminished vii⁰ chord is essentially absent from pop music.

implied harmony happens somewhere between the ages of five and seven in typical Westerners.

Change of key

Until around the middle of the nineteenth century, harmonic progressions in Western classical music tended to be rather formulaic and conservative. They typically involve transitions from one chord to another that is closely related, so that the change isn't jarring. For a major key, this generally means a movement from the tonic to one of the six related triads on the diatonic scale shown in Figure 6.6 (Figure 6.18*a*). Pop and rock music has inherited much of this tradition, so that its chord transitions show a rather similar pattern (Figure 6.18*b*) – although as we'll see, it introduced some new moves too. These progressions may lead to a true modulation, where the key itself is altered. For example, the sonata form generally involves a modulation from the major tonic to the dominant (V) in the 'exposition' section at the beginning, which introduces the main themes of the music. If the sonata is in a minor key, the modulation here is often instead to the relative major – for example, C minor to E♭ major.

In principle, even these conservative progressions and modulations can carry one stepwise to rather remote keys. In the Classical and Romantic traditions, this was typically done via the cycle of fifths. A change from C major to A major sounds rather abrupt and disconcerting if done suddenly, but not if it is 'prepared' by first modulating to the dominant (G) and from there to D, the dominant of G. Then A is within easy reach as the next new dominant. Here the transition is effected by means of intervening 'pivot' chords that are related to both the original and the new key.

Note that these modulations aren't simply a matter of playing a new backing chord. Rather, a true modulation has only occurred when chords *shift their functions*: when we hear a G no longer as a dominant chord but as a tonic, say. This might sound odd – how can we 'hear' a chord differently, if it contains exactly the same notes in both cases? But remember that in music, context is everything: notes and chords convey information and meaning because they appear in relation to other chords and notes.

A modulation, then, consists of a change in the perceived *root* of the music. You can stay all day on the fifth floor of a tall building and never lose the sense of it being a fifth floor rather than a redefined ground level. That's because the ground level can't miraculously rise up to the fifth floor. But in music it can: a modulation has taken place when the fifth floor becomes the new ground level, and that changes all the other levels too. It is hard to describe, but easy to experience. Take 'Fool to Cry' by the Rolling Stones: the end of the song modulates from the major (F) to the relative minor (D minor), and you know it when it happens even if you couldn't have designated it that way. You feel the tonal centre of gravity move, so that there is a new resting place for the harmonic progression.

Carol Krumhansl and Edward Kessler have found that it typically takes about three chords, which might occupy the space of a few seconds, for a simple modulation to become firmly established in our minds – that is, for us to conclude that we must mentally relinquish the old tonic centre. One way of signalling this shift is to alter the key of the melody so that it uses the scale of the new tonic. For a shift from C to G, for example, this means that we throw out F natural and start using F♯. A move to the key of D then means sharpening the C to C♯ too.

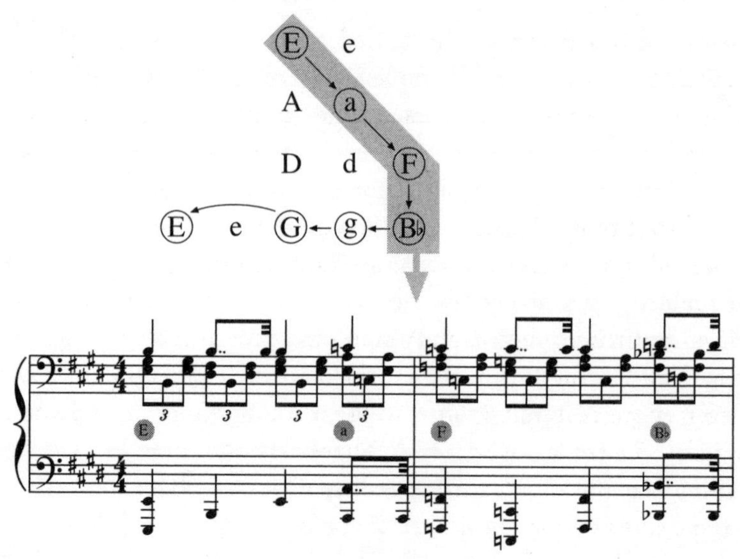

Figure 6.19 The harmonic progression in the third phrase of Chopin's Prelude in E major, Op. 28 No. 9. Here I show the trajectory within the 'harmonic space' described later (p. 193), in which related chords are grouped together. This progression makes use of pivot chords to achieve transitions between keys that are rather remote from one another, such as E major→A minor→F major. The full score for the shaded part of the progression is shown in the lower frame.

Yet we shouldn't imagine that music is like chess, in which only certain moves are permitted. The 'rules' are largely a matter of convention and experience: we come to accept certain moves as legitimate merely because we have heard them before and become accustomed to them. While music students were taught that the way to modulate was via the cycle of fifths and pivot chords, anything else sounded odd. But one of the characteristics of classical music from Beethoven onwards is that it began to take more risks and detours with modulation. Chopin's changes of key are sometimes rather striking, even though he generally achieves them via the conventional means of pivot chords. For example, in his Prelude No. 9 in E major, Op. 28, there is a dramatic modulation from E major to F major, which is achieved via the stepping stone of A minor – this is both the vi chord of E major and the iii chord of F major (Figure 6.19). Moreover, whereas in Mozart's music modulation not only tends to follow well-established patterns but also happens rather quickly, as though it is something to be got over with at the earliest opportunity, later composers prolonged these transitions and thereby lengthened the periods during which the listener couldn't be sure of the key and was left struggling with the ambiguity. Composers such as Chopin and Liszt began not only to take unexpected, meandering routes in harmonic space, but to use chords that didn't obviously belong to any tonality – a technique refined to literally epic proportions by Wagner.

The rules of classical modulation and harmonic progression were well and truly shredded by modernists such as Prokofiev, whose music is filled with unexpected and unprepared modulations, such as the move from IV to ♭III (F to E♭) at the start of his Piano Sonata No. 5, or even more blatantly from i to ♭I (E minor to E♭) in the waltz of his ballet *Cinderella* (Figure 6.20). These shifts are in many ways more startling than the angular motions of atonal music, because Prokofiev's appropriation of traditional tonal forms and phrases lulls us into a false sense of security.

As well as using classical progressions, pop and rock music has created its own set of permissible moves, which break classical rules but which are now so well established that they don't sound strange at all. For Baroque audiences the transition I-♭VII (C to B♭, say) would have seemed decidedly odd, yet for rock audiences it is utterly unremarkable – for example, in the Who's 'My Generation', or reversed in the Kinks' 'Your Really Got Me'. The progression I-♭VII-♭VI-V is also a pop staple, notably used in 'Hit the Road Jack', made famous by Ray Charles. Another classic example is the semitone rise in key: a gear change that injects fresh energy from the jolt. There are countless examples – Stevie Wonder's 'You Are the Sunshine of My Life' and Abba's 'Money, Money, Money', say. Indeed, the trick is overused to the point of cliché, becoming a way for the songwriter to snatch unearned impetus. Equally common is the whole-tone modulation, found for example in the Beatles' 'Penny Lane' (where the key changes from A to B). Typically these shifts come shortly before the end of a song, just when the momentum is starting to flag. In classical terms they are hardly modulations at all, but merely transpositions of key – what comes before bears none of the theoretical relationship to what comes after that true modulation entails, but is merely repetition in a new key. We are not taken on a journey in harmonic space, but simply have the coordinate grid redrawn under our feet.

Figure 6.20 Unprepared modulations in the music of Prokofiev: the Piano Sonata No. 5 (*a*) and the waltz from the ballet *Cinderella* (*b*).

Harmonic cartography

However bold, trajectories in harmonic space can be no more arbitrary than those in melodic space. To anyone brought up in the diatonic tonal tradition – and that means fans of Michael Jackson as much as Michael Tippett – there needs to be a logic to modulations if they are to be comprehensible. New paths can be invented, but they still have to get somewhere. This means that we need some sense of what harmonic space looks like.

Just as music theorists and psychologists have attempted to map out the relations between notes in diatonic and chromatic scales, so they have attempted a cartography of chords and keys. The earliest attempts to do so were based on musical theory, such as the relationships between a tonic triad and the various triads derived from its major scale. We might conclude from these that any major chord I is in some sense 'close' to two other major chords (IV and V) and three minor chords (ii, iii and iv). A harmonic progression from the tonic to any of these chords will 'sound right', although not necessarily to the same degree (as we'll see shortly). But associated with a major chord there is also its own minor (i) – the transition between the tonic major and minor is also an 'acceptable' move. It is extremely common in classical music, for example in the well-known Rondo alla Turca

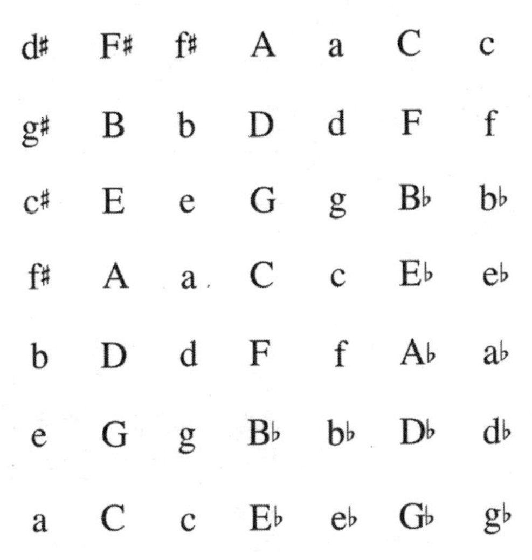

Figure 6.21 A two-dimensional representation of harmonic space. The grid continues indefinitely.

movement of Mozart's Piano Sonata No. 11, K331, the Rondo of Beethoven's Piano Sonata No. 19, Op. 49 No. 1, or Chopin's Waltz in B minor, Op. 69 No. 2. A minor chord, meanwhile, has associated minors v and iv, as well as its relative major rooted on the minor third (for example, c-E♭).

All this means that one plausible way of representing the harmonic space of keys, which places related ones close by in a two-dimensional grid, is that shown in Figure 6.21). Arnold Schoenberg drew a map like this in his 1954 book *Structural Functions of Harmony*. Movements vertically step through the cycle of fifths, while horizontally they progress alternately between major/minor and major/relative minor. You'll see that the pattern repeats vertically (but shifted horizontally) every three rows. And allowing for the fact that in equal temperament F♯ is equivalent to G♭ and so on, it also repeats horizontally every eight columns. The harmonic universe is closed, curling back on itself in a shape which is actually that of a doughnut or torus. Conventional modulations in music theory are those that move in small steps across this grid, whether vertically, horizontally or diagonally.

So much for theory – but does a map like this bear any relation to the way we actually perceive harmonic space? Carol Krumhansl has tried to answer that. One way of judging the relationships

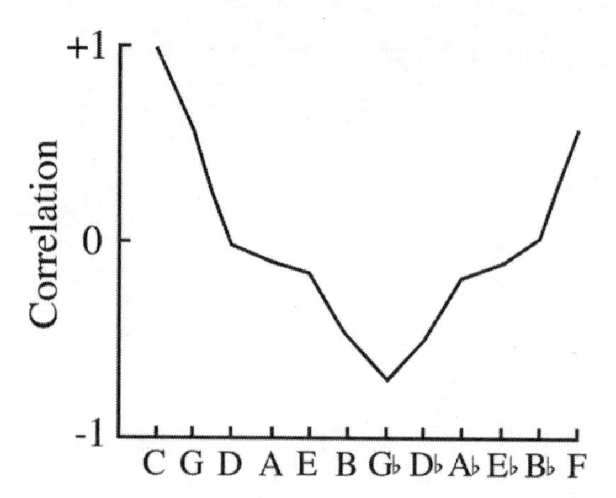

Figure 6.22 One way of gauging the 'relatedness' of keys is to look at how statistically correlated their tonal hierarchies are (see p. 103). Here are the results for C major compared against the other major keys, progressing around the cycle of fifths.

between keys is to look at how closely correlated the tonal hierarchies (p. 103) are for each pair in turn. Recall that these tonal hierarchies are measures of how well each note in the chromatic scale is judged to fit within a specified tonal context. We can compare one graph against another and see how well they overlap. When Krumhansl and Edward Kessler did this, they found that it vindicated musical intuition: the further round the cycle of fifths you go, the less related two keys seem to be, until you come round the other side and the similarities start to increase again (Figure 6.22). A similar pattern is found for minor keys, with the difference that the U shape of the correlation is skewed and that there is a hump corresponding to the relative minor, where the degree of overlap is boosted.

But harmonic space is more complicated than this, because it involves relationships not just of a tonic chord to other chords, but of those other chords to one another. Is a ii chord (D minor in the key of C) perceptually closer to a IV chord (F) or a iv chord (A minor)? In Figure 6.21 the latter two are equivalent neighbours. But how do we actually judge their proximity? There are many possible permutations of relationship here, and the resulting harmonic space is therefore multidimensional. It's a little like trying to map out our social network of friends. Joe and Mary might both be very good friends of mine, and yet have hardly ever met. So if I show them on the map

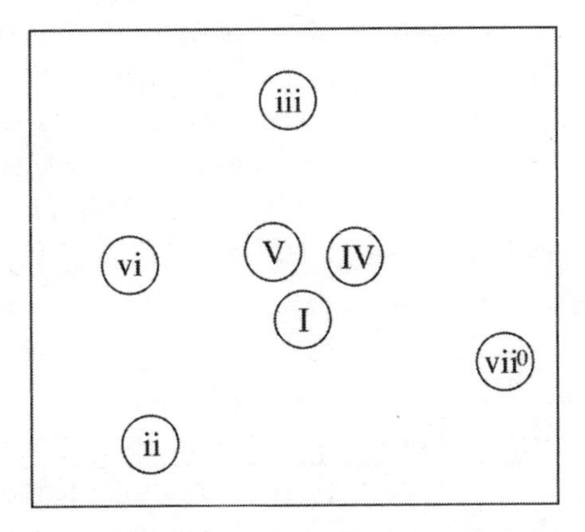

Figure 6.23 The harmonic space of tonic chords as deduced from 'probe-tone' experiments. The distances between chords correspond to their perceived degree of relatedness.

as both being close to me, they're surely going to be close to each other, which gives a false impression of intimacy to their relationship. How do we depict the space in a way that gets round these problems?

Happily, when Krumhansl and her co-workers evaluated the relationships with the 'probe tone' (or here, 'probe chord') method – playing subjects a scale to establish a tonal context and following this with two chords whose fittingness was to be assessed within that context – they found that the results could be pretty well represented on a two-dimensional map (Figure 6.23). You could say that, just as in the social analogy two good friends of yours are pretty likely to know one another, similarly if two chords are both felt to be closely related to a third, they'll probably be perceived as closely related to one another too – regardless of the tonal context. The multidimensional space can thus be fairly effectively collapsed into a flat one that we can draw on paper.

There are no big surprises in the resulting map: the tonic, dominant and subdominant all cluster closely together, the iii, vi and ii chords are all roughly equidistant from this cluster, and the diminished vii⁰ chord is a little more distant still. A music theorist might have guessed something of this sort.

But what determines the pattern? It might be something to do with how many notes the chords share in common, or how consonant they sound together. Or it might be something we have simply learnt by association. Krumhansl and colleagues found that, as with the tonal hierarchy, learning probably predominates: there is a match between the judgements of fit and the frequency of chord use for classical music in the respective key for music of the eighteenth and nineteenth centuries up to 1875. In other words, the more often chords are used together, the better they are deemed to fit. There was far less of a match with the amount of sensory consonance between the respective chords.*

This, however, probes only a small part of the harmonic terrain. What if the context is not a major key but a minor one, for example? And how are major and minor keys themselves related? Is C major *cognitively*, as well as theoretically, close to C minor? How do the related chords fit around them? Krumhansl and Kessler conducted a series of psychological experiments to map out the space of all major and minor keys, and found that again the results could be represented in two dimensions (Figure 6.24). The map is basically the same as that postulated above from theory (Figure 6.21).†

One of the most remarkable findings in musical neuroscience is that this map is *imprinted in a musician's brain.* No one expected that. It is one thing to possess a mental representation of this abstract space, just as we might visualize the layout of our neighbourhood, or indeed the faces of our loved ones. But it is quite another thing for such a representation to be literally projected on to our brain tissue. We don't, for example, seem to have clusters of neurons that represent 'our house', and others for the post office, the bank, the grocery shop, all in the right spatial locations to reproduce a local map in grey matter. Yet that is what happens for harmonic space.

*This test is far from perfect, since it would only really apply if all the subjects had gained their musical experience from listening to classical music of the period concerned. It would be far better to have looked for the amount of correlation with chord use in the music that the subjects actually listened to. But as we've seen, the harmonic conventions of this period of classical music are broadly similar to those of much popular music.

† There seems to be a little bit of corrugation in the rows, so that for example E minor looks closer to C than does G. But this is an artefact of drawing the map in two dimensions. Because the real space has a torus shape, the actual distances are slightly distorted on the flattened-out map, just as they are in a Mercator projection of the world.

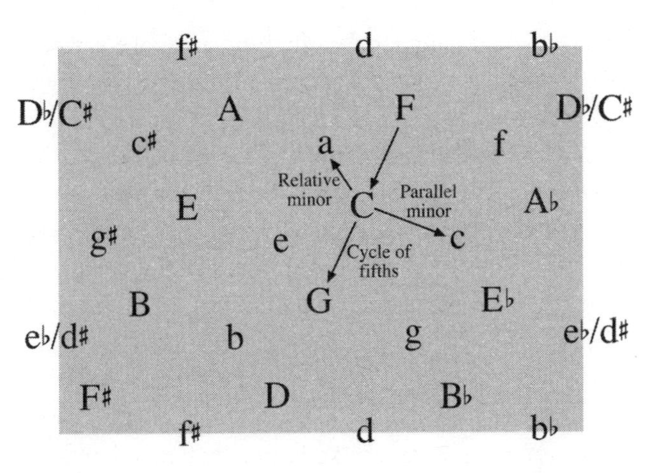

Figure 6.24 The harmonic space of all major and minor keys as deduced from psychological tests. Note that this has basically the same form as Figure 6.21.

This was discovered by neuroscientist Petr Janata of Dartmouth College in Hanover, New Hampshire and his co-workers in 2002, using the technique of magnetic resonance imaging. As I explain in Chapter 9, this method reveals which regions of the brain become activated during a cognitive processing task: it detects the increase in blood flow that accompanies neural activity. Janata and colleagues looked at changes taking place in subjects' 'harmony-processing centre' while listening to music. This brain region, situated in the so-called prefrontal cortex, acts as a juncture of several different modes of processing: purely cognitive (concerned with the 'hard facts' of the stimulus, such as pitch), affective (concerned with emotions) and mnemonic (concerned with memory). It is also associated with the assessment of consonance and dissonance. In short, it is well placed to integrate all the key elements of musical experience. By scanning the brains of eight musically experienced people while they heard a tune that systematically modulated through all twelve major and minor keys, Janata and colleagues found that different regions of this cortical area – distinct populations of neurons – were activated by different keys. One area lit up for C minor, say, and another for A♭ major.

This might naïvely be interpreted as saying that the harmonic map is biologically imprinted, and therefore somehow 'natural'. But that's not really the case. For one thing, the map is 'redrawn', with

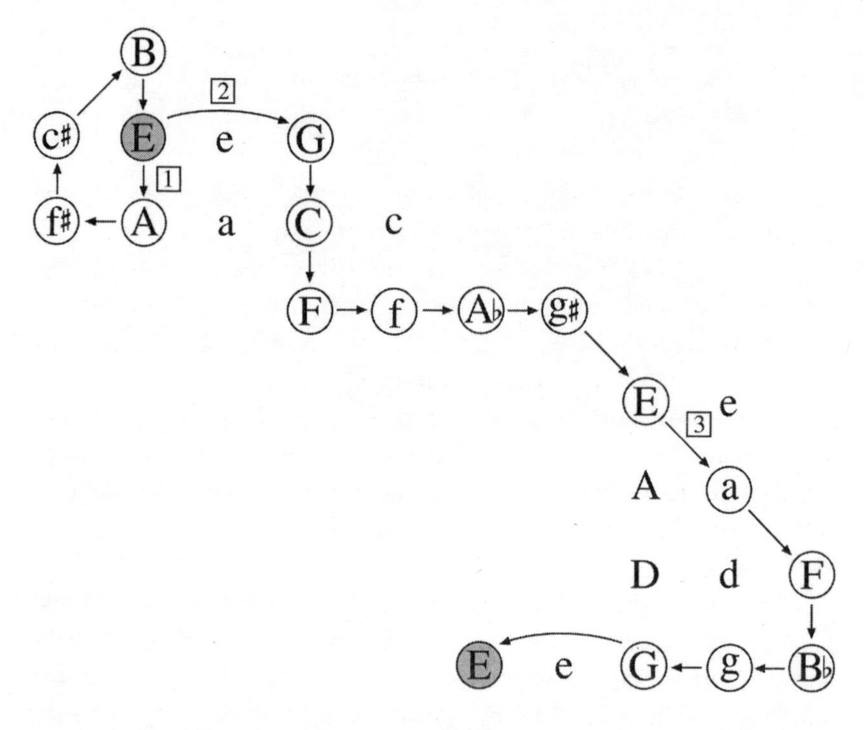

Figure 6.25 The full trajectory through harmonic space during Chopin's Prelude in E major. The three phrases of the piece (1: bars 1–4; 2: bars 5–8; 3: bars 9–12) are marked with numerals, and the starting and ending chords (both tonic) indicated in grey. Notice that here the basic repeating unit of Figure 6.21 (or equivalently, Figure 6.24) has been repeated several times, although only the relevant portions are shown.

an arbitrary orientation, on each separate occasion that this region is activated by music. It's as if we have a neuronal blackboard for drawing the map, but we wipe it clean when the music is over and then draw it again, at a random angle, the next time we need it. And the relationships themselves are most probably learnt, and recalled from memory each time, rather than being hard-wired. But that is in a way equally striking, because it means that we can abstract this map from music itself, where it is encoded only in a very indirect way. With sufficient musical experience, we come to 'know' the map of harmonic space by heart even though we've never actually seen it.

Lost in space

Let's now look more closely at the kinds of journeys music can offer. Chopin is one of the most inventive tour guides: in his E major Prelude mentioned earlier, we range far afield in the space of just twelve bars (Figure 6.25). The first phrase (bars one to four) stays securely in the vicinity of E major, until we move to G major in bar five, and from there to C major in bar six. Here G acts as the pivot chord. Then we get to A♭ by bar eight – a long way from E major, but less so from C.* From here Chopin jumps back to E major in bar nine, and then deftly moves to A minor, which, as we saw earlier, paves the way to F and thence to B♭. From here it's a short step to G minor (bar eleven), and from there to G major. Finally we return, via an ephemeral B major, back to E.

This is a remarkable voyage, but every step has a logic to it, with pivot chords being used to achieve long jumps in a short space of time. This piece illustrates how modulations and harmonic progressions became blurred in the nineteenth century: changes of tonic centre happen so rapidly that it is hard to keep pace. In a typical piece by Mozart, harmonic progression is quite conservative and can alter a local sense of tonality without our losing sight of where the overall phrase is rooted – the return to the tonic usually happens quite quickly. But at various points in the Chopin prelude, it's doubtful that any listener would still have a strong sense that the piece is nominally in E major.

This movement in harmonic space allows composers to create quasi-narratives in musical form. When Chopin unleashes one of his trademark chromatic progressions with all its ambiguity of tonality (Figure 6.26), it is as if we have left the open plain and entered a thicket in which all direction is confused – only to emerge eventually into a new tonal vista where at once the path becomes clear again. In Chapter 13 I explore how some composers seem to have used this sort of harmonic

* A♭ is itself identical to a G♯, which is the major third of E. Yet Chopin writes it as A♭, not G♯, in bar eight. Why? Because A♭ is reached via C and F minor, where the A♭ note is heard as a flattened A, not a sharpened G. As we saw earlier, just because a sharp and its corresponding flat *sound* the same in equal temperament, this doesn't mean they serve the same musical *function*.

Figure 6.26 A chromatic progression in Chopin's Nocturne in G minor, Op. 15, No. 3. Notice that this eventually resolves itself in a key (D♭) close to the one from which it began (G♭).

trajectory to represent quite specific 'meanings'. One's sense of these journeys is dependent on how clearly one can perceive the map: where naïve listeners might sense only a constant shifting of direction, more sophisticated listeners who have a good sense of harmonic relationships will be able to hold on to their bearings and are more likely to develop predictions and expectations about where the path is taking them.

The steps along the way don't have to start and finish on specific keys: our perception can dither in the space between them. Classical composers often create a sense of transition begun but not completed by using modified chords whose tonic root is not clear. There's one on the fourth beat of bar seven in Chopin's E major Prelude: B♭, D♭ and G set against an F♭ bass.

You can find this between-ness even in the basic twelve-bar blues progression, where the switch from the tonic chord (C, say) in the first four bars to the subdominant (F) in bars five and six is often presaged by a flattened seventh chord in bar four: C, C, C, C^7, F, F, C, C, . . . The flattened seventh here adds B♭ to the major triad, and it creates a sense of impending change because B♭ is in the major scale not of C but of F. So are we in the key of C in bar four, or of F? We probably feel we are somewhere in between.

Krumhansl and Kessler have studied how people perceive changes of key centre in painstaking experiments using a series of chord changes that move through harmonic space. The researchers played subjects the sequence repeatedly but stopping after two, three, four chords and so on. In each case they used the 'probe chord' method to figure out where the listeners felt they were in relation to each nearby chord. This was extremely laborious, but it enabled Krumhansl and Kessler to pinpoint the precise location of the listeners' sense of tonality on the map of harmonic space (Figure 6.27). Some of the chord sequences stayed close to the original tonic, whereas others modulated to a new key. This was reflected in the wandering trajectories of the subjects' perceptions.

It seems that we're happy for our perception of harmonic centre to wander by small degrees, but resist sudden big shifts. When a modulation in these experiments involved a big jump in harmonic space, listeners appeared to resist it for as long as they could: they tried to fit what they were hearing into the tonal context of the original tonic. When that was no longer tenable, their perception made an abrupt jump over to the vicinity of the new tonic. In other words, we're comfortable with lingering between nearby keys, but not between remote ones.

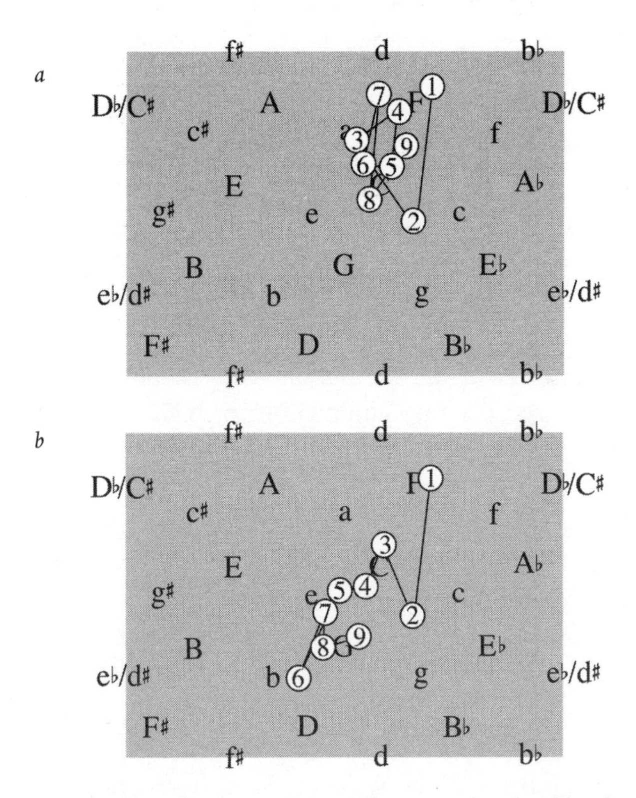

Figure 6.27 Journeys in harmonic space. These are the chord-by-chord locations perceived by listeners during two nine-chord progressions. That in (*a*) was designed to create a sense of remaining in the key of C major, whereas that in (*b*) suggests a modulation to G major.

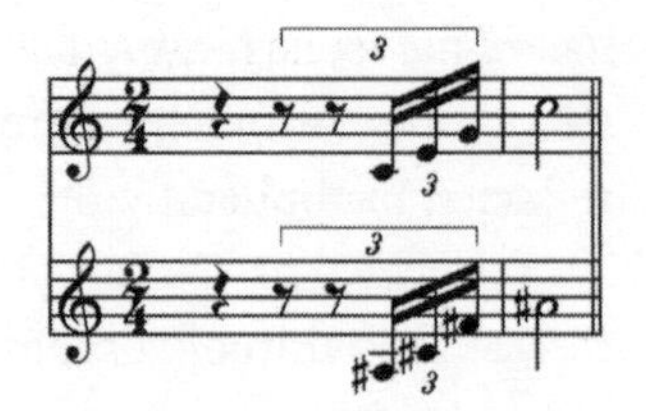

Figure 6.28 Stravinsky's 'Petrushka' chord – a juxtaposition of the C major and F♯ major triads.

This is a more general characteristic of how we listen to music. We've seen now that we are always seeking some familiar framework or structure on which to order what we hear. We do that by making best guesses, comparing what we have heard with what we have learnt from experience. Each new bit of auditory information requires us to revise and update this guess. But each guess is tenacious – once we think we have the right structure, we don't give it up lightly but only once it becomes obviously implausible.

Rather than hovering between keys, is it possible to hear two different keys at the same time? Modernist innovations such as the dual orchestras of Charles Ives make this more than a theoretical question. Most famously, Stravinsky superimposed the arpeggiated chords of C and F♯ major in the 'Petrushka' chord that appears in the second tableau of the ballet of that name (Figure 6.28). It gives this passage what you might call a crooked quality – it's not exactly dissonant, since both of the superimposed chords are consonant in their own right, but there is a jangling sensation to it that perfectly suggests the disjointed movements of the eponymous puppet. The tonalities of C and F♯ are as far apart as they could be – a tritone, on opposite sides of the cycle of fifths, with only two notes of the major scales in common.

This aspect of Stravinsky's work has been called polytonal, meaning that the music retains tonality but uses more than one tonic at the same time. Polytonality has been used by several composers: Darius Milhaud, for example, juxtaposed a right hand in D major against a left hand in G major in his piano dance suite *Saudades do Brazil*. But is the Petrushka chord truly bitonal, simultaneously in C and F♯? Stravinsky himself thought so, but other musicologists have suggested that the C/F♯ combination can be rationalized in a single, non-diatonic scale: the eight-note octatonic C, C♯, D♯, E, F♯, G, A, A♯. Is this scale instead the organizational structure that a listener uses to interpret the chord?

Krumhansl has investigated that question experimentally by asking which notes of the chromatic scale seem to fit most comfortably with the context supplied by the Petrushka chord. Do listeners pick out the diatonic scales of C and F♯, or the octatonic? The answer is neither! The tonal hierarchy obtained from these perceptual tests is better matched with a scheme proposed by musicologist Pieter van den Toorn, according to which the octatonic collection of notes is partitioned hierarchically, like the notes of the major scale: C and F♯ are the most salient, followed by the groupings (C-E-G) and (F♯-A♯-C♯), and then all eight notes.

This suggests that listeners, faced with these unusual harmonies, seem able to construct some way of making sense of them that doesn't really feature in the conventional theory of Western music. Hardly anyone knows about van den Toorn's exotic octatonic hierarchy, and yet it seems that people subconsciously abstract this organizational scheme from the music – on the spot, and for a section of music lasting only several seconds! This recalls the way listeners deduce new tonal hierarchies to make sense of unfamiliar scales of other cultures (p. 106). It brings home again how little formal training or theoretical knowledge we need to interpret, and perhaps to enjoy, new musical experiences.

The big picture

Krumhansl's harmonic space, or something like it, seems then to be a fair representation of the mental map we use for negotiating harmonic progressions. But it is still just an approximation, a sketch of a slice through a multidimensional space. Musicologist Dmitri Tymoczko of Princeton University has used musical and mathematical theory to trace out a more rigorous and exhaustive plan of musical space, by considering the formal relationships between all possible notes and chords. The result is rather frightening, for this convoluted, high-dimensional space is challenging even for mathematicians to visualize and comprehend.

And yet at root this is an exercise in simplification. If we consider all the notes on a piano keyboard and all the ways they can be combined, sequentially or in chords, into musical sequences, the possi-

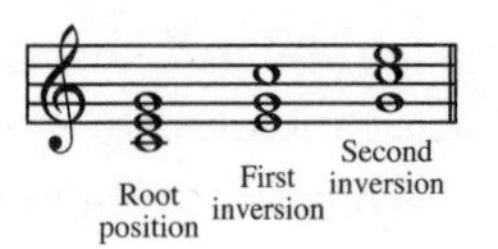

Figure 6.29 The inversions of a C major triad.

bilities are astronomical. No one can hope to understand such an abstract space. Without musical concepts such as scales, chords and keys, we would be utterly lost.

What these concepts do is organize certain groups of notes into *classes*. The note groups C-E-G and B♭-D-F are equivalent in the sense that both are major triads – the latter can be derived from the former by moving each note down one tone. Similarly, C-E-G is equivalent to E-G-C and G-C-E in that all are C major chords, the latter two being inversions of the first: they share the same pitch classes, but with different bottom notes (Figure 6.29). C-E-G is also the same kind of chord regardless of which octave it is played in. Tymoczko, with colleagues Clifton Callender and Ian Quinn, has looked for ways to represent geometrically all such equivalences that musicians recognize between different groups or sequences of notes. These relationships can be described mathematically as *symmetry* properties, akin to the way an apple reflected in the mirror is still recognizable as an apple. By enumerating all the symmetries, the immense number of ways to arrange notes into melodies and chord sequences can be collapsed into much more compact subspaces. It's rather like the way different objects can cast shadows that look identical: by 'projecting' these shapes from three to two dimensions, we find a symmetry that makes them all look the same (Figure 6.30). Projections in other directions might reveal different sets of relationships.

The researchers say there are just five common kinds of transformation that are used in judging equivalence in music, including octave shifts, reordering of notes (for example, in chord inversions), and duplications (adding a high E' to the triad C-E-G, say). These equivalences can be applied individually or in combination, giving thirty-two ways in which, say, two chords can be considered 'the same'.

Such symmetries 'fold up' the vast space of note permutations in particular ways. The geometric spaces that result may still be complex, but they can be analysed mathematically and are often intuitively comprehensible. In these folded-up spaces, classes of equivalent musical objects – three-note chords, say, or three-note melodies – can each be represented by a point. One point in the subspace that describes three-note chord types – a subspace that turns out to be a cone-shaped surface – corresponds to major triads. Another corresponds to augmented chords, in which some notes are sharpened by a semitone, and so on.

Musical pieces may again be seen as paths through this space. But because it is a more comprehensive map, new relationships can become clear. For example, Tymoczko says his approach reveals how a chord sequence used by Debussy in *Prélude à l'après-midi d'un faune* is related to one used slightly earlier by Wagner in the prelude to *Tristan und Isolde* – something that isn't obvious from conventional ways of analysing the two sequences. Needless to say, Debussy had no intimation of this formal mathematical relationship to Wagner's work. But Tymoczko says that such connections are bound to emerge as composers explore musical space. Just as a mountaineer will find that only a small number of all the possible routes between two points are actually negotiable, so musicians will have discovered empirically that their options are limited by the underlying shapes and structures of musical possibilities.

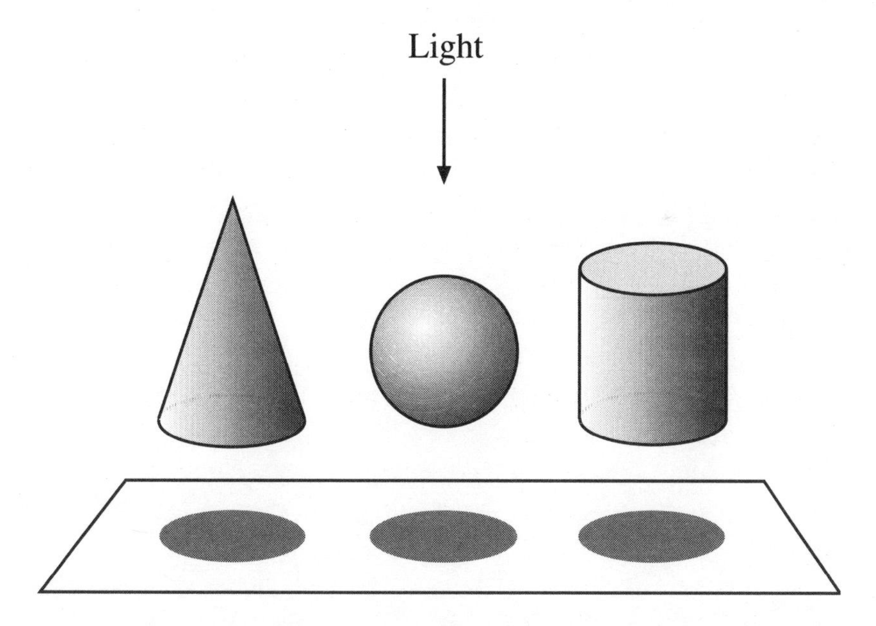

Figure 6.30 Symmetry properties can reveal correspondences between groups and sequences of notes, simplifying musical space, just as different three-dimensional objects can be 'collapsed' into a single class by projection onto two dimensional space.

For example, early nineteenth-century composers such as Chopin began to look for short cuts to the traditional ways of moving between two keys, which they often negotiated through a high degree of chromaticism. Music theorists have sometimes criticized these experiments in harmony as whimsical and unprincipled. But in Tymoczko's big picture they can be seen to take advantage of certain geometric features of chord space, constrained by definite rules that the composers intuited even if they had no sense of their mathematical grounding. Gottfried Leibniz once wrote that musicians use mathematics without being aware of it; but nineteenth-century innovators in harmony were in effect exploring through music geometrical spaces that were beyond the comprehension of their contemporary mathematicians.

7
Con Moto
Slave to the Rhythm

What gives music its pulse?

Imagine that you have never heard any jazz. (If you don't have to imagine that, go at once to www.youtube.com/watch?v=wrTrk WJNyOY and www.youtube.com/watch?v=ukL3TDV6XRg, and then come back.* See now what you've been missing?) Now imagine having to learn about jazz rhythm from sheet music alone. It's not going to work, is it? Elvis Costello was overstating the case when he said that writing about music was like dancing about architecture, but when it comes to rhythm, you can see his point.

Perhaps it sounds a bit silly to suggest that someone might consider learning about jazz rhythm purely from transcriptions. But Igor Stravinsky attempted just that when he was writing his opera-cum-ballet *The Soldier's Tale* (1918). Stravinsky was marooned in Switzerland by the First World War, and jazz had not yet permeated far into European culture. He'd heard about this exciting new music, but there was no YouTube for him to consult. However, his friend Ernest Ansermet, who later conducted the premiere of *The Soldier's Tale* in Lausanne, had picked up some sheet music of jazz scores during a recent American tour. From this alone, Stravinsky tried to imagine what jazz would sound like.

It doesn't sound much like *The Soldier's Tale*. With its shifting and complicated time signatures (5/4, 5/8, 7/16), the piece seems to confuse mere irregularity with jazz's use of irregular rhythms set against a steady beat. That doesn't detract from the many pleasures of Stravinsky's composition, but if a composer with his rhythmic sophistication can be found wanting, we surely have to concede that

* Curious? These are video clips of Ellington's 'Cottontail' (with fabulous dancing), and of Charlie Parker in a quintet with Marquis Foster on drums, playing 'Groovin' High'.

rhythm is a subtle business. On paper just a regular subdivision of time, in practice it lives or dies as a 'felt' quality.

At the same time, however, few aspects of music can seem simpler. I defy you to watch those jazz clips without twitching. Music is often said to find a short cut to the heart, but the right sort of music takes a quick route to the legs. There's nothing that compares with rhythm for turning us into puppets on a string, powerless to resist the tugs and jerks. (Nevertheless we sometimes must, poor things, be bound by the conventions of the concert hall. Children know better.)

What is it about rhythm that literally moves us? How can it be instantly catchy yet so hard to capture in notation? What, indeed, *is* rhythm, really?

Where's the beat?

Not all music possesses rhythm. Some compositions by György Ligeti (like that in the hallucinatory climax of Stanley Kubrick's movie *2001: A Space Odyssey*) or Iannis Xenakis are more or less continuous skeins of sound, hardly articulated with any discernible pulse. At the other extreme, Karlheinz Stockhausen's electronic work *Kontakte* (1958–60) seems to be made up of more or less disconnected aural events, defying the imposition of a temporal grid. And even though Samuel Barber's *Adagio for Strings* has an underlying rhythm, we barely hear it – the strings seem to slide continuously between their heady, lamenting chords. Music for the Chinese fretless zither (*qin*) has rhythm in the sense of notes with different lengths, but these are not arranged against a steady underlying pulse – the musical notation specifies not how long a note should be, but the manner in which it should be plucked.

These are relatively rare cases, however. A quasi-regular pulse pervades most of the world's music. The songs of the Australian Aborigines, for example, may be accompanied by the steady clicks of a pair of rhythm sticks, or the clack of boomerang 'clapsticks', or simply by handclaps (Figure 7.1). We might imagine that here the percussive sounds are what supplies the rhythm. But that's not quite right. Rhythm is a familiar concept, but surprisingly hard to define. It is easily confused with the *metre* of music – the regular division of time into instants separated by equal intervals, providing what we

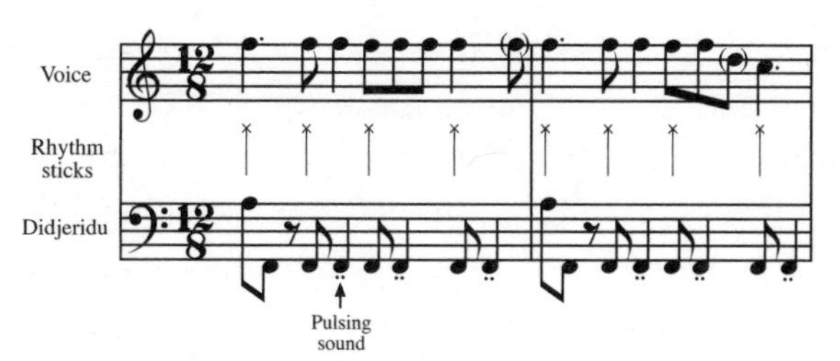

Figure 7.1 This Aboriginal song is accompanied by a steady beat on the rhythm sticks.

colloquially call the 'beat'. The notes or sounds themselves don't have to coincide with those instants: sometimes they fall off the beat, sometimes they are sustained over the beat, sometimes they crowd the gaps between beats.

Even the metre or beat is not as straightforward as it sounds. A piece of music can have regularly spaced notes and yet lack a true metre – Gregorian chant is an example. To create a beat from a regular pulse, some pulses need to be emphasized over others. In general this is done by making them louder: we count ONE two three four ONE two three four . . . But our minds tend to impose such a differentiation of pulses even where none actually exists: play someone a series of identical pulses, and they'll typically hear it as divided into such groups, most probably to create a duple rhythm *da*-da-*da*-da-*da*-da. The basic grouping instinct may be innate, being found even in young babies, but it seems to be shaped by culture – English and Japanese people divide up simple regular sequences of tones in different ways, for example, probably reflecting the patterns learnt from the respective languages. The tendency can be lent a helping hand by other musical factors: pitch patterns, for example, may suggest rhythmic groupings by the repetition of notes, contours and phrases (Figure 7.2).

The grouping of pulses defines the music's *metre*. Western music uses mostly simple metres: recurring groups of two, three or four pulses, or sometimes six. There is a two-pulse metre to 'Twinkle Twinkle Little Star', and a three-pulse metre to waltz time (think of Johann Strauss II's *Blue Danube*).

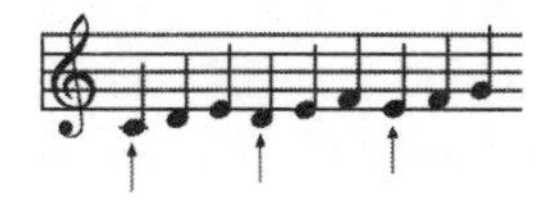

Figure 7.2 This pitch pattern encourages us to hear the series of notes as rhythmic groups of three, with strong accents on the notes arrowed.

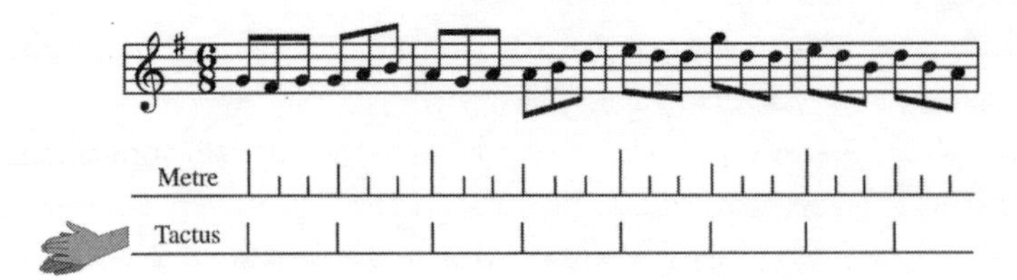

Figure 7.3 Metre and tactus for the 'Kesh Jig'.

Related to but different from the metre is the *tactus*, the beat we would clap out while listening to music (Figure 7.3). The tactus may be culture-specific: Western and South American audiences, for example, clap along to some dance music in sequences displaced from one another by half a beat. The two groups hear the stressed accents of the rhythms in quite different ways. And we tap out a slower pulse to familiar than to unfamiliar music, since in the former case we're better able to discern larger-scale rhythmic structures.

The *rhythm* of music is the actual pattern of note events and their duration, and so tends to be much less regular than the metre or tactus. Music that aligns its notes precisely with the metre is rather rare, and apt to be dull: even 'Twinkle Twinkle' takes a break with a long note after every six short ones. Everyday language fails to respect this particular connotation of rhythm, however, and I will be sometimes forced, for want of a better word, to talk of a 'rhythmic' signal or sense in relation to the production or perception of a periodic stimulus.

Typically, a rhythm is created by elaborating the periodic beats in some way, as for example the didjeridu does in the Aboriginal song above. It adds a second pulse after each beat, subdividing the pulse into a short (beginning) and long (end) note. And the large pitch jump every fourth click of the rhythm sticks supplies another, longer division of time (which supplies the justification for writing the song this way in Western notation, which is needless to say not how the Aborigines would annotate it). Subdivisions and stresses superimposed on a steady pulse give us a sense of true rhythm: they help us to locate ourselves in an otherwise homogeneous train of pulses, just as the tonal hierarchy of scale notes enables us to find our position in a pitch space that would otherwise be just a series of pitch steps. This sort of orderly and hierarchical structuring of time is found in the rhythmic systems of many musical traditions: different gongs in a Javanese gamelan ensemble, for example, are used to designate subdivisions of the basic unit of musical time, the *gongan*.

The metre is often portrayed as a kind of regular temporal grid on which the rhythm is arrayed. But the real relationship between the two is considerably more complex: musicians tend subconsciously to distort the metrical grid in small but significant ways so as to bring out the accents and groupings implied by the rhythm. Paradoxically, this stretching and shrinking of metrical time actually helps us to perceive both metre and rhythm, since it slightly exaggerates and accentuates the patterns they contain. Music psychologist Eric Clarke revealed musicians' elastic approach to time in experiments in which he made very accurate measurements of the timing and duration of notes played by several skilled pianists. Clarke gave them all a single melody to play, but placed it both in two different metres (2/4 and 6/8) and in different positions within a metre (Figure 7.4). The musicians' timings deviated from strict regularity in distinct and repeatable

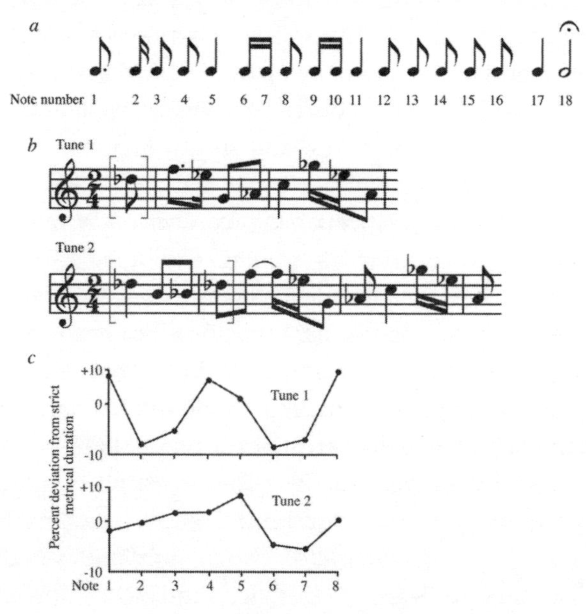

Figure 7.4 The 'same' rhythm is played in different ways depending on how it fits with the metre. Here a melody with the basic rhythmic pattern in (*a*) was placed in ten different metrical contexts, of which two are shown in (*b*) (with additional notes in square brackets). Performers' deviations from strict metrical time depended on this context (*c*) – in Tune 1, for example, the second note was significantly shortened, while in Tune 2 it was played pretty much exactly to time.

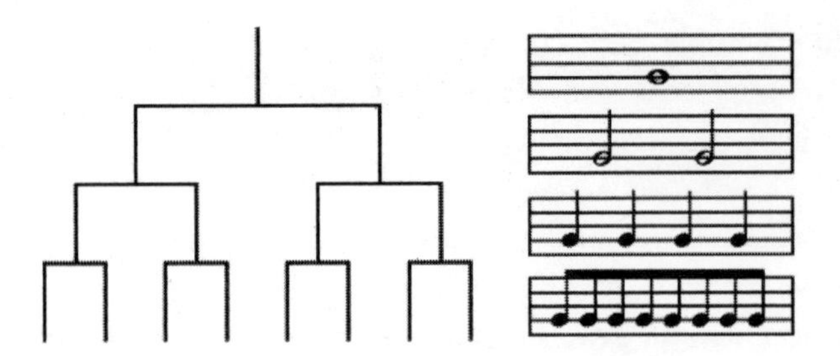

Figure 7.5 In Western music, metre and rhythm are generally constituted by binary subdivision, evident in the way that note durations are dictated by a succession of halvings.

ways, depending on the way the rhythm of the melody was superimposed on the metrical grid. Metre and rhythm, said Clarke, are linked in a network in which changes to the onset and duration of one note create changes in the notes all around.

Music in Western Europe has traditionally chopped up time by binary branching. A song's melody is typically broken down into phrases grouped in twos or fours, and each of these is divided into bars, again typically two, four or eight. There may be four beats to the bar, and each beat could consist of notes arrayed on a grid of two half-notes, or four quarter-notes, and so on (Figure 7.5). Of course, non-binary divisions of the phrase or the bar are possible, particularly threefold. But the dominance of binary division is reflected in the designation of note durations: a semibreve (whole bar of four beats), minim (two beats), crochet (a beat), quaver (half-beat) and so on.

The Eastern European musical tradition doesn't use this hierarchical splitting, but instead builds up phrases from less symmetrical sequences of 'chunked time'. Indeed, Slavic and Balkan music doesn't really recognize metre in the same sense as the West; the basic divisions are instead into groups of two and three beats. Combinations of these typically give rise to odd numbers of beats per bar: of the five common forms of Balkan folk dance, only one is in duple (2/4) time. The others are 9/16, 7/8 (twice) and 5/4.* The assumption that two- and four-beat metres somehow stem from the natural

* Bruno Nettl remarks that in much Balkan music the number of beats in a bar is a prime number: 3, 5, 7, 11, 13. This is probably not significant, however, since all the low-valued odd numbers except 9 are primes.

Figure 7.6 Transcription of a Romanian carol by Béla Bartók (1935). Notice how Bartók struggles to adequately convey the grace notes (shown in brackets) – see p. 304.

rhythms of marching and dancing in bipedal humans is thus groundless: Greeks have no problem dancing to seven beats, and these irregular metres are used not as a show of virtuosity but because they seem natural within those cultures.*

This additive, rather than top-down hierarchical, approach means that the 'metre' of Eastern European song may change constantly, sometimes almost every bar (Figure 7.6).† This irregularity of song rhythm might be due at least in part to the different rhythmic structure of this region's poetry and song. Whereas in the West it is not uncommon for the number of syllables in a line to vary, with only the number of *accented* syllables remaining the same (for instance, 'For he's a jolly good fellow/And so say all of us'), in Eastern Europe the total number of syllables is the same in each line, meaning that the accents may fall in different places between one line and the next.

Western musicians tend to struggle with anything outside of duple or triple time, so that attempts like those of Dave Brubeck to broaden the rhythmic palette, as in his 'Blue Rondo à la Turk' (in 9/8 time), can sound awfully laboured. The rhythmic instincts of Western musicians are particularly exposed as four-square when it comes to improvisation: both in Brubeck's 'Blue Rondo' and in Pink Floyd's 'Money' (with its surging 7/4 riff), the solos retreat to the safety of 4/4 time.

* While the prevailing view is that such distinctions are cultural, it's nonetheless intriguing that in 1999 a team of Japanese neuroscientists found evidence that rhythms based on simple durational ratios (1:2:4, 1:2:3) are processed in different parts of the brain – with a different hemisphere predominating – than those in more complex ratios of 2:5:7. However, this study didn't involve subjects whose musical culture commonly uses the latter ratios.

† Western musical notation is biased towards regular rhythms. Music like that in Figure 7.6 looks, in this system, extremely complex, and yet that is not likely to be the way Romanians perceive what is after all a Christmas carol. Arguably, this notation system actively discourages rhythmic complexity.

Hop and skip

It is not just the stressing of some beats but an *asymmetry* of events that creates a true sense of rhythm and avoids monotony. The same is true in language, where the stream of spoken syllables is typically broken up into something that skips rather than plods. It's telling that the way we 'speak like a robot' is to make all syllables equal in strength and duration. In contrast, spoken English often uses a rhythmic structure called the iamb, which consists of a short syllable followed by a long one: da-DAH. It is familiar from the iambic pentameter of poetry:

When forty winters shall besiege thy brow
And dig deep trenches in thy beauty's field

This is the rhythmic pattern of the Christmas carol 'I Saw Three Ships' (Figure 7.7). The reverse of the iamb is the trochee: a long pulse

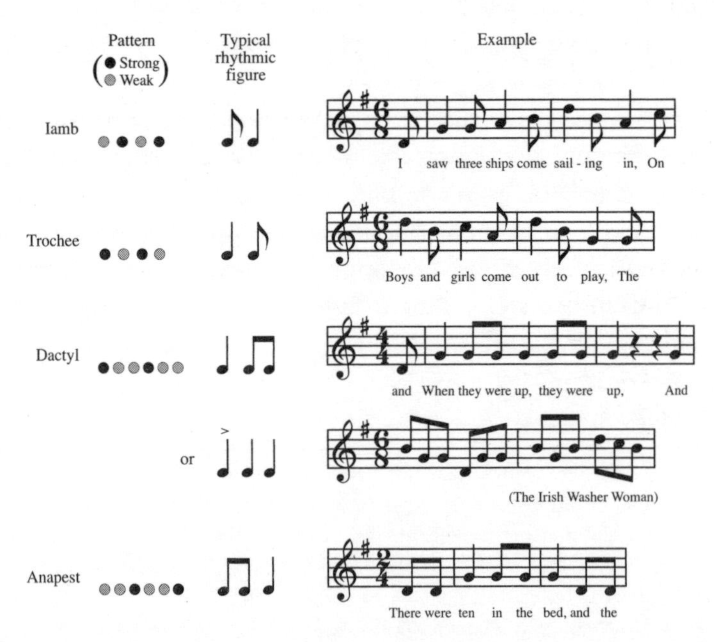

Figure 7.7 Basic rhythmic figures: the iamb, trochee, dactyl and anapest. These are exemplified in, respectively, 'I Saw Three Ships Come Sailing In', 'Boys and Girls Come Out To Play', 'The Grand Old Duke of York', and 'Ten In the Bed'.

followed by a short one. Trochaic metre is also widely used in poetry, and in children's rhymes too:

Boys and girls come out to play
The moon does shine as bright as day

The asymmetry is evident in both of these structures: the two notes or syllables are of unequal duration. Another kind of asymmetry comes from dividing groups of two equal notes into one long one and two short ones. In the dactyl, the long note comes first; in the anapaest, it's the short notes that come first (Figure 7.7). These patterns can also be defined by stress patterns rather than durations: a strong syllable followed by two unstressed ones is also a dactyl, as in the rhythm of the word 'happily', or Tennyson's 'Half a league, half a league'. That's often the form in which they are found in music: groups of three equal notes with the first one stressed, as in triple metres.

These figures are the atoms from which we build a sense of rhythm. They suggest to us how to interpret and make sense of the stream of musical events: how to apportion them into coherent temporal units. For that coherence to be felt, the units have to be repeated. Needless to say, some music is more rhythmically regular than others. The jigs and reels of traditional British, Irish and Scandinavian folk music, mindful of the need to give dancers clear guidance, tend to fill up every beat or half-beat with notes (Figure 7.8*a*). This regularity is common also in music of the Baroque and Classical eras, some of which also derives from dance forms (the gigue, courante, allemande and so forth) (Figure 7.8*b*). We are quick to identify rhythmic patterns, and tenacious in our choice – as musicologists Grosvenor Cooper and Leonard Meyer say, 'Once a rhythm becomes established, it tends, if possible, to be continued in the mind of the listener – it tends, that is, to organize later patterns in its own image. This is true even where the natural grouping would seem to be otherwise.'

In making these assignments of rhythmic pattern, we draw on information of various kinds – not just the duration of notes (which might, in very regular music, be mostly equal) but on the nature of the melody, the phrasing, dynamics, harmony, timbre. For example, consider the version of 'Twinkle Twinkle' in Figure 7.9*a*, which has been written here with all notes of the same duration. How do we group them into rhythmic atoms? The natural impulse is to form groups of two notes in a trochaic pattern – a stressed and weak beat. But a variation of the melody shown in Figure 7.9*b* produces an iambic grouping: weak-strong. Then the stresses fall on relatively stable notes in the tonal hierarchy – all the out-of-scale sharps are weak – while notes that are close in pitch are grouped together. Making some notes last longer, as in Figures 7.9*c* and *d*, produces an anapaestic grouping – weak-weak-strong – in both cases. But the first one of these sounds a bit odd and strained, because the strong beats then come on less stable notes (**6**, **4**, **2**, as opposed to **5**, **3**, **1** in *d*). In such ways, composers can direct and manipulate our sense of where the rhythm lies. Hadyn was particularly adept at using rhythm to elaborate and enliven rather mundane melodic material.

Confusing the issue

Composers generally want us to know which rhythm they intend, and so they may use other factors like these to reinforce it. Stressed notes are usually planted as firm, unambiguous signposts: they are rarely split into notes of small duration. While stresses can fall on silent

Figure 7.8 Rhythmic regularity in traditional jigs and reels (*a*) and in Mozart's Gigue in G, K574 (*b*).

beats, they rarely fall *within* long notes. Even composers such as Erik Satie who dispensed with time signatures and bar divisions altogether were not thereby eliminating any sense of metre.

On the other hand, composers may seek to confuse our expectations in order to introduce tension and emotion into a piece, an idea that I explore in Chapter 10. Compared with melody and harmony, rhythm is a particularly potent tool for this sort of manipulation because the expectation is so transparent: it is very easy to hear when a beat is disrupted. And so composers toy with our instinct for finding grouping rules, giving us conflicting or ambiguous signals that suggest alternative ways of grouping so as to sharpen our attention and create a lively stimulus.

Beethoven does so at the start of his Fifth Symphony. I doubt that most people actually hear this, perhaps the most famous theme in Western music, as they 'should'. That's to say, we hear the first note as falling on a strong beat (*da*-da-da *dah*!), whereas in fact the first downbeat of the bar is a rest (Figure 7.10). Yes, Beethoven's Fifth begins with a silence! (So, in fact, does the theme of the Sixth, although the clue

Figure 7.9 The role of melody in determining rhythmic grouping patterns. The basic melody (*a*) implies a grouping based on the trochee: strong – weak. But altering the tune as in (*b*) makes the preferred grouping iambic instead. And altering the note durations (*c, d*) creates anapests. We may not start to hear these groupings straight away (the initial note isn't always included in them), but once they are established we expect the pattern to continue.

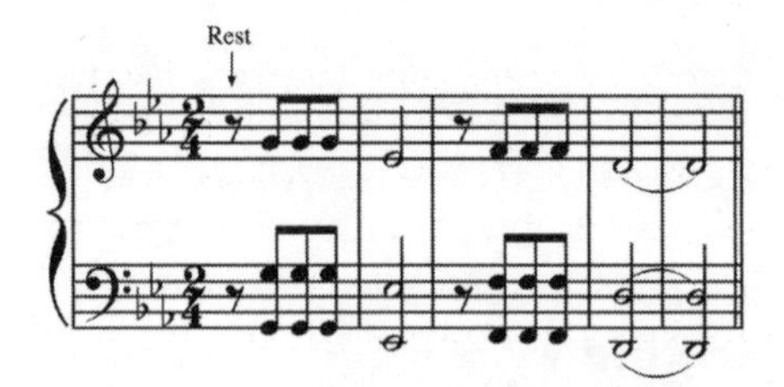

Figure 7.10 The opening of Beethoven's Fifth Symphony. To hear the rhythm correctly, one has to infer the initial rest.

there is a bit clearer.) Beethoven delights in these tricks of rhythmic perception. In his Piano Sonata No. 13 in E♭, he sends out contradictory messages through rhythm and pitch, in which the melody descends in groups of four semiquavers before jumping back up each time, while the metre actually demands groupings of six (Figure 7.11). And Leonard Bernstein does something similar in the theme of 'America' in *West Side Story*, which is nominally in 3/4 time but could be heard as 6/8 (Figure 7.12). Here the conflict is fostered by using pitch to encourage a different way of grouping the notes than is implied by the metre: the initial two groups of three identical pitches imply the 6/8 rhythm.

Popular music abounds with these rhythmic elisions. The Rodgers and Hart song 'Getting to Know You' from *The King and I* has a deceptively simple rhythmic figure laid out at the beginning in the title phrase – a triplet followed by two longer notes – that repeats subsequently at a quite different starting point in the bar, making this an unexpectedly difficult (and in my view, a rather cumbersome) song to sing (Figure 7.13). In Led Zeppelin's 'Nobody's Fault But Mine',

Figure 7.11 Beethoven's rhythmic ambiguity in his Piano Sonata No 13 in E♭. The brackets show the way the pitch structure encourages us to make rhythmic groupings, contradicting the metre.There's a more obviously deceptive example of Beethoven's rhythmic trickery on p. 288.

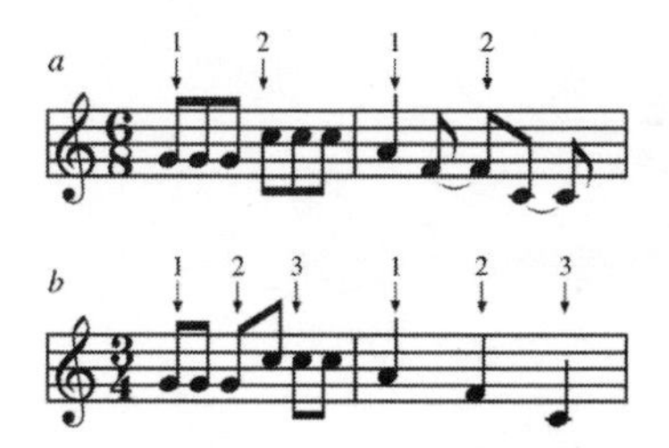

Figure 7.12 The ambiguous rhythmic interpretations of 'America' in Leonard Bernstein's *West Side Story*. The theme could be 'read' as either in 6/8 (*a*) or 3/4 (*b*) time, with two or three stresses per bar respectively; the first bar seems to imply the former, but the second bar suggests the latter.

the deceptively simple oscillating octave of the rhythm section defies you to locate it within the metre (Figure 7.14*a*), while the riff of 'Black Dog' delights in its twists and turns. The latter doesn't actually do the violence to the beat that it pretends to: the pattern shifts by half a beat while the drum remains steadily four-square beneath, nicely illustrating the difference between metre and rhythm (Figure 7.14*b*). According to bassist John Paul Jones, this ambiguity was quite conscious: he says the riff came from one used by Howlin' Wolf, 'a blues lick that went round and round and didn't end when you thought it was going to'. Bob Marley's 'Is This Love?' is another adroit example of how timing, phrasing and rhythm can be used subtly to hint at parallel, alternative structures.

As well as creating tension through ambiguity, composers may introduce quite unambiguous disruption of the metre, creating not the sensation of being tantalized and held suspended but the jolt of being proved wrong. In pop music, dropping or adding a beat is a common trick for adding a little kick to a song (see p. 290). It's often implied that much modernist music lurches from one such jolt to another, but the fact is that the impact is far greater in music that has a more traditional metre and beat than in music that is constantly shifting: we can only have our expectations violated if we feel confident enough to have expectations in the first place. A famous example is the 'Dance of the Adolescents' section from Stravinsky's *The Rite of Spring*, where the pulse is very clear and regular but the accents fall in a different place every time (Figure 7.15*a*): we're constantly being caught out,

Figure 7.13 A repeated pattern with shifting rhythmic emphasis: 'Getting to Know You', from *The King and I*.

Figure 7.14 The rhythm section baffles our sense of regularity in Led Zeppelin's 'Nobody's Fault But Mine', in spite of the regular metre (*a*). And the riff of 'Black Dog' sounds more complex than it is because it shifts its starting point by half a beat (*b*). Here the repeating rhythmic phrase in bars 4–8 is bracketed so that the shift relative to the bar lines is clearer.

but only because the pulse compels us to expect regularity. Contrast this with the 'Evocation of the Ancestors', where the metre itself is dislocated (Figure 7.15*b*).

Our sense of metrical regularity is not as strong as we might suppose, unless it is reinforced by the music itself. People without musical training will distort metre without being conscious of it – happily switching from waltz time (three beats in the bar) to quadruple time (four in the bar), or inserting extra beats and bars at random. Early rural blues musicians might often play an occasional thirteen-bar segment, or just omit a beat here or there (something that their white imitators would laboriously mimic). Literal transcriptions of this music can seem as complex as anything served up by Bartók, with quintuplets and septuplets to challenge our sense of orderliness. The same is true of other folk styles; the Australian Percy Grainger, an early ethnomusicologist as well as a composer, transcribed some of the odd excursions in rhythm and metre practised unconsciously by rural English folk singers (Figure 7.16). These encouraged Grainger to include irregular metres in his own works. One such, called 'Sea Song' and composed in 1907, has such a complicated series of time-signature changes that it was almost unplayable, and arguably demolishes any sense of metre at all. The first thirteen bars run like this: 1/4 | 7/23 | 3/32 | 5/64 | 5/16 | 3/8 | 7/64 | 3/32 | 5/64 | 9/32 | 3/8 | 7/64 | 5/16 | . . .

But there was in a way nothing new in this. Western music in the Middle Ages and early Renaissance often did not have any all-embracing

Figure 7.15 The steady pulse of the 'Dance of the Adolescents' in Stravinsky's *Rite of Spring* is undermined by the erratic stresses (*a*). This confusion of regularity is rather different to the fragmentation of metre evident in the 'Evocation of the Ancestors' (*b*).

metric beat. It became common to overlap voices with very complex crossed rhythms, although this wasn't made explicit in the metre-free notation of the time. In the fifteenth-century example shown in Figure 7.17, the phrases can be considered to switch repeatedly between 6/8,

Figure 7.16 Percy Grainger's annotation of the folk song 'Bold William Taylor' as sung by Mr George Gouldthorpe at Brigg, Lincolnshire, in 1906 is full of metre changes. Note also the ambiguities of pitch, to which we return in Chapter 10.

Figure 7.17 The rhythmic complexity of polyphonic music in the Renaissance: there are no bar lines, and no real metre.

3/4 and 2/4 time, all calculated to converge simultaneously in the final cadence. We don't really know how the composers thought about this rhythmic interplay, because there is no documented theoretical discussion. Perhaps they relied on intuition alone to create a lively pattern of accents.

Some traditions seem almost wilfully to delight in confounding our grouping instincts. Cross-rhythms are common in African, Indian and Indonesian music, where they might be used to trick the other musicians into losing the beat. This playfulness is a strong feature of Indian music – here the American music critic Winthrop Sargeant saw a similarity with jazz, in which drummers often play 'across the beat'. With the help of Indian music specialist Sarat Lahiri, Sargeant wrote that

> It often happens among Indian musicians that a *vina* player and a drummer will engage in a friendly contest to see which can confuse the other into losing track of the *sam* [the beat] . . . The *vina* player delights in apparently losing himself in the most abstruse counter-rhythms, leaving the listener with a sense of utter bewilderment, only to issue forth triumphantly at the *sam* again without a hair's breadth of inaccuracy, and with a sparkle of obvious satisfaction. The effect, to one who is accustomed to this idiom of expression, is that of being hurled through chaos, and then suddenly landing right side up on terra firma again with no bones broken and a feeling of intense relief.

Western classical music has traditionally made rather little use of cross-rhythms, because its traditional focus on the 'vertical' harmonic relationships between notes has tended to demand that the different voices and emphases coincide. But these techniques have been particularly exploited in the hypnotic 'minimalist' compositions of Steve Reich and Philip Glass. Reich often uses repetitive riffs played by different instrumentalists at slightly different tempos, so that they form nearly interlocking patterns that are constantly shifting: a process he calls phasing, since it is like the movement in and out of phase of two waves of similar frequency. Reich first developed this method with taped music or voices, and he initially imagined that it would be too difficult for live performers to supply the delicate gradations in tempo that it required. But on trying it, he found it was surprisingly easy, and in 1967 Reich wrote *Piano Phase* for two pianists, playing the same phrases (of twelve, then eight, then four notes) at slightly different speeds.* Reich used the phasing technique in several other works, one of which (*Drumming*, 1971) made an explicit connection to the polyrhythmic tradition of Africa, which Reich visited around this time.

What strikes one most about these pieces is that they are not *heard* in quite the manner they are composed or performed, that is, as separate voices playing at slightly different speeds. Rather, the voices interlock into distinct rhythmic patterns at different stages, which then dissolve and crystallize into new ones. We hear discrete jumps from one pattern to the next, falling suddenly into place with the pleasing sensation of an object abruptly slotting into the right-shaped hole. Again, then, we form an interpretation of what the rhythmic pattern is, only to have to keep revising it as a new structure emerges. Some find the result banal (and it's true that the melodic themes are in themselves of rather little interest); others are thrilled by this constantly shifting texture. The appropriate visual analogy here is not the Necker cube but the Moiré patterns produced by two identical grids when one is rotated relative to the other (Figure 7.18). As the rotation proceeds, we see a series of regular geometric patterns formed from the movement of grid lines in and out of phase with one another. But the abruptness and clarity with which we hear these changes in rhythmic patterns reveal a mind

* In virtuoso displays of metrical technique, some pianists have played *Piano Phase* as a solo, each hand playing one part on different pianos – an extraordinary demonstration of Reich's discovery that our rhythmic abilities are deeper than we might imagine.

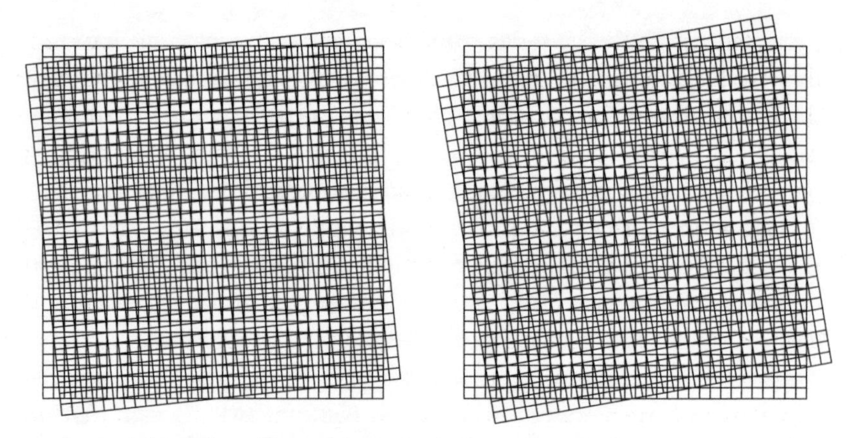

Figure 7.18 The shifting 'phase relationship' of repetitive rhythmic patterns in the music of Steve Reich creates ephemeral larger-scale regularities in a manner similar to the Moiré patterns generated by overlapping grids.

imposing its determination to perceive structure, seizing on a hint of regularity and doggedly hanging on to it until it becomes untenable. The brain's intolerance of ambiguity seems here to orchestrate the primary musical experience.

Born to boogie

We seem to have an innate capacity to discern auditory regularity: to pick out a repetitive pulse in a series of sounds. This is *not* the same as a capacity to discern rhythm or metre, but it seems very likely to be a precondition for it. If you can't identify a regularly repeating sound, you're unlikely to be able to make sense of the patterns of accents and sub-pulses that rhythm and metre weave around it.

The crucial thing about this sense in humans is not our ability to produce repetitive sounds or movements – many animals do that – but the capacity to *entrain* our actions to some external pulse. That's to say, we can 'get in time' with a rhythmic signal. Curiously, this sort of entrainment seems to be a very primitive characteristic of cells, and yet appears most uncommon in higher organisms. In 2008, a team of Japanese researchers found that the single-celled slime mould amoeba *Physarum polycephalum* can learn to anticipate beats in a periodic stimulus. They subjected the cells to a series of 'shocks' at regular intervals, exposing them to blasts of dry air that slow down their speed of movement. After three episodes of dry air in regular succession an hour apart, the slime mould apparently came to expect more: it slowed down when a fourth pulse of dry air was due, even if none was actually applied.*

This kind of rhythmic sense can be understood using the mathematical theory of so-called 'linked oscillators', like an array of pendulums that 'feel' each other's oscillations. Two pendulums attached to a single support will eventually fall into step because of the vibrations transmitted to one another through the support. The Dutch scientist Christiaan Huygens discovered this in 1665 when he noticed that the pendulums of two clocks on his mantelpiece always ended up beating in unison (actually, in synchronized but opposite motions). Like all living organisms, slime moulds have built-in biochemical oscillators, and it seems likely that the versatile rhythmic sense of *Physarum* stems from many biochemical oscillators in the colony pulsing with different frequencies, giving the cells an ability to pick up and anticipate a wide range of pulse rates. A comparable kind of entrainment is found in fireflies, which can synchronize their flashing when many are gathered together. This isn't, however, like a human 'group drumming' session, not least because fireflies can only synchronize over a rather limited frequency range and because they can only keep a strictly periodic pulse – nothing complex or syncopated.

Until very recently, no higher organism was ever found unambiguously to move in time to a musical beat. Asian elephants can *make* a beat: they can strike out an extremely regular series of bangs on a drum with a mallet held in the trunk.† But they don't adjust their beat to that of another drummer. It was widely suspected that this might be a uniquely human activity, but in 2008 Aniruddh Patel and his colleagues studied the 'dancing' behaviour of a sulphur-crested

* One of the most surprising findings in this study was that the slime mould retained a memory of the beat for several hours. A single renewed shock after a 'silent' period left the mould expecting another to follow in the rhythm it learnt previously. This led the Japanese researchers to speculate that the results 'hint at the cellular origins of primitive intelligence'.

† This ability is exploited in the Thai Elephant Orchestra, an ensemble of elephants at the Thai Elephant Conservation Center near Lampang, which play on huge percussive musical instruments. You can, if you wish, buy CDs of their 'music'.

male cockatoo called Snowball, who was kept in a bird sanctuary in Schererville, Indiana. Snowball had become a minor celebrity after videos of him dancing to pop music were posted on YouTube. Although it looked as though Snowball was genuinely moving to the music, it was hard to be sure until Patel and his co-workers showed that the cockatoo could adjust his movements to match different tempos, and that the episodes of apparent synchrony were intentional and not due to pure chance.

Snowball seems unlikely to be unique among dancing birds. Can we say, then, that cockatoos are 'musical'? On the contrary, the finding implies that a capacity for rhythmic synchronization is *not* a specifically musical adaptation, because animals have no genuine music. That's significant for the notion that music originated as a group activity which, being reliant on a steady rhythm, encouraged coordination and social cohesion. The cockatoo's ability to move to a rhythm doesn't exactly undermine that idea, but it hints that music became possible because our ancestors *already* had a sense of rhythm, rather than that a sense of rhythm evolved from a need for music.

A conclusion of that kind presupposes that beat entrainment in cockatoos has the same origin (in neural and evolutionary terms) as that in humans. This may or may not be true. Certainly, it seems that we're not born with rhythm: in Western cultures, the ability to synchronize movements to a beat doesn't happen until around the age of four. Before that, children dance to a tempo of their own making (which has its own charm, of course).* And while four-year-olds can hold a steady rhythm in song for short periods of time, they don't tend to keep it up for long. The ability to sustain a rhythm tends to emerge a year or so later, as does the ability to clap along to a tactus rather than simply to mimic note durations.

This doesn't mean that very young children can't *discern* a beat; it may simply be that they haven't acquired the motor coordination skills to move to it. We're certainly sensitive to rhythm from a very young

* Children actually tend to dance spontaneously to music progressively less readily between the ages of two and five, by which time they are more likely to remain motionless while listening. This might be because, as they grow older, they demand some 'model' that shows them 'what they are supposed to be doing', rather than following their own impulses.

age: babies two to four months old can detect a (relatively modest: fifteen per cent) change in the tempo of a beat and can tell apart different simple rhythmic patterns, while seven- to nine-month-olds can identify rather subtle changes in rhythmic patterns that transform a pattern with a regular beat to one without.* (As with other infant studies, these relied on indicators of attention changes such as head-turning.) Infants seem to prefer steady rhythms to complicated, less regular ones.

So we can't say for sure whether or not we have an innate predisposition for extracting regular beats from a quite complex sequence of sound pulses. And the popular notion that infants acquire sensitivity to rhythm from the maternal heartbeat is sheer speculation. Even if that were so, the ability to hear a steady pulse is not in itself a *musical* attribute. Acoustic regularity is too common a feature of the world for this to be a music-specific adaptation.

* Aniruddh Patel points out that these latter studies are open to criticism, because only a single irregular pattern was used in the switch from regular beats.

8

Pizzicato
The Colour of Music

Why do instruments sound different,
and how does that affect the music?

The most objectionable aspect of orchestral versions of rock music is not the patronizing implication that this is after all 'real music', nor that the rudimentary harmonizing of rock music doesn't lend itself to polyphonic orchestration. No, it's the fact that rock music without a rock timbre is like vegan ice cream. Why try to make a dish when you have proscribed its central ingredient? You might as well play 'Purple Haze' on the xylophone.

Timbre is arguably the most personal characteristic of music. When it comes to singing, timbre often holds the key to our preferences. Billie Holiday, Frank Sinatra, Nina Simone, Bob Dylan, Tom Waits, Robert Plant, David Bowie – all have vocal timbres that are instantly recognizable, and it is this, more than choice of song, that is the defining quality of their sound. The emotive power of timbre is tremendous. When jazz organist Jimmy Smith flips the switch that puts his Leslie loudspeakers into rapid rotation, the timbral change (a quickening of the pitch modulation or vibrato) sets pulses racing too.

Violinists are obsessive about timbre, and will pay a small fortune to obtain the right instrumental tone. But unlike fine wine, the quality of timbre isn't necessarily proportional to investment or finesse. Most people will agree that a concert Steinway sounds nicer – has a better timbre – than the old upright in the village hall, but there's no consensus about whether Sinatra's golden croon is 'better' than Tom Waits' guttural rasp. And there are times when a badly amplified Fender Stratocaster can be much more eloquent than a forty-piece string section. Timbre has no absolute measure of quality, but is all about context.

This chapter on timbre is rather short, because timbre is one of the least studied of basic musical attributes. And the reason for *that* is *not* that timbre is a relatively unimportant feature of music, but that it is one of the hardest to pin down. No one is quite sure what timbre is. The definition given by the American Standards Association is basically just an indication of what it is *not*: specifically, it is that attribute of a sound signal that enables us to distinguish it from another of the same pitch and loudness. In other words, if they sound different even though they are of the same pitch and equally loud, the difference is down to timbre.

Even this waste-bucket description isn't enough, because not all sounds have a well-defined pitch. There is a timbre to the sound of gravel being shovelled, of a jet engine or a falling tree. Albert Bregman suggests that the definition of timbre be phrased a little more honestly: 'We do not know how to define timbre, but it is not loudness and it is not pitch.'

It is, however, utterly a part of music. All the tomes of intricate musicological analysis of the works of Mozart or Stravinsky could be read (a little perversely, it is true) as implying that it doesn't much matter if we hear the notes on a piano or a trombone or an electric guitar. You can say what you like about the beauty of the melodies or the exuberance of the dynamics, but music will leave us indifferent and unmoved if the timbre isn't right. Equally, a change of timbre may transform the meaning. Is 'Mack the Knife' the 'same song' when sung by Louis Armstrong or by Ella Fitzgerald or Nick Cave?

The real mystery of timbre is that, despite it being such an elusive, ill-defined musical characteristic, the human mind is so astonishingly attuned to its nuances. Conventionally, timbre is attributed to the precise mixture of frequency components in the acoustic signal. But we can play havoc with that signal – broadcasting it through a cheap, tinny transistor radio, say – and yet still remain able in an instant to discriminate between the sound of a saxophone and a trumpet. Why is that?

Instrumental personalities

We saw earlier that musical instruments create complex tones that contain overtones of the fundamental pitch with whole-number multiples of its frequency, arrayed in the harmonic series. The acoustic strength of an overtone – how loudly it sounds in the mix – varies from one kind of instrument to another, and, to a lesser degree, from one particular instrument to another. These differences account for a substantial part of the timbral distinctions. The astringent sound of a clarinet, for example, relies on the strong presence of odd-numbered harmonics (Figure 8.1): the third, fifth and so on. Bright-sounding instruments such as trumpets are rich in high harmonics. The timbre of a violin can be altered by specialized bowing techniques that excite different harmonics. The same is true of a piano as the keys are struck with different degrees of force.

Not all overtones sit within the harmonic series. In particular, I explained earlier that percussive instruments tend to generate 'inharmonic' overtones that are not pure harmonics of the fundamental, giving them an ambiguous pitch and creating the 'metallic' timbre of bells and gongs. This has posed difficulties when these instruments are used in ensembles, because the inharmonic overtones can produce dissonant clashes with other instruments even when the fundamental

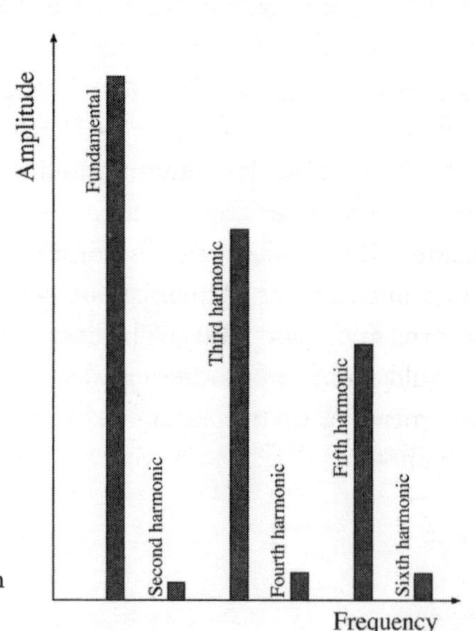

Figure 8.1 The overtone spectrum of the clarinet.

tones are consonant. The carillon, an array of bells sounded by a keyboard, has for over 300 years relied on the rather mysterious art of bell-tuning to suppress these mistuned overtones. But however skilfully they are prepared, the fact remains that traditional bells have a third overtone with a frequency 2.4 times that of the fundamental – a minor third, which is potentially dissonant. Acoustic scientists have now designed bells that generate only pure harmonics, obviating these clashes. Australian composer Ross Edwards makes use of such 'harmonic bells' in his Third Symphony (1998–2000).

Timbre is also crucially influenced by the way these blends of overtones change over time as an instrument is sounded. The quality of a sound is particularly dependent on its onset or 'attack': how the sound intensity rises in the first fractions of a second. Recorded notes played on different instruments with this initial attack clipped off can be hard to identify or even to tell apart. A piano recorded and played backwards doesn't sound like a piano at all, even though the overtone content is unchanged.

Timbre somehow emerges as an integrated quality of these static and dynamic features of the sound. What we perceive is not a particular blend of harmonics modulated by an envelope of rising and falling loudness, but instead a single perceptual phenomenon, a gestalt that is instantly recognizable as a 'piano' or 'marimba' or 'trumpet'. There are good evolutionary reasons why we should be adapted to unify the contributing aspects of timbre into a lone sonic entity, since this helps us to attribute sounds to specific sources even if these abstract characteristics of the signal vary: to distinguish the voices of friends and kin, say, or to recognize warning sounds and animal calls. If this is the adaptive function of our timbral sense, it stands to reason that timbre should be so emotionally charged.

If the sound quality of different instruments is indeed the result of distinct blends of harmonics, then it should be possible in principle to 'synthesize' complex sounds from their single-frequency components. This is essentially what organ stops do: they specify particular combinations of pipes, each pipe creating (in theory) a single tone determined by its size.* These amalgams simulate a whole range of instruments, from woodwind to brass and strings.

* There are actually two types of pipe: flue pipes are tubes with a mouth like that of a whistle or recorder, while reed pipes have a resonating reed, like a clarinet. Neither produces a pure, single-frequency tone.

But it's obvious to any ear that pipe organs don't serve up anything more than a very crude approximation of the designated instrument, often barely identifiable as such. Early electronic synthesizers sought to mimic instrumental timbres more accurately, both by using electronically generated pure sine tones as the ingredients and by applying loudness envelopes to modify the attack and decay. The results were still disappointing: brass and woodwind could be passably mimicked, but synthesized pianos and violins sounded wholly unconvincing. It became plain that we are extremely demanding in our perception of timbre and won't be easily palmed off with cheap imitations. Early synthesizer music, such as Wendy Carlos' seminal 1968 album *Switched-On Bach*, gained its popularity more from the novel, 'electronic' character of its sound than from any fidelity to the acoustics of traditional instruments. Modern synthesizers do a much better job by abandoning this 'bottom-up' approach to creating mimetic sound. Instead, they typically use either digitally sampled recordings of the actual instruments (reminiscent of the cumbersome magnetic-tape loops of the Mellotron, the staple of progressive rock) or so-called physical-modelling approaches in which mathematical equations and algorithms are used to reproduce the acoustic features of the instrument.

The timbres of instruments give them distinct personalities, which some composers have exploited in literal fashion to emulate personified 'voices'. Perhaps the most famous example is Prokofiev's *Peter and the Wolf*, where the high, pure warbling of a flute represents a bird, the flat sound of the oboe depicts a duck, the cat is a jaunty clarinet and the wolf a strident trio of French horns. Ravel's orchestration of Mussorgsky's *Pictures at an Exhibition* gives distinctive voices to the two bartering Jews in 'Samuel Goldenberg and Schmuyle', using timbre to convey their personalities where Mussorgsky's piano score was limited to purely dynamic contrasts.

Timbral 'personality' is a defining stylistic feature for some jazz and rock instrumentalists. Aficionados can tell in a blink the difference between the fluid wailing guitar of Jimi Hendrix and the full-bodied, mellow purity of Jeff Beck. Eric Clapton produced his trademark 'woman tone' by using a wah-wah pedal, a device that manipulates timbre by filtering the frequencies of the guitar signal via a moveable foot pedal. The era-defining theme from the 1971 blaxploitation movie *Shaft* makes timbre the driving force of the music – the wah-wah sound used here by Isaac Hayes' guitarist Charles Pitts became ubiquitous in 1970s soul. The wah-wah and fuzzbox, memorably introduced by Keith Richards for '(I Can't Get No) Satisfaction', were among the earliest of a vast array of timbre-altering effects pedals that guitarists now deploy to carve out their own timbral niche in the rock pantheon.

The Russian composers of the late nineteenth and early twentieth centuries – Mussorgsky, Rimsky-Korsakov, Prokofiev, Stravinsky – were masters of musical 'tone colour' derived from the judicious choice and blending of timbres. Rimsky-Korsakov listed what he considered to be the expressive qualities offered by woodwind timbres, for example:

	Low register	*High register*
Flute	Dull, cold	Brilliant
Oboe	Wild	Hard, dry
Clarinet	Ringing, threatening	Piercing
Bassoon	Sinister	Tense

But timbre acquires perhaps its most elevated and contentious status in Western instrumental music from the violin. The eighteenth-century instruments made by the Cremonese masters Antonio Stradivari and Giuseppe Guarneri del Gesù now change hands for millions of dollars – a Stradivarius auctioned in 2006 went for $3.5 million. These prices reflect a conviction that the instruments have timbres that are literally incomparable. Countless theories have been proposed to explain that, ranging from the Cremonese luthiers' 'secret formula' for the varnish (although this seems likely to have been no different to that used by contemporary furniture-makers) to variations in wood density or special chemical treatments of the wood.

Are the tones of these instruments really unique and unsurpassed? Absolutely, say the top players. According to distinguished concert violinist Ara Gregorian, 'Every violin has its own voice or tone.' If that's so, you'd expect it to be measurable. But acoustic scientists have long sought in vain for the characteristic sound quality that the musicians' ears pick out in an instant. There's now some evidence that the truly greatest instruments have a more even tone across their entire pitch range, and that they generate more volume in the lower registers than cheaper instruments. Yet these timbral distinctions are subtle at best, and it is hard to understand why on this basis they have come to command such extraordinary price tags.

It is all too common to ascribe the alleged superiority of tone to the instrument alone, overlooking the person playing it. That error was derided by the great Russian violinist Jascha Heifetz. He was once allegedly approached after a concert by a female fan who complimented him breathlessly on the 'beautiful tone' of the del Gesù violin he had played that night. Bending down to put his ear close to the violin lying in its case, Heifetz remarked, 'I don't hear anything.'

And the adoration of the Stradivarius sometimes seems more like a fetish, a danger that deters some top violinists from craving one. The young American virtuoso Hilary Hahn plays a nineteenth-century violin made by Jean-Baptiste Vuillaume and insists she doesn't want a 'greater' instrument. And some even dare to acknowledge that many of the best modern instruments are comparable, perhaps even superior, to some of the old Cremonese ones. All the same, no one can quite put their finger on where these distinctions reside at the acoustic level, and the art of making violins remains very much that, aided rather little by acoustic or cognitive science.*

How many directions does timbre go in?

Despite how readily we hear differences in timbre, it's difficult to say what they are. When we hear a flute and a clarinet, we don't think 'Ah, one has more odd-numbered overtones', but instead perceive the distinctions in a qualitative manner: we might say that the flute sounds 'pure' and 'soft', the clarinet 'hard' and 'brilliant'. Cognitive scientists have sought to identify the basic perceptual categories we use for judging timbre. Are they universal and well defined, or idiosyncratic and fuzzy? This is how Albert Bregman puts the problem:

> If you take each moment of each frequency component into account, you realize that sounds can differ from one another acoustically in an astonishingly large number of ways. Does the auditory system deal

* It's said that the conservatism of many luthiers makes matters worse. The idea of blind testing is anathema here.

with this complexity by collapsing the differences into a small number of dimensions?

In the 1970s, musicologist John Grey suggested that there are just three key 'dimensions' to timbre: three characteristics (other than pitch and loudness) that make one sound different from another. Roughly speaking, these are brightness (the strength of high overtones), attack (how the volume of each harmonic rises initially), and the loudness profile of the harmonics over time, or how they sustain and decay.*

In contrast, music psychologist Gerald Balzano argues that we interpret timbres more materialistically, by translating them into the physical processes deemed to be causing them: blowing, plucking, whistling, striking and so forth. Thus, he says, we hear a change in the timbre of a gong in terms of a change in the force with which it is struck.

There is no consensus on this issue. We don't really know if the mind has a concise way of classifying timbres that we have yet to identify, or if it handles timbre on an ad hoc basis, making comparisons within the immediate acoustic context without regard for any 'global' filing system. Partly for this reason, some intriguing questions remain unresolved. Are there 'metameric' timbres that are perceptually indistinguishable despite being made up of utterly different acoustic frequencies, just as there are metameric colours of which this is true for the frequencies of the constituent light rays? What are the emotional connotations of timbre? How does timbre affect our perception of harmony (or vice versa), and how easy is it to prise them apart? Are there cultural preferences for different timbres, and if so, why? The answers, if they are ever found, won't just be interesting in their own right, but might open up new possibilities for composition. For already there are many examples of experiments in timbre that have led to exciting new music.

* I say 'roughly' for good reason. For example, the last of these properties was said even by Grey to be hard to summarize, but relates to the changes over time in the 'spectrum' of harmonics.

Phantasmagoric instruments

The artificial combination of harmonics to manipulate timbre did not begin with electronic music. During the Classical era of Mozart and Haydn, a flute often shadowed a violin melody, adding high harmonics that the violin lacked in order to brighten the tone. Later composers, particularly since Berlioz, began to use the same principles to make timbre a 'composed' element of classical music, rather than something imposed by the physical characteristics of the instruments. As instruments became increasingly standardized and could be guaranteed to give fairly constant and predictable timbres, composers employed them as the basic 'sound sources' of orchestral synthesizers, which they blended into composite 'sonic objects' with new timbres – or, as Pierre Boulez has called them, 'phantasmagoric instruments'.

In this ensemble timbre, the individual instruments are inaudible as such: they fuse into hybrids. For timbre offers a category of perceptual similarity that can, like pitch and rhythm, be used by composers and musicians to produce gestalt grouping. Arnold Schoenberg even wondered whether timbre might supplant the function of pitch in melody: instead of a melody comprising a succession of pitches, it might be made from a succession of timbres. But he did not know how to construct such music, and neither does anyone else, because our ignorance of the theoretical and perceptual principles of timbre means that we have no idea what the timbral equivalents of scales, keys and metre might be. We don't even know whether the human mind would classify them in the same way, so that, for example, a 'melody' that steps between tones of different brightness would be perceived as 'getting anywhere' and as having an organized form at all. My suspicion is that such an analogy doesn't really exist.*

Nonetheless, in the twentieth century some composers began to produce a kind of timbral music in which a melody was parcelled out

*North Indian tabla music, which is percussive and non-melodic, does contain something like a timbral system of organization, although this is rare. Furthermore, psychologist Josh McDermott and his colleagues have shown that we do seem to recognize contours of brightness in music and can use them to identify familiar melodies. So it is possible that there is indeed more that might be made of timbre as an organizing feature of music.

between several different instruments. Schoenberg and his pupil Anton Webern wrote musical compositions of this type which they called *Klangfarbenmelodie* or 'sound colour melody'. In one example, Webern arranged a Bach ricercar for orchestra in which the theme was divided up between several instruments (Figure 8.2). The result is extraordinary – quite different from the interweaving of voices in Bach's original, and more like a delicate, pointillist blend of hues, with a beauty that is wholly original. Nowhere does the chromatic analogy seem more appropriate.

This dissection of a melody into short snatches differentiated by timbre gives the music a constantly shifting quality, a kind of tonal shimmer or sparkle. It can be regarded as a descendant of the compositional technique known as hocket (the word possibly stems from the Latin *ochetus*, hiccup), in which small groups of notes in a melody are apportioned between different instruments. Hocket was practised since around the thirteenth century, and Beethoven used it in some of his late string quartets. It is also common, and rather sophisticated, in the music of Indonesia and Africa.

Figure 8.2 A part of Webern's *Klangfarbenmelodie* orchestration of Bach's ricercare.

It can be argued that, in Western music, timbre found its place as a means of musical expression only in the twentieth century. The styles of Baroque and Classical composers were generally defined by how they made use of melody and rhythm, rather than in the 'colours' of their compositions. But beginning with Berlioz and particularly with Debussy, timbral texture, as much as the notes themselves, became a composer's trademark. Stravinsky's spiky sonorities – muted trumpets and caustic clarinets – mark him apart from Mahler's muscular mush. With the recorded tapes of Edgar Varèse and Karlheinz Stockhausen and the wailing ondes martenot of Olivier Messiaen (featured in his *Turangalîla-symphonie*), entirely new timbres began to enter the classical repertoire.

Changes in compositional style helped make room for these new tonalities, textures and timbres. In the pan-tonality practised by György Ligeti, the aim was to create a more or less saturated harmonic space where many different pitches overlapped. But this stacking up of vast numbers of traditional instruments turns out to produce a sound that is not just more complex but qualitatively different. The massed, discordant strings of *Atmosphères* no longer sound like strings at all, but like some cosmic resonator played by God, reverberating through the universe. American composer Glenn Branca achieves a similar result with massed electric guitars, mostly each playing a single note amplified to distortion, which blend in his Sixth Symphony into an exhilaratingly convincing imitation of a 'devil choir at the gates of hell'. The Beatles brought similar textural innovation into pop music with the chaotic orchestral crescendo that appears twice in 'A Day in the Life'; but Branca moulds that sort of chromatic morass into extended musical forms. The sound is scary and exciting partly because it foils our timbral gestalt, presenting us with a wholly unfamiliar soundscape.

Other contemporary classical composers weave timbres into dense, dizzying textures that Albert Bregman compares to a 'spraying shower or flowing hair in vision' – what we hear are not individual melody lines or trajectories but a general sensation of movement and 'granularity'. Such music demands a new way of listening, in which we must be prepared to relinquish many of the expectations and emotional triggers acquired from exposure to tonal music. One can argue that this is also the best way to appreciate the extreme serialism of Pierre Boulez: to regard it not as a dissonant and jumpy 'melody' but as a

Pollock-style splatter-painting of notes. Their precise sequence in time is then all but irrelevant; what matters is the unfolding sonic canvas. No doubt Boulez would grumble that this misses the point entirely; but in any event, the usual armoury of musical expectations and organizing principles is useless here.

As I said at the outset, it is in the arenas and sweatpits of rock music that timbre takes centre stage. It's widely acknowledged now that the experiments in sound creation and manipulation pioneered in recording studios by the Beatles, the Beach Boys and Pink Floyd were as innovative as anything Varèse achieved (and considerably more popular). And anyone who sniffs at the musical crudeness of the MC5, the Stooges and punk rock has failed to see that, sociology aside (and let's face it, that is ultimately the key), this music needs to be understood not in terms of chords and melodies but timbres. The point of the rock band's Big E chord, as conjured from a guitar by the windmilling arm of Pete Townsend, is not that it is a major triad but that it is a sonic entity, in its way as redolent with musical potential as Wagner's famous Tristan chord. Some of the most compelling rock music of recent years, from Sonic Youth to Ministry, has taken that raw timbral idea and shaken it furiously to see what comes tumbling out.

9

Misterioso
All In the Mind

Which bits of the brain do we use for music?

Which is your baby's favourite composer: Mozart, Beethoven or Bach? Don't tell me you're just singing 'Humpty Dumpty' to him/her? Don't you know that there's now a miniature library of recordings available to 'build your baby's brain' through classical music? Most offer Mozart, but you can also try Beethoven ('Brain-training for little ones') or compilations of Handel, Pachelbel, Vivaldi and others. There are selections of Mozart especially for newborns, or for every one of your infant's moods 'from playtime to sleepytime'.

I suspect these recordings might bring genuine benefits, for if I were a parent with a mind in tatters from sleepless nights and the primal struggle of attending to the unceasing needs of an infant, the Adagio of Mozart's 'Hoffmeister' String Quartet might well be just what I'd need. My baby might even like it too. But that is not, of course, quite how this music is supposed to be doing its magic. Rather, the idea is that your little one will be having its brainpower boosted by the 'Mozart Effect', which (the story goes) makes kids smarter.

We should hardly be surprised that recording companies spotted and exploited this marketing opportunity, especially given the connection to the emotive issue of child development. But there is simply no evidence at all for a Mozart Effect. The German research ministry was even forced to commission a special report on the matter – which came to this negative conclusion – after it was flooded with funding requests for work on music and intelligence. All this stemmed from a short paper published in *Nature* in 1993 which transparently failed to show – indeed, did not even investigate – any enhancement of intelligence either by Mozart's music in particular, or in children.

I'll come back later to the spurious Mozart Effect. For now, it's enough to recognize that this claim did not come out of nowhere. As the authors of the 1993 study pointed out, there is a long history to the idea that music stimulates the brain in ways that assist with other cognitive tasks. Previously that notion was based on little more than anecdote. But now it's a different matter, because new technologies such as magnetic resonance imaging (MRI) are enabling neuroscientists to see exactly what the brain is up to when it processes music, and to investigate whether there is overlap with other mental functions. We'll see some examples of this in the later chapters.

'Exactly' is perhaps putting it too strongly, however. Right now the technology outstrips our ability to interpret what we are seeing: we do not yet understand very well what all the parts of the brain do or how they interact with one another. The pictures look impressive, but they often don't tell us much more than that the brain uses one region both for *this* and for *that*. We don't actually know what the cognition of *this* or *that* entails.

And as far as music is concerned, MRI studies are something of a mixed blessing. For whereas many cognitive tasks, such as vision or language, have fairly well-localized centres of brain activation, music does not. To put it crudely, when we listen to music, all the lights are apt to come on at once. Pretty much the whole brain may become active: motor centres that govern movement, the 'primal' emotion centres, the language modules that seem to process syntax and semantics, the auditory highways . . . Unlike language, say, music has no dedicated mental circuitry localized in one or a few particular areas: it is a 'whole brain' phenomenon. On the one hand this makes it immensely challenging to understand what is going on. On the other hand, it shows why music is so fundamentally important: no other stimulus comparably engages all aspects of our mental apparatus, and compels them to speak with one another: left to right hemisphere, logic to emotion. And that's why we do not need any Mozart Effect to validate the importance of music in development, cognition, education or socialization. It is quite simply a gymnasium for the mind.

A tour of the grey matter

Here's the basic problem for neuroscience: how on earth can the extraordinary richness of human experience be summed up by some bright spots on a brain map, where the wizardry of nuclear magnetic resonance has spotted an increase in blood flow? Or by some electrical impulses sensed by electrodes taped crudely to the cranium? How can we ever hope to forge an equation that links Bach and Picasso to this grapefruit-sized mass of jelly?

If I sound somewhat dismissive about the current techniques of neuroscience, it is simply because of the magnitude of this problem. But even the crude generalizations we can now make about the workings of the brain, thanks to these blurry, absurdly coarse-grained maps of cogitation, are of tremendous worth. By identifying which regions of the brain we use for different tasks, we can form a picture of how the brain classifies and interprets the very nature of the cognitive demand: which parts of its wet hardware it selects for the job, and thus which jobs apparently share processing skills in common. This is a large part of what the neuroscience of music is about: spotting when the brain uses circuitry whose purpose we already know, or suspect, from its involvement in other functions.

It's simplistic but largely unavoidable to portray the brain as a vast bureaucracy of specialized departments, some in constant communication and others barely aware of their mutual existence. Most prominently, the brain is divided into two hemispheres, on the left and the right of the cranium. One of the virtues of musical neuroscience is that it undermines the tiresome trope of 'left brain' and 'right brain' processing, which popular belief labels as logical or analytical, and emotional or intuitive, respectively. That's not a totally false picture – the hemispheres do show some tendency towards specialization, musical tasks included. Pitch perception, for example, seems to be mostly (but not entirely) localized in the right hemisphere. But the full picture is more complicated; for example, while left-hemisphere processing seems dominant for positive emotions, the right hemisphere becomes engaged for negative ones.

Most of the brain's volume is taken up by the cerebral cortex (often simply called the cortex), which is divided in each hemisphere into four lobes: the frontal, temporal, parietal and occipital (Figure 9.1). These

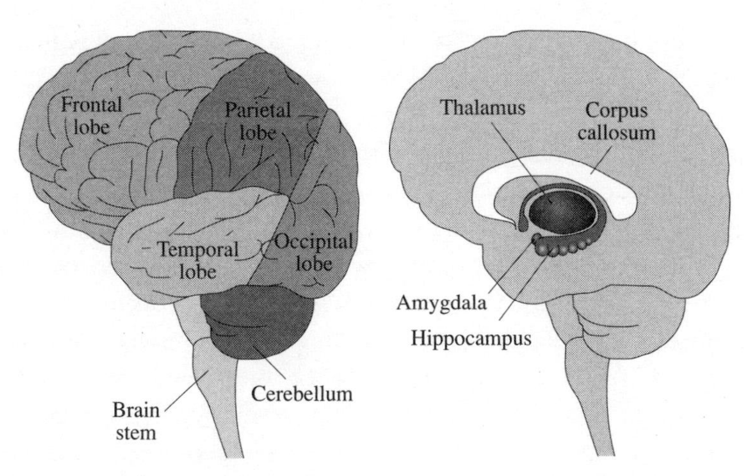

Figure 9.1 The anatomy of the brain.

have quite generalized functions. The frontal lobe is involved in planning and the organization of perception, and also (to the rear) in motor and spatial skills. The temporal lobe deals with memory and hearing – it houses the hippocampus, the seat of long-term memory, as well as the primary auditory cortex, where sound stimuli coming from the ear are first processed. It also has a role in handling the semantics of speech. The parietal lobe integrates various kinds of sensory information, for example those governing our sense of space. And the occipital lobe specializes in vision processing. Below the temporal lobe and close to the brain stem is the cerebellum, the oldest part of the brain, which governs our emotional responses as well as coordination and motor control. But there is also another important 'emotion centre' in the temporal lobes of the cortex, called the amygdala.

When we hear music, information is sent from the cochlea in the ear, via the brain stem (where some initial pitch processing seems to take place both in music and in speech), to the primary auditory cortex.*

* Of course, we can 'hear' music in the brain without any auditory input at all. Oliver Sacks describes people who, owing to brain damage, 'hear' such music all the time, involuntarily and sometimes at great volume, to their immense discomfort. Not much seems known about this 'imaginary' perception of music, although Petr Janata and colleagues at the University of California at Davis have found that the electroencephalograph signals detected from the brain of people listening to and just imagining music are all but indistinguishable.

From there it gets sent out for processing to many different regions of the brain, some of them performing overlapping tasks. For example, pitch intervals and melody are processed in the lateral part of a domain called Heschl's gyrus, within the temporal lobe, which has circuitry involved in pitch perception. These two things are also handled by the planum temporale, a domain that deals with rather sophisticated attributes of audition such as timbre and the spatial location of sound sources, and by the anterior superior temporal gyrus, which deals with streams of sound, including spoken sentences. From what we've seen so far, it should be no surprise that pitch and melody have to be, if you like, approached by the brain from several different angles. For example, perceiving pitch involves an analysis of the structure of harmonics (the brain can 'fill in' a fundamental pitch even if it isn't sounded, so long as the higher harmonics all fit together). And we need to separate out pitch streams into different voices, as in polyphonic music or for multi-instrumental ensembles. Then we need to turn a succession of pitch steps into a coherent melody – to switch from a local to a global view. Neuroscientist Isabelle Peretz and her co-workers at the University of Montreal have suggested that the right hemisphere discerns the global pattern of pitch contour while the left hemisphere fleshes this out with detailed aspects of pitch steps. What's more, pitch is obviously central to harmonic processing. Even rhythm and note duration provide clues about the significance we should attribute to particular pitches. The analysis of pitch generates expectations about melody and harmony that are processed in the right-hemisphere analogue of a left-hemisphere language-processing region called Broca's area. In short, pitch sequences encode a whole range of musical dimensions, and to some degree the brain assigns distinct modules to process each of them.

Needless to say, the brain's response to music is not simply a matter of coldly dissecting patterns and regularities of pitch and rhythm. As soon as the primary auditory cortex receives a musical signal, our 'primitive', subcortical brain kicks in at once: the cerebellum's timing circuits fire up to pick out the pulse and rhythm, and the thalamus takes a 'quick look' at the signal, apparently to check for danger signals that require any immediate action before more complex processing occurs. The thalamus then communicates with the amygdala to produce an emotional response – which, if danger is detected, might be fear. Only after this primeval scan to warn of hazards does the detailed dissection of the sound signal begin. We call on the hippocampus to supply memories, both of the immediate past course of the music and the more distant associations and similarities it evokes. The prefrontal cortex does some sophisticated work on anticipation and expectation, while Broca's area itself, associated as I say with language processing, seems to be enlisted to deal with high-level 'syntactic' aspects of the music. A performer might require the visual cortex to read music and to watch the conductor or colleagues, while the sensory cortex enables them to feel the instrument beneath their fingertips. And our rhythm-processing circuits seem to fire up motor functions not only to produce rhythm but also as part of simply hearing it – which might go a long way to explaining why it is hard to sit still when listening to James Brown.

Music can trigger physiological processes apparently far removed from the purely cognitive. It can, for example, affect the immune system, boosting levels of proteins that combat microbial infection. Both performance and listening can also regulate the body's production of mood-influencing hormones such as cortisol, showing that there is a sound biochemical basis for the use of music in therapy.

This could all seem bewilderingly complex. But it basically entails reducing the cognitive task to a series of some more abstract processes, which operate in a cascade of increasingly sophisticated functions. The brain's initial aim is to identify the basic acoustic building blocks, such as fundamental pitch frequencies and harmonics, note durations, and loudness. These must then be separated out into individual instruments or melodies. The results are then compared with stored musical memories and experience – for example, our implicit and explicit knowledge of harmonic relationships and cadences, of genre and style. Typically this emerging musical landscape is informed by information from other cognitive domains, for example if we are listening to lyrics with semantic content. At the same time, each of these aspects of analysis and synthesis is engaging our emotions, and eventually the emotional, associative and syntactic information is combined to stimulate some kind of behavioural response: we are aroused, soothed, moved, annoyed.

This processing involves somewhat general-purpose neural circuitry that is brought to bear on the musical context of the stimulus – for example, pitch and rhythm processing are general features of audition that are needed to interpret speech and ambient sound too. Are there, however, any brain regions that are dedicated *specifically* to musical tasks? This is a profound question. If there are, it would suggest that musicality is innate, engineered by evolution – which in turn would imply that it serves an adaptive function, that musicality somehow helped our ancestors to be reproductively successful. The search for specialized music modules in the brain has therefore sometimes tended to become an emotive mission, because it might pronounce on the question of whether music is fundamental to human nature or merely parasitic on other cognition functions. But that quest has so far failed to produce definitive results.

When things go wrong

One of the most fertile approaches to understanding how the brain deals with complex cognitive and perceptual tasks involves studying people who have selective impairments in those areas. Damage to particular parts of the brain can weaken or destroy very specific functions while leaving others intact. Loosely speaking (and this is not always a totally safe assumption), if a person loses the ability to do one task while retaining an ability to do another, it is likely that they are processed in different parts of the brain.

In music processing, just as in other cognitive tasks such as vision and memory, the consequences of these selective dysfunctions can be bizarre, and also sadly very distressing – although they might alternatively involve little more than mild inconvenience.

Studies like this might tell us whether there are genuine 'modules' in the brain devoted only to music.* But any evidence of that is so far ambiguous at best. The only *widespread* specific deficit identified in music processing is in the perception of pitch relationships, which is faulty in about four per cent of the population. These people are genuinely tone-deaf, being unable or barely able to judge things such

* Modularity of processing doesn't necessarily imply that the functions are physically localized in the brain – although the two may go together, and indeed the fact that localized lesions may create very particular processing deficits suggests as much.

as whether one pitch is higher or lower than another. Surprisingly, this faculty seems to be independent of the perception of individual pitches, which is why it is possible simultaneously to possess the seemingly exclusive conditions of absolute pitch and tone-deafness. And even being tone-deaf doesn't in itself make someone unmusical – it doesn't mean that they can't play an instrument, for example, although most people who are severely tone-deaf don't do that, probably because they have decided (or have been told) that music 'isn't for them'. Moreover, neurological studies have shown that even tone-deaf people do hear 'wrong notes' in some sense: these elicit a specific kind of electrical brain signal just as they do in other people, but the incongruity doesn't seem to filter through to conscious awareness.

The incidence of genuine tone-deafness is considerably smaller than the number of people who will call themselves 'tone-deaf'. One study by Canadian psychologists found that seventeen per cent of a sample of students at Queen's University in Kingston considered that they belonged to this unfortunate category, whereas very few of that group fared significantly worse than anyone else in tests of music perception. It seems likely that most people who think of themselves as unmusical do so for reasons unconnected to their actual musical potential – maybe they are embarrassed by the belief that they can't sing, or have been told by others that they lack any musical talent, or were simply never encouraged to take any interest in or enjoyment from music. It's hard to know whether to be heartened or dismayed by these findings. On the one hand they imply that nearly all of us have what it takes to be at least musically competent. On the other hand, many of us are somehow receiving the message that we haven't.

If music merely draws on cognitive abilities that evolved for other purposes, how to explain, for example, the musical precocity sometimes found in people with autism, who don't seem to obtain any more general cognitive benefits from those abilities? This by no means settles the argument, but it's puzzling.

Conversely, there are many examples of people with cognitive defects that seem to hamper only the way they process music, and nothing else. These musical dysfunctions can be remarkably particular: one doesn't necessarily just become 'unmusical'. Recall, for example, Oliver Sacks's highly musical patient who lost the ability to hear harmony, while being able to hear the separate 'voices' perfectly (p. 143). And one professional violinist who suffered brain damage reported a seemingly arbitrary list of impairments, including an inability to read sequences of notes and to transcribe music, and an inability to recognize familiar music and to identify melodic intervals. Yet he retained acute pitch discrimination and was able to name single written notes and to identify major and minor scales.

More generally, any of the traditional attributes of music can become impaired while leaving others intact: the conditions of 'dysmelodia', 'dysrhythmia' and 'dystimbria' are all recognized. Particularly poignant are people who remain able to perceive music perfectly while losing any emotional response to it. In one case, neuroscientist Jason Warren of London's National Hospital for Neurology and Neurosurgery and his co-workers described a former radio announcer who, having developed damage to the amygdala, found that the 'chills' he had always experienced when listening to Rachmaninov's Preludes had vanished. And Isabelle Peretz's group found that damage to a part of the amygdala left some of their patients unable specifically to identify 'scary' film music as such, while having no effect on their perception of 'happy' and 'sad' music.

Conversely, and perhaps even more strangely, emotional receptivity to music can remain even in the face of a complete inability to process melody. This is what Peretz and her colleagues observed in a patient called 'IR', who suffered damage in both hemispheres when brain surgery conducted for other reasons developed complications. IR could make perfect sense of speech, and could discriminate between different sounds in her environment, but she couldn't tell one tune from another, or distinguish melodies altered to become dissonant from the original versions. But she claimed still to find music listening rewarding, and could classify melodies as 'happy' or 'sad' almost as reliably as unimpaired listeners. It's possible that IR was able to make use of emotional cues unrelated to pitch, particularly rhythm. We'll explore how these might work in the next chapter.

Music as brain food

Musicians' brains are not like everyone else's. Just as physical exercise changes the shape of the body, so too it seems that *musical training alters the brain*. For example, music processing tends to become more analytical in musicians. The conventional view that labels this more 'left brain' is not amiss here: whereas melody is primarily handled by the right hemisphere in non-musicians, the left hemisphere is more involved for musicians.*

But this *doesn't* mean that musicians replace a feeling response with a thinking response – and given the emotional investment that most musicians make in music, it would be absurd to think otherwise. Musicians, especially those whose musical training began before about the age of seven, appear to have an enlarged corpus callosum, the brain region that connects and integrates the hemispheres. Neuroscientists Christian Gaser and Gottfried Schlaug have found a host of differences in brain functions for musicians (specifically, keyboard players) and non-musicians, including improved motor, auditory and visual-spatial skills. Moreover, musicians who use their fingers for performance have an enlarged representation of the hand in their cortex: you could say that they are more in touch with their fingers. And musicians also show boosted development in a part of the auditory cortex used for processing pitch. In general, these anatomical changes are more pronounced the longer the musician has been trained, suggesting that they are effect rather than cause: a systematic product of musical learning rather than being innate differences that make a musical career more likely.

Mightn't all this mean that music really does make us smarter? That there *is* a Mozart Effect?

What neurobiologist Frances Rauscher and her co-workers at the University of California at Irvine reported in 1993 was that college students showed slightly better results in spatial-reasoning tests after listening to ten minutes of Mozart (Sonata for Two Pianos, K488) than after either listening to a 'relaxation tape' or sitting in silence. The tests involved analysing patterns and figuring out how a piece of paper folded according to a diagram will look when one corner is snipped off and it is then opened up. The differences in performance were

* There are probably differences between the sexes too. It's generally recognized that, in cognitive tasks, women's brains apportion the mental processing less specifically to left or right hemispheres than do men's brains: women are said to be less lateralized. Men and women show some differences in how well they can discriminate similar melodies when delivered just to the right or the left ears, which tends to confine the processing to those hemispheres.

small but significant: Mozart boosted IQs by eight or nine points relative to the other two cases.

Surely an extra eight or nine IQ points are worth having? Well, let's think about it. First, 'IQ' here was being derived from just three tests of spatial reasoning – a pretty limited measure of intelligence. (The researchers provided no indication of the error bars in the values they measured.) Second, the alleged Mozart Effect was so transient that it was seen only in the first of the three tests, having apparently faded away by the time the subjects carried out the second and third. And there was no reason to believe (nor was any claimed) that this was an effect specific to Mozart's music – the researchers didn't test any other music apart from the 'relaxation tape', which was presumably pretty mind-numbing stuff.*

The *Nature* paper stimulated a flurry of attempts to verify the claims. Some reported similar enhancements; others, often using different tests of cognitive ability, found none. One follow-up study made a particularly big splash: in 1996 Susan Hallam from the University of London's Institute of Education collaborated with the BBC to test more than 8,000 ten- and eleven-year-old British schoolchildren. The study found that listening to Mozart gave the children no advantage in two tests of spatial and temporal abilities. Yet in 1999, K. M. Nantais and Glenn Schellenberg at the University of Toronto at Mississauga saw an enhancement in listeners' performance in a paper-folding and cutting test after listening to either Mozart or Schubert, relative to a preparatory period of silence. And they saw a similar improvement when the subjects first listened to a *story*. In other words, there was nothing 'musical' about the effect at all: it had something to do with the complex quality of the stimulus. Then Nantais and Schellenberg asked the participants which they preferred listening to: Mozart or the story. They found that those who preferred Mozart performed better after hearing it, and likewise with the story: the improvements reflected what the subjects *liked*.

* One follow-up study by Rauscher and colleagues found a result that will please critics of minimalism: it replicated the improvements in performance for Mozart, but found none after listeners heard music by Philip Glass. Any triumphant proclamations of the superiority of the old maestros should be squashed, however, by the fact that a separate study also found improvements after exposure to the music of the Greek New Age composer Yanni.

They reasoned that the Mozart Effect therefore had something to do with how the stimulus affected arousal and mood in the listener: that is, whether it interested them, and whether it made them happy. When listeners heard Albinoni's Adagio, a stereotypically 'sad' piece, they performed *worse* than they did after hearing nothing.

So if our spatial reasoning receives a temporary boost – at least for some types of spatial task – from hearing music we like, we'd expect children to respond best not to Mozart but to their favourite kind of music. This hypothesis led Schellenberg to team up with Hallam and reanalyse her 1996 data on schoolchildren. As well as listening to Mozart, the children had also been tested after listening to pop music from BBC's Radio 1 station, on which they heard three contemporary songs including 'Country House' by Blur. Schellenberg and Hallam discovered that the children did better in one of the two tasks (again, a paper-folding problem) after they heard the pop music. In other words, they said, there was a 'Blur Effect' too.*

So music *does* appear to induce a small and temporary improvement in some kinds of cognition – but the mechanism has nothing 'musical' about it. Rather, we perform better if the music makes us feel good. That is broadly in line with what psychologists have long known: cognition is dependent on mood and arousal, and stimuli that we find interesting and pleasant give us a lift. Mozart has no intrinsic, mysterious power to make your child smarter.

This episode is salutary for several reasons. It stemmed from a long-standing suspicion that music boosts intelligence, but it is also a sad reflection on our modern response to that possibility. If this is true of music, we seem to be saying, then we want a quick fix: a CD-sized Mozart pill.

This is all the more unfortunate because it seems clear that music *can* enhance intelligence. Many of the past claims of that sort have been simplistic, consisting of little more than the observation that children who take music lessons tend to do well in other, non-musical areas of learning, performance and intellectual development. It's not

* Why was there an improvement for just one of the tasks, and not the other? Schellenberg and Hallam speculated that it might have been because the folding task was the hardest of the two: that's to say, perhaps the boost in motivation is more apparent when the challenge is greater.

hard to imagine why this should be so. Learning to play and read music demands focus and attention, self-discipline and good memory, all of them handy attributes for other areas of learning. Musical training could help these qualities to develop in a child, and equally, children who already possess them are more likely to persist with learning music. Not only is cause and effect hard to unravel, but one might expect similar developmental advantages from any subject that requires concerted study. What's more, children who take music lessons are more likely to come from an environment conducive to intellectual development: to have better-educated, affluent parents.

We can't conclude that music per se improves development and IQ, then, without a more careful look at other confounding factors. One study in 1999 found that, when equivalent groups of children were or were not given piano lessons from age three, the piano learners were better at spatial tests for the first two years, but that any benefit disappeared thereafter. Glenn Schellenberg has conducted one of the most careful studies so far. He recruited 144 six-year-old children for 'free weekly arts lessons' by placing an advertisement in a local newspaper, and divided the children into four groups. One group received keyboard lessons, another got singing lessons using the Kodály technique, developed specifically for children and involving movement-related exercises such as clapping and hand signs. A third group was given drama lessons; the fourth received no lessons at all during the one-year period of the study (but they were given keyboard lessons the following year, so as not to have been recruited under false pretences). All the children received a full IQ test after the first year. The groups were randomized according to precise age and parental annual income. The study was set up, therefore, to search for any IQ effects due specifically to music.

All four groups showed an improvement in IQ, but the increase for both of the music groups was significantly larger (about seven points) than that for the other two (about four points). In short, said Schellenberg, 'the results provide evidence of relatively modest but widespread intellectual benefits from taking music lessons.' There was, however, a caveat. When tested for 'social behaviour', the drama group showed significantly more improvement than the other two. If music makes you smarter, it doesn't necessarily make you any easier to get along with.

If you think that the cost of music lessons is worth three extra points of IQ, that's fine. But while there is value in establishing quantitative measures of the intellectual stimulation that music affords, this should never be more than a small component of any argument for why music is the food of the mind as well as of the heart. When you see the range of cognitive functions that music engages, you can hardly be surprised that some abstract measures of intelligence are among the attributes it enhances. Some, perhaps most, of the real benefits for the mind probably lie beyond the reach of tests. Learning to be discriminating in music and to listen well (or to know why it is at least sometimes worth doing that), learning *how* to let music move you, and ideally learning how to make your own music, no matter how crudely or haltingly – these are mental skills that will serve you far beyond the pleasures they provide from music itself. As ethnomusicologist John Blacking puts it:

> The development of the senses and the education of the emotions through the arts are not merely desirable options. They are essential both for balanced action and the effective use of the intellect.

And while it is true that sitting at a piano learning scales may not do much for your social skills, once you have played or sung in an ensemble, or danced a 'Strip the Willow', or have laughed or wept or sat shell-shocked at the passions and ideas that someone else has put into musical form, you will have discovered a direct route to the core of our shared humanity.

This is the reason why music must not be an optional extra in the curriculum. This is the real reason to play your baby Mozart, or Muddy Waters or *Revolver* or Transylvanian airs or whatever it is that makes you feel that way.

10
Appassionato
Light My Fire

How does music convey and elicit emotion?

Do you recognize this?:

> Suddenly [I] experienced a tremendously strong feeling that was felt both in my body and in my head. It was as if I was charged with some kind of high tension, like a strong intoxication. It made me ecstatic, inconceivably exhilarated, everything concentrated to a single now. The music flowed as [if] by itself. I felt like I was penetrated by Bach's spirit: the music was suddenly so self-evident.

Or this?:

> I was filled by a feeling that the music started to take command of my body. I was charged in some way . . . I was filled by an enormous warmth and heat. I swallowed all tones . . . I was captured by each of the instruments and what they had to give to me . . . Nothing else existed. I was dancing, whirling, giving myself up to the music and the rhythms, overjoyed, laughing. Tears came into my eyes – however strange it may seem – and it was as a kind of liberation.

If you have never had such experiences in response to music, you might be feeling now like the gloomy outsider at a party, excluded from the drunken euphoria all around. But I suspect you have. Or that, if you haven't, then at least you *could*. Intense emotional responses to music are not unlocked by any master key – the first description above is by a young performer rehearsing Bach, the second by a woman listening to Finnish tango in a pub. What is striking, however, is that the experiences themselves are often so similar, to the extent that it can be hard to describe them without recourse to cliché – 'the music swept me away' and so on. But they are clichés because that is what it feels like.

Although transcendental ecstasy is a response to music found in many cultures, intense musical experience is not always so blissful. Here is someone recalling a performance of Mahler's Tenth Symphony:

> [There was] a chord* so heart-rending and ghost-ridden that I had never experienced before . . . My brother and I reacted the same: we were both filled with such a primitive horror, almost prehistorical, that none of us could utter a single word. We both looked at the big black window and both of us seemed to see the face of Death staring at us from outside.

It seems astonishing, culturally and psychologically, that anyone would wish to compose music that has such an effect. (One might debate whether that was what Mahler intended, but there are plenty of examples of musicians explicitly aiming to unnerve or distress their listeners.) It also seems bizarre that someone would willingly expose themselves to it (again, even if this listener didn't do so intentionally, others surely do). And perhaps most strikingly of all, it seems extraordinary that an arrangement of musical tones can create such responses in the first place.

While the question of *how* music does this must be framed by its cultural and historical context, the capacity of music to move us is surely universal. When Tolstoy wrote that 'music is the shorthand of emotion', he was apparently expressing approval and admiration of this trait. But to St Augustine in the fifth century, the ineluctable emotionality of music was a worry. He loved music, but fretted over the thought that worshippers might be 'more moved by the singing than by what is sung'. Medieval clerics recognized with dismay that music was as apt to incite lust and desire as it was piety, and this was a central consideration in the efforts during the Counter-Reformation to expunge sacred music from the contaminating influences of the secular. For churchmen, John Dryden's remark cut both ways: 'What passion cannot Music raise and quell?'

* The chord appears in bars 203–6 of the first movement, piling up dissonant minor thirds. Listen if you dare.

Why are we moved by music? For any attempt to understand how our minds grasp and process music, this is perhaps the hardest question of all. There is something uniquely intangible about the way music works its alchemy. Many great paintings literally depict feelings, whether in faces, gestures or circumstances. Even abstract art can evoke associations with its shapes and colours: Yves Klein's celestial, infinite azure, Mark Rothko's sombre horizons, Jackson Pollock's spattered traces of frenzied intensity. Literature stimulates emotion through narrative, characterization and allusion, and even if the 'meaning' can be somewhat elastic, the limits are relatively narrow: no one suggests, say, that *Great Expectations* is about coal mining. But music is invisible and ephemeral: it sighs and roars for a moment, and then it is gone. Save for the occasional moments of intentional mimicry, it refers to nothing else in the world.* Particular phrases and tropes had conventionalized 'meanings' in the Classical era, and some have argued that Western instrumental music more generally has specific meanings that can be objectively decoded. But those arguments always run aground in arbitrary assertion. The answers that they offer are too easy, too superficial. For it is hard enough to understand why we should be able to make the slightest sense of these eliding acoustic signals, let alone why they should move us to tears and laughter, make us dance or rage.

It is only rather recently that the question of emotion has been raised at all by cognitive scientists and musicologists. Eduard Hanslick, whose 1854 book *Vom Musikalisch-Schönen* (translated in 1891 as *The Beautiful in Music*) was one of the first serious modern studies of musical aesthetics, complained that previously music had tended to be discussed either in the 'extremely dry and prosaic language' of technical theory or by aesthetes 'wrapped in a cloud of high-flown sentimentality'. His focus was not on emotion per se but on the aesthetic effect, particularly the sense of beauty, that arises where

* Clearly I am not talking here about opera. In opera, as in film, the music enhances the emotion that is already there in the storyline, and then the same question arises: how can music effect this enhancement? Indeed, music is often the primary agent of affect in operas, many of which would otherwise have banal plots and stodgy dialogue (all too often woodenly acted). Opera is not music, but an art form that makes use of it. One can enjoy opera while being indifferent to music (many do). And vice versa.

feeling and intellect meet. In the eighteenth and nineteenth centuries, said Hanslick, it was all but taken for granted that music is a crafting of sound with the primary purpose of expressing and exciting passions. And yet he doubted that there was any real correspondence between a musical composition and the feelings it elicits. After all, Beethoven was considered in his time to be passionate in comparison to the cool clarity of Mozart, yet Mozart was likewise once deemed fervent in comparison to Haydn. 'Definite feelings and emotions', Hanslick claimed, 'are unsusceptible of being embodied in music.'

The question of whether specific emotions can be conveyed by music is a complex one to which I'll return later. But no one can doubt that some music is capable of exciting some emotion in some people some of the time. Indeed, many might still agree with the old view that music's *raison d'être* is to inspire emotion. The question is how.

Hanslick insisted that this will always be mysterious. 'The physiological process by which the *sensation* of sound is converted into a *feeling*, a *state of mind*, is unexplained, and will ever remain so,' he wrote. 'Let us never appeal to a science for explanations which it cannot possibly give.' That was unduly pessimistic (or, one suspects, hopeful). And yet someone coming to the psychology of music hoping to learn how it inflames the heart may well be disappointed. At present, such studies often seem woefully inadequate, even simple-minded, on the topic of emotion. When neuroscientists examine how and when people classify particular musical extracts as 'happy' or 'sad', the music-lover might reasonably see that as traducing the emotive qualities of music – as though we sit through a piano concerto either beaming or moping.

All the same, the issue of emotion has become central to the field of music cognition. One happy consequence is that an atomistic dissection of music – for a long time the preferred approach – must be laid aside in favour of examining how people respond to the sort of music they actually listen to. For no one is likely to feel much emotional response to a monotone sine wave from a tone generator. This has meant in turn that the 'test material' is more diverse. Someone contemplating the field in its early days could have concluded that there were just two broad categories of music: Western classical, and 'primitive' music from preliterate cultures whose function was primarily sociological. It is now permissible to ask why and how people listen to, say, the Eagles or Grandmaster Flash, while the canonical corpus for exploring musical responses has shifted from Mozart to the Beatles.

Some people might react with horror to the thought that the emotive effects of music can be explained and classified with the cold instruments of science. Doesn't this threaten to reduce the composer to a technician manipulating the responses of a passive audience, just as Kandinsky hoped the abstract artist might deploy colour and shape systematically to play the keyboard of the soul? It might even conjure up the spectre of music being used for social manipulation and behavioural control: muzak making us bovine, or Wagner's 'Ride of the Valkyries' turning us homicidal. But one needn't look very deeply for these worries to recede. For one thing, although music probably can, in the right contexts, encourage violence or calm, so can language, drama, perhaps even colour or the weather. We do not need cognitive psychology to tell us what kinds of music tend to relax or arouse people, nor is it at all likely to show us how to make music that achieves these things significantly more effectively than is already possible. And music is not a pill that, when swallowed, inevitably produces a prescribed state of mind. Music psychologist John Sloboda warns that psychological studies of musical emotion run the risk of seeming to reduce much-loved pieces 'to a set of "effects" such as might similarly be realized by the colour of paint on a wall, or the administration of caffeine'. He calls this the 'pharmaceutical model' of music.* Sloboda says he regularly receives requests along the lines of 'Can you tell me which music will have such and such an effect?', or 'What is the best soundtrack for a marriage proposal?' Music, he tries gently to tell his correspondents, doesn't work that way.

So Hanslick was doubtless right to assert that no amount of scientific and musicological study will tell us all there is to know about the link between music and emotion. Yet neither must it remain inscrutable. We have good reason to think that there are some general

* That will do nothing to stem the flood of compilations of both classical and popular music marketed as 'Music for Relaxing', 'Music for Lovers', 'Music for Concentration' and so on.

underlying principles of musical affect, and that it is by no means some magical property beyond the reach of rational analysis. I doubt that any composer or musician has ever truly believed that.

How do you feel?

Before we can even start to ask how music creates emotion, we must wrestle with the conundrum of what 'musical emotion' means. Hindemith scoffed at the notion that composers actually feel, while they are composing, the emotions they strive to convey. Tchaikovsky agreed:

> Those who imagine that a creative artist can – through the medium of his art – express his feelings at the moment when he is *moved*, make the greatest mistake. Emotions – sad or joyful – can only be expressed *retrospectively*.

Performers too need not feel emotions in the course of expressing them, and often – in rehearsal, say – they will not. But in what sense does the listener 'feel' anything? Very often, we may recognize music as having a particular emotional quality without it actually awakening that emotion in us. Hindemith believed that 'the reactions which music evokes are not feelings, but they are the images, memories of feelings'. Philosopher Peter Kivy agrees: we might recognize these emotions, he says, but not necessarily feel them. Mozart's 'Jupiter' Symphony sounds happy to me even if I am feeling rotten, and even if it doesn't make me feel any better. It's the difference between *inducing* and *expressing* emotion, which Arthur Schopenhauer proposed in 1819 as a distinctive feature of music.

Western musicians and theorists have persistently exhibited a certain vagueness about the distinctions between the emotions music is supposed to express and those it seems to induce. We are on firmer ground with the former, because then we are at least always referring to the same thing: the music itself, and not the experience. It generally makes little sense to say that such-and-such a piece of music 'makes me feel X'. Is that equally true if you hear it during a dinner party, on top of a mountain, or blasting from your neighbour's house at four in the morning? It's obvious that our responses to music don't depend just on the music itself. A melody might move us to tears on one day and leave us cold on another. These responses are surely conditioned at least in part by context, and this has to make us wonder what to make of any experiment in which listeners are placed in booths with a pair of headphones. In some non-literate cultures (and not those alone), the context of music might be the key to any emotional response: the point is not what the music itself 'does', but the fact that it is known to be a birth (joyous) or funeral (mournful) song.

Our response to music is in any case not always congruent with what we would objectively characterize as its emotional content. Sad music can make us cry, but frankly (in my experience) so can happy music. And sad music presumably offers more pleasure than pain, or we would not listen to it so much (and country & western would not exist). Clearly, we can't equate sensuous pleasure in music solely with happy or pleasant attributes, such as beauty. Music can arouse all manner of passions, from anger to despair, even disgust, while remaining music to which we choose to listen. Where is beauty in Bartók's *The Miraculous Mandarin*, or Mussorgsky's 'Gnome' from *Pictures at an Exhibition*, or the prelude to the churchyard scene in Act Two Scene Three of Stravinsky's *The Rake's Progress*? This is music that is deliberately grotesque and ugly, and yet we rightly award it high aesthetic value.

We also need here to distinguish emotion from *mood*, although admittedly it is not always easy to do so. A piece of music can offer a more or less consistent mood while creating constant variations in intensity or complexion. (When we are genuinely listening to music, mood seems far too passive a word: our response involves something like a dialogue with the acoustic stimulus.) In contrast to moods, emotions are typically of shorter duration, often accompanied by distinct facial expressions, and most importantly, evoked by identifiable, specific stimuli rather than by a generalized ambience. A piece of music doesn't necessarily retain any single emotional quality for long. Some leap from grief to joy to calmness within minutes, and for our emotions to follow this in any serious way would, as Hindemith remarked, demand something approaching a mental disorder. Yet it is precisely this polyvalent aspect of music that may provide its attraction.

Swedish music psychologists Patrik Juslin and Daniel Västfjäll argue that many of the ways in which musical emotions are aroused share psychological mechanisms with more everyday emotions. Our pulses might be quickened by the basic acoustic properties of the sound, such as sudden loudness. Haydn's Symphony No. 94 in G major, nicknamed the 'Surprise' Symphony, is the most famous example of this, springing a loud chord in the tranquil Andante movement – apparently an example of Haydn's notorious delight in musical jokes. Such aural jolts are like a shriek in a library, activating primitive alarm reflexes hard-wired into the brain stem which serve to alert us to danger – the roar of a predator, the crack of a falling tree. Such surprises can be enjoyed when they are revealed immediately to pose no real threat – but, being instinctive, they are also barely dulled by familiarity, since they kick in before they can be suppressed by slower cognitive reasoning.

Some researchers have attempted to identify a limited number of 'basic' emotions, such as happiness, sadness, anger, fear and disgust. Others argue that emotions can be mapped on to a multidimensional space in which the coördinates are attributes such as valence (good/bad), activity and potency. Serenity, say, would have a strong positive valence but moderate activity, while fear is rather strongly negatively valenced and also highly active – it makes the heart pump faster and can prompt us into flight. Or perhaps emotions exist in a kind of hierarchical, tree-like relationship: scorn and fury might be branches of anger, which is itself on the negative branch of a basic good/bad bifurcation. All this seems a rather faux-scientific attempt to define mental states that are truly too complex to be broken down this way: after all, the same event can make us simultaneously happy and sad, and these states may change over time. That's not to say that such classifications aren't useful – they may even be essential in researching emotion – but they shouldn't be mistaken for representations of reality.

Yet it's by no means clear that all musical emotions, or their triggers, are of the 'common or garden' variety: happy, sad and so forth. For example, people often report a feeling of 'chills' in response to music, but these don't necessarily come with any emotional *valence* – that's to say, the feeling is neither 'good' nor 'bad'. Bach's music can make me tearful without being either happy or sad. (I'll look later at how in fact we might begin to unpack such feelings.) Steve Reich's music might make me feel at the same time relaxed yet alert: a clear emotional state, but not one I can articulate very coherently. Swiss music psychologists Klaus Scherer and Marcel Zentner suspect that basic emotional states might not be of much use for describing what we feel in response to music. 'It may be necessary', they say, 'to focus more strongly on terms for more subtle, music-specific emotions (such as longing, tenderness, awe, activation, solemnity) that seem to describe the effects of music more readily than anger, sadness, or fear.' This sounds likely, but it still insists on using terms that are in fact not music-specific at all, but more nuanced forms of everyday emotion.

Might there be a mode of emotionality inherent to music for which, as Aaron Copland put it, 'there exists no adequate word in any language', and which American philosopher Diana Raffman calls ineffable? Another philosopher, Susanne Langer, agrees with Copland that music may induce emotional states that have a quality of their own, for which we have no suitable vocabulary. Her stimulating idea, proposed in the 1950s, is that music doesn't so much *represent* emotion as *mimic* it: the ebb and flow of music is analogous to the dynamics of emotion itself. As the American psychologist Carroll C. Pratt put it in 1931, 'music sounds the way emotions feel.' 'The forms of human feeling', wrote Langer,

> are much more congruent with musical forms than with the forms of language . . . There are certain aspects of the so-called 'inner life' – physical or mental – which have formal properties similar to those of music – patterns of motion and rest, of tension and release, of agreement and disagreement, preparation, fulfilment, excitation, sudden change, etc.

Some composers and music theorists seem to feel much the same way. Roger Sessions wrote in 1950 that music expresses 'the dynamics and the abstract qualities of emotion', while Michael Tippett calls music 'a significant image of the inner flow of life'. To the influential German theorist Heinrich Schenker, 'music mirrors the human soul in all its metamorphoses and moods'. The philosopher Ernst Cassirer argued that this is true of all art: 'What we feel in art is not a simple or single emotional quality. It is the dynamic process of life itself.'

No one has clearly identified a 'music-specific' emotion. But how would we know one when we saw it, and what would we call it? In many studies, the emotions are prescribed by the experimenters, who often ask listeners to tick off which emotions they experienced in a predetermined list. Without such an emotional menu, the responses can be so disparate and vague as to be beyond interpretation. We know what 'happy' and 'sad' mean, and can also identify more subtle emotional states such as 'serene', 'optimistic' or 'anxious'. But might we, in describing musical experience, merely grope for these words because there seems to be nothing better?

Another problem in studying emotion in music is that it is generally easiest to investigate the more extreme responses, which can create a focus on the atypical. While it is valuable to identify the kinds of musical events that create intense feeling such as chills, it's not clear that the experience of hearing music consists solely of such occasions more or less amplified. Besides, when music psychologist Alf Gabrielsson solicited the quotes with which this chapter begins, by asking people to describe 'the strongest, most intense experience of music that you have ever had', few were able to supply any direct link to specific *musical* events. Rather, they seemed to be more generalized accounts of intoxicating, overwhelming, blissful or elated experiences that happened during a music performance. It is important, indeed essential, to recognize that music is capable of producing such intense experiences – but the study shows just how hard it is to figure out *why*. In most of these cases, the triggers seemed to have more to do with the general environment, perhaps relating to the psychological feedback that exists in group activities or the way music may, in a rather unspecific manner, heighten emotions already present in the listener. Even in the one case cited where a particular musical stimulus could be identified – the utter dread and 'primitive horror' caused by a massive chord in Mahler's Tenth Symphony – it seems that this was just the trigger for the listener's latent feelings of despair at the death of a loved relative.

This raises a critical but often overlooked issue: music doesn't need much in the way of sophistication to arouse emotion. A soppy song or a surge of schmaltzy strings at the cinema can make us come over all weepy. Many of rock music's harmonic structures and rhythms can be considered crude by comparison to other music (not just classical), and yet, my goodness, they can hit the mark, stimulating all kinds of extreme passion. One could make a case that, as far as emotion is concerned, less is often more: a bluesman with a dobro on his knee can exude buckets of raw emotion that is absent from a bebop combo improvising on the same twelve-bar progression. An aesthete like Hanslick would no doubt look on this kind of emotional (some will say sentimental) response with disdain, and indeed that is perhaps why he was careful to explain that he was concerned not with emotion but with aesthetics.* Yet we should not fall into the trap of imagining that the emotiveness of music is somehow a measure of its quality. Some great music is rather emotionally barren, and plenty of 'low' music is emotionally rich. And as we will see, there are some quite mechanical 'tricks' for eliciting emotion that can be exploited quite shamelessly, for we will fall for them whether we like it or not.

There is another odd thing about the emotionality in music: it's not clear what the emotions are *about*. When we see a sad film or read a sad book, we might be moved from empathy and identification by the experiences of the characters. But if music can be said to be 'sad' (itself a simplistic description), in what does the sadness reside? It is hard to see that the sadness can be anywhere but *in the music itself*. We may or may not feel sad in response, but most people can recognize a quality of sadness in the music. Yet if it is in the music, from what does it arise? Emotions can usually be attributed to particular events: we feel sad when a friend dies, say. Even if – in fact, especially if – we can pinpoint the sadness in a piece of music to a particular phrase, it sounds a little absurd to put it that way: 'this descending F minor scale is sad'. How can notes be sad? It seems a little like saying that a tree or an ocean is 'sad' – not that we feel sad at the tree's plight, but that it contains sadness. What can this mean?

One seemingly obvious candidate for the location of the emotion – albeit one that we might generally reject – is the performer. If we say that a musician gave an emotional performance, we don't generally mean that we assume the musician experienced those emotions.

*It should be said this somewhat elitist attitude, which dominated musicological studies until rather recently, has tended to overlook a wealth of sophistication that can be present, outside of matters of pure form, in allegedly 'simple' music – as we'll see later.

Even in opera we would usually imagine that a 'moving performance' involved an excellent representation of emotion – perhaps the singer genuinely felt some of those things, but we don't know and probably don't care whether they did or not. This is a deeply complicated issue. When I watch my friend the British jazz singer Barb Jungr perform, I have no doubt that she is using her emotional as well as her technical resources. When Billie Holiday sings 'Strange Fruit', I don't think any of her audience could imagine, even if listening to a mere audio recording, that her emotions are anything less than fully engaged. But it is anyone's guess what the intensity of Glenn Gould's performances reveal, if anything, about his hermetic inner life, and this doesn't seem terribly important anyway. C. P. E. Bach asserted that a musician cannot move the listener if he is not himself moved – but it's not straightforward to interpret that statement given the emotionally codified context of the Baroque. In any event, today's classical musicians generally agree that some emotional detachment is needed in performance, and that it is not necessary to feel emotion (in this kind of music) in order to convey it.

A universal language of emotion?

Even if we don't necessarily feel the emotion that we imagine a piece of music seeks to express, most people will agree that music *does* sometimes move them. This need not be (as Hanslick implied) because they are determined it should do so, in some egotistical display of 'sensitivity'. It's frequently quite the opposite – who hasn't struggled to hold back embarrassed tears when the manipulative soundtrack swells in some corny movie? It's almost disturbing how susceptible we are to this emotional compulsion. 'Music operates on our emotional faculty with greater intensiveness and rapidity than the product of any other art,' wrote Hanslick. 'The other arts persuade us, but music takes us by surprise.' I had listened to *The Well-Tempered Clavier* many times before I read somewhere that the F♯ minor Prelude in Book I was particularly beautiful. So I sat down to play it – which of course means attending to it properly – and within a few bars, out of nowhere, there were tears in my eyes. How right neuroscientist Isabelle Peretz is when she says, 'My intuition is that musical emotions typically occur without consciousness or willingness.' *This* is the central mystery we need to explain.

Tears are an obvious sign of emotion, but people listening to music show other well-established physiological indications of genuine arousal too, such as sweating and changes in heartbeat rate. But these are not universal signals – some people can experience emotion without them. Neither are they diagnostic: different types of emotion may provoke the same physical symptoms. Carol Krumhansl has found that certain physiological responses could be reproducibly elicited in subjects listening to excerpts of classical music judged to reflect specific emotional states, and that different extracts drew recognizably different responses. But it wasn't clear that these responses could be reliably linked to those found for the corresponding emotional states induced by other means – that there is, say, a set of 'sad' bodily indicators that can be aroused and identified both for a mourning person and someone hearing Barber's *Adagio for Strings*. What this means is that in general we have to rely on what people report about their emotions. Even brain imaging might be of limited value in making emotional testing more objective. For what does it mean, philosophically and neurologically, to insist that someone is sad when 'sad' brain centres light up, unless they confirm that they truly *feel* sad?

With that proviso, we can begin to look at how music might create these feelings. I suggested above that it seems meaningless to say that a tree is sad. But of course some trees *are* described as being sad – everyone from Billie Holiday to Ella Fitzgerald and Nina Simone has sung Ann Ronell's great 1932 song 'Willow Weep for Me'. And willows 'weep' because they look like a weeping person, with a slouched, drooping physiognomy. It seems likely that some broad characteristics of musical emotion stem from a similar source: they mimic in music the physicality of that emotion in people, particularly in speech and gesture. When most listeners agree that music is sad, it is likely to be slow and soft – rather like the way sad people tend to move and speak. This doesn't mean that all slow, soft music is sad, but few people would find Japanese Kodo drumming or Handel's *Music for the Royal Fireworks* sad. Similarly, music commonly designated 'happy' tends to be relatively fast-paced and jaunty, and at a moderately loud volume. Such purely acoustic properties allow simple emotional messages to be easily discerned on the 'surface' of the music. Early opera tended to work in this way: the music was deemed to be emulating and thus heightening the emotions in the drama.

Plato seems to have regarded the link between musical modes and moods as a type of mimesis, saying that one Greek mode was fortifying because it resembles the cries and shouts of brave warriors.* The English biologist and social theorist Herbert Spencer proposed in the nineteenth century that there might be common emotional cues in speech and music. There is some evidence that the acoustic indicators of core emotions, such as happiness, sadness, anger and fear, are common to the speech patterns of many cultures, perhaps because (as Spencer speculated) of the physiological effects that these emotions have on the vocal apparatus. For example, angry speech seems rather universally to be fast and loud, sad speech to be slow and quiet, happy speech to be fast but of medium volume.

Music is not alone in possessing the capacity to mimic emotions in speech and gesture. When ancient people witnessed a storm, for example, they might regard it as a sign of divine anger. Of course, a storm can wreak damage as an angry person can, but the widespread connection with 'anger', as opposed to ebullience, say, also seems to imply that some human qualities (darkening, sudden outbursts, loudness?) were read into the meteorology. Again, these characteristics seem to apply across cultures, and there is good reason to think they are an innate result of our anthropomorphic instincts, our tendency to project human characteristics on to non-human things: John Ruskin's 'pathetic fallacy'.

In music, the mimetic features are rather simple. As well as having moderately fast tempo and moderately high volume, 'happy' music tends to use a high pitch register and relatively simple melodies (it stands to reason that we are hardly likely to feel uplifted if we are struggling to decode some complicated tune). Sadness retains simplicity but brings down both the tempo and the register. Anger is conveyed

* This reiterates the question of what the Greek modes actually were (see p. 50), since it isn't clear that a mere scale of small pitch steps will resemble anything. In one Socratic dialogue of his *Republic*, Plato has the protagonist Glaucon say that 'Of the harmonies I know nothing, but I want to have one warlike, to sound the note or accent which a brave man utters in the hour of danger and stern resolve.' Socrates replies that he therefore needs the Dorian mode. The implication seems to be that, as some have suggested, these modes had specific kinds of melody associated with them.

by complex rhythms and frequent changes in tempo or volume;* anguish is invoked by slow tempos coupled to melodic complexity, and, in Western music, plenty of chromaticism. While it remains hard to separate the role of convention, you can see how these qualities have a physiological flavour: an anguished vocalization, say, is typically drawn out and of awkward, ragged contour. String instruments seem particularly suited to imitating vocalized sighs of despair and heartache, while saxophones and trumpets suggest this with wailing glissandi. At the other end of the emotional spectrum, the muted trumpets of Ellington's big band might offer a braying, exuberant laugh, while Jimi Hendrix and imitators such as Steve Vai make their guitars seem literally to talk. Some acoustic cues of emotion seem to be preserved and recognized across cultures in both speech and music – in this respect, the brain might experience music as a kind of 'super-expressive voice'.

These qualities may offer a universal musical vocabulary of 'basic' emotional states. Patrik Juslin and his colleague Petri Laukka have found that, at this level, people are rather good judges of what quality is intended in the music of other cultures. Western listeners have been found able to distinguish whether pieces of music from unfamiliar cultures – Kyrghistani, Hindustani and Navajo – were meant to convey joy or sadness. And likewise, the Mafa people from a remote part of Cameroon, who have never been exposed to Western music, identify 'happy', 'sad' and 'scared/fearful' extracts of this music more accurately than can be explained by chance alone. Tempo seems to be the main clue here: happy or joyous music is faster.

One of the most detailed cross-cultural studies has been carried out by music psychologists Laura-Lee Balkwill and William Forde Thompson of York University in Canada. They asked north Indian musicians to play to listeners – fifteen Western men and women, and four experts on Hindustani music – extracts of ragas that were intended to convey joy (*hasya*), sadness (*karuna*), anger (*raudra*) and peacefulness (*shanta*). (Each raga has an associated mood or *rasa* which is believed to act on a specific energy centre or *chakra* of the body to

*As an illustration of how crude these generalizations can be, consider the Sex Pistols' 'Anarchy in the UK': melodically and rhythmically simple, but you'll probably agree that anger is a better descriptor here than joy.

instil one of the nine moods recognized by Hindustani music theory.) The listeners had to associate each piece with one of these four emotions. In addition, Balkwill and Thompson wanted to identify what musical parameters the performers were using to convey emotion. So they asked the listeners to rate, in each piece, five candidate vehicles of emotion: tempo, melodic complexity, rhythmic complexity, pitch range and timbre.

The subjects were pretty good at identifying the 'sad' and 'joyful' ragas, but were less able to spot anger and especially peacefulness (curiously, given that north Indian people regard music as inherently peaceful by default). What's more, the 'naïve' listeners made similar ratings to the experts, although the latter made rather more clear-cut distinctions between moods. Again, the key clue for both listening groups seemed to be the tempo, as well as melodic complexity, and timbre in the case of anger.

We must be cautious about what was truly being probed in these experiments, however. In the West, emotion tends to be understood today as 'what one feels'. This isn't the universal view, and indeed wasn't held even in the West before around the early nineteenth century. When Indian musicians talk about the *rasa* of a raga, they don't mean whether it makes the listener feel a particular way. The word *rasa* refers to an impersonal emotion, something outside the self: an emotional state viewed, in one apt description, as though through a plate-glass window, which keeps it apart from the viewer and thereby removes any unpleasant sensations. Personal emotion, in contrast, is referred to by the word *bhava*: a kind of distillation of disembodied emotion, which can transform consciousness and give access to a spiritual state of being. (Medieval sacred music had a similar purpose.) In the same way, while types of sadness may be expressed in Javanese music, Javanese culture does not regard such mental states as having the 'negative' valence they are attributed in the West, but rather, as having a neutral quality. The Western listeners in the Canadian experiment may have sensed the correct emotional flavour of the ragas, but not the manner in which that flavour was to be tasted.

Emotions may not only be viewed in a different framework in different cultures, but also be aroused and expressed according to different social codes. I can appreciate why the qawwali music of

Sufism might have transcendental effects, but I suspect I am unlikely to feel them, not just because I lack fluency in the musical language but because I am outside of the traditions within which the ecstatic effects take place. The ecstasy is not a spontaneous response to the music but one that is created in a highly prescribed and ritualistic way, involving a gradual intensification of signal and response between the listeners and the musicians: as the former give signs of being moved, the musicians identify and intensify the corresponding stimuli. Here and in other cultures – the religious intoxication of Pentecostal churches, say – 'emotion' is not a personal matter but a communal event that unfolds in a somewhat formalized way, and which may have a predominantly social function.

Mimetic qualities often account only for relatively superficial aspects of our emotional response to music. They set the mood, but offer little nuance. And they can be overridden: sad music can be animated, for example (listen to Irish folk music), and while most people would assume that a buzz-saw noise is intrinsically unpleasant, Glenn Branca has used a similar timbre to create some of the most jubilant music I have heard. Moreover, we seem able to recognize only a very small number of basic emotions this way – what, for example, would we expect in the 'physiognomy' of music that sounded 'hopeful' or 'resentful'? Here, at least, Hanslick's lofty pronouncements have some validity: 'That which for the unguarded feelings of so many lovers of music forges the fetters which they are so fond of clanking, are the *primitive* elements of music – *sound* and *motion*.' If that's all there is to it, he sniffs, then there is no such thing as musical appreciation or judgements of quality, but only crude acoustics. 'If to a musician who considers the supreme aim of music to be the excitation of feelings we present several pieces, say of a gay and sprightly character, they will all impress him alike.' If you want to feel happy, any old bit of jaunty Mozart, or almost any early Beatles song, will do. One might as well get one's pleasure or relaxation from a good cigar or a warm bath, Hanslick scoffs. He says that the much-quoted assessment of music by the duke in *Twelfth Night* betrays him as a person of this sort: 'If music be the food of love, play on, give me excess of it.' It doesn't much matter what music it is, the duke implies – I just want a nice big slice, like a piece of cheesecake, to make me feel good.

There's no denying that many people *do* use music like this. I suspect, in fact, that at times we all do, even the most refined of aesthetes: we don't think 'Oh, I feel like *appreciating* the Temptations', but rather, 'I want a bit of uplifting music – ah, here's "My Girl"'. John Sloboda and his co-workers find that people often use music as a mood modifier in their daily lives: a 'means of enhancing everyday activities', of basking in good memories (or wallowing in sad ones), of winding down or winding up. Hanslick (and others today) might respond that these people aren't truly listening to the music at all, but are merely indulging superficial characteristics devoid of aesthetic worth. That may or may not be true – we shouldn't underestimate the factors that, say, allow a person to select precisely the right music for the target mood. But in any case, the condemnation of experts in 'music appreciation' risks merely alienating: as Juslin and Sloboda put it, 'many people feel they are being told "the thing that strikes you most immediately and powerfully about the music you hear, the thing which keeps you engaged with it and wanting to hear more of it, is really totally irrelevant to a proper appreciation and understanding of it."' One can hardly pretend to be interested in the emotional effects of music while dismissing most of the ways and situations in which people are aroused by it. It might be disheartening that great art is sometimes treated as a kind of self-therapy, but there seems little point in pretending otherwise.

What's more, the rather rough-and-ready, basically acoustic mechanisms by which music can stimulate the listener are germane to the ways it fulfils its function in many cultures. And even the great Western composers themselves have not always been terribly sophisticated about emotion. During the Baroque era it was thought that music should achieve stylistic unity by aiming to represent only a single emotional quality throughout. This was considered to be an objective, rationalistic property of the music: it was expected to sound 'gay' or 'solemn' to all listeners, and was not in any sense an expression of the composer's own feelings. This attitude encouraged the use of conventional designations rather than spontaneous forms of self-expression: the composer assumed that the audience knew how to 'read' the signs, rather than that the music would awaken such feelings of its own accord. In this regard, 'emotion' becomes more a matter of musical *meaning* than a question of cognition, and as such is explored in Chapter 13.

All in the music?

Most people would call Louis Armstrong's 'What a Wonderful World' a happy song – a little wistful, maybe, sentimental undoubtedly, but who can doubt that the words express joy and delight at life itself? Yet I can't hear it without tears welling up, because it was played at a friend's funeral.

According to one school of thought, all emotion invoked by music has something of this character: the feeling is not in the music, but in what it reminds us of. Psychologist John Booth Davies has called this the 'Darling, they're playing our tune' theory. One might call it rather more loftily the Proust effect, for Proust found that music (and not just the taste of madeleines) could stir vivid recollections:

> Taking advantage of the fact that I still was alone, and drawing the curtains together so that the sun should not prevent me from reading the notes, I sat down at the piano, turned over the pages of Vinteuil's sonata which happened to be lying there, and began to play . . . I was carried back upon the tide of sound to the days at Combray – I do not mean at Montjouvain and along the Méséglise way, but to walks along the Guermantes way – when I had myself longed to become an artist.

The power of music to evoke association is illustrated in the movie *Casablanca*, where Rick forbids his bar pianist Sam ever to play 'As Time Goes By' because the feelings of sadness and loss that it awakens in him are unbearable.

Beyond question, this sort of thing happens all the time, if not necessarily with the acuteness that Bogart's Rick feels. Equally, no one can seriously suggest that this is all there is to it, or we'd never agree on the broad emotional tone of a piece of music (and we'd expect most pieces to have none at all).

This 'emotion by association' can also encompass responses that don't refer to any non-musical events or characteristics but which have become ingrained by acculturation. Song lyrics and music for film and television send children a clear message about what 'happy' and 'sad' music sound like, which they have thoroughly assimilated by the age of ten if not considerably sooner. These associations may not all be arbitrary – as we've seen, they might draw on 'physiognomic' parallels – but in any event they come to serve as simple labels of musical intent. It is even possible that music's emotive power is itself in part merely a tacit cultural consensus: as John Sloboda and Patrik Juslin put it, 'One answer to the question "why is music so emotionally powerful" is that we have decided to construe it thus. At one very important level of analysis, music is not inherently emotional. It is the way we hear it that makes it so.'

Yet learnt correlations alone don't seem to account for the immense emotional nuance possible even in relatively 'unsophisticated' music. Moreover, we can wholly misunderstand the music of other cultures, missing all the important allusions and structures within it, while still deriving from it intense and emotionally charged enjoyment.

In contrast to the 'emotion by association' or so-called referentialist position, the absolutist school insists that whatever music conveys is intrinsic to the music itself and not dependent on what lies 'beyond' it. Hanslick is often regarded as one of the originators of this view, although that is somewhat to miss his point, for as we've seen, his focus was not on emotion but on aesthetics and value – what matters, he said, is not whether music is 'gay or mournful', but whether it is good or bad. He saw it as wrong to suppose that the purpose of music is to arouse or represent everyday emotions, and although he didn't deny that it does so, he considered this a mere side effect, of no real artistic or aesthetic relevance. That formalist position was echoed by the English psychologist Edmund Gurney in his 1880 book *The Power of Sound*.

Formalists insist that musical affect is primarily *intellectual*: one appreciates the formal relationships between notes, harmonies, rhythms and so on, just as the knowing observer might derive satisfaction from a well-played chess game. In this view, any emotional effects are coincidental: desirable, maybe, but largely unintentional. Stravinsky thought about music this way, and the sparse clarity of his compositions might be seen to reflect that. But of course Stravinsky's music also offers abundant reason to reject a purely formalist approach: one can't imagine for a moment that the excitement we feel at the abandoned stomp of *The Rite of Spring*, the tense unease of the 'Petrushka' chord or the triumphant joy of the *Firebird*'s closing march are an unintended product of a formal exploration of harmonies and rhythms.

There are more -isms within -isms in this territory. In contrast to formalists, an 'expressionistic' absolutist position would say that music *does* convey genuine emotion, but that it may do so without reference to non-musical factors or learnt responses. Something in the pitch, rhythm and combination of notes has an intrinsic capacity to move us.

That would be remarkable. How can a series of sounds at different acoustic frequencies, with no obvious resemblance to human cries or sighs or so forth, and which we cannot easily connect with similar patterns we have heard before – how can such a thing be transformed into emotion? Yet the dominant theory of musical affect for the past half-century assures us that it can be, and I suspect it is correct to do so.

The truth is that one can't carve up musical emotion as neatly as all these labels and schools would have us believe. Like most -isms, they offer useful polarities rather than descriptions of the world. To imagine that we can hear music without any reference to what we've heard before – without any degree of referentialism – is absurd. Moreover, the formalist position seems to insist that music cannot be appreciated without being 'understood' in theoretical terms, which is self-evidently false. And it now seems that anything resembling expressionistic absolutism relies mostly on learnt patterns rather than innate tendencies, although this learning process is unconscious and more or less independent of formal training. That places limits on how universal an 'emotional language' music can be. When students and teachers at a school in Liberia were played Western classical music in the 1930s, they didn't recognize any emotion in it. Villagers were even more direct in their response: they became restless, and some simply got up and left.

Nonetheless, we've seen that mimetic devices can offer a rather coarse and somewhat cross-cultural lexicon of musical emotion. The more subtle signals are coded in scales and modes, rhythms, harmonies and the other building blocks we've encountered already. To what extent are these just arbitrary conventions? Many people in the West are convinced that there must be something inherently cheerful in the major scale and inherently sad in the minor. Peter Kivy suggests that the 'pulling down' of the major third to the minor creates a kind of sinking, plaintive effect (a more extreme view held that this stimulates a Freudian castration complex). Kivy argues that when *both* intervals of the major triad sag by a semitone, producing a diminished chord, the result is even more anguished.

But there's no reason to believe that people will inevitably experience the minor and diminished chords as sad and anguished, if they have not learnt those associations. I should be rather surprised if it were otherwise. For one thing, how was this (admittedly crude) aspect of Western music's emotional tenor to be gauged before these modalities were clearly established during the early Baroque? And let's not forget that just about all emotional expression at that time was codified in a standard vocabulary of musical figures and devices: a series of signs that were no more universal than the words used to denote the respective emotions.

This failure to distinguish between what is learnt by convention and what is innate pervades musicologist Deryck Cooke's dogmatic arguments for why major and minor modes are intrinsically happy and sad. Cooke considered it sufficient to pile up examples from Western music in support of that contention, as well as of a host of other assertions about the emotional 'meanings' of musical figures. Look, he said: sad lyrics are almost always set to a minor melody. And look: the ascending major triad **1-3-5**, or the full scale figure **1-2-3-4-5**, is regularly used 'to express an outgoing, active, assertive emotion of joy'. This may be true, and recognizing it can help us see why (mostly classical) composers have made some of their melodic choices. But it says precisely nothing about why music evokes emotional states beyond the fact that Western composers have decided on a convention. And even then, well-informed listeners merely acknowledge the link: 'Ah, Mozart wanted to sound a happy note here.' They don't necessarily feel that way themselves.

Cooke's 1959 book on the alleged universality of musical emotion, *The Language of Music*, was influential and widely praised at the time. But his arguments sometimes become contorted to the point of (prejudiced) bathos. He tells us that the medieval Church tried to exclude the major third from sacred music not just because it was associated with secular music but because the happiness it evoked conflicted with the pious attitude of life as a 'vale of tears'. But don't some cultural traditions, even in the West, use a minor third in music that is far from sad and sombre – Slavic and Spanish folk music, for instance? Ah, says Cooke, but this is because those rustic, unsophisticated cultures have never assimilated the humanistic belief in a right to happiness – they remain 'inured to a hard life'. That's the kind of arrogance with which Western musicians and musicologists have long sought to assert their own principles as the 'natural' ones against which other cultures must be measured.

The fact is that there is no fundamental reason to suppose that minor-key music is intrinsically sad. Christian sacred music in the Middle Ages commonly used modes that now sound to us like minor keys, but it wasn't seeking a 'sad' ambience: it just so happened that those were the notes the mode contained. Minor-key compositions with relatively upbeat connotations aren't rare in the Western canon: Bach's Double Concerto for Violin, say, or the Badinerie in his Second Suite for Orchestra, where he even made his jocose intentions explicit (*badiner* is French for 'to jest'). There are plenty of examples in popular music too, such as Van Morrison's 'Moondance'. And while some studies have claimed to find the major-happy/minor-sad association in children as young as three, apparently it doesn't get firmly established until the age of seven or eight, despite all the conditioning children must receive.

Similarly, we should be wary of suggestions by Cooke and others that there is an intrinsic emotional context to the rising and falling of pitch: upwards to heights of despair, joy or heavenly bliss, downwards to calm, stability and mundanity, or further still to darkness and the inferno. It is easy to find strictly symbolic examples in Western music: the ascending **1-2-3-4-5** of Beethoven's Gloria, for example, or the **1-3-5** cries of 'Triumph' in the chorus' greeting to Tamina and Pamina from Mozart's *The Magic Flute*. But there is no reason to suppose that high notes automatically connote heightened emotion, or that rising melodies create rising spirits. We've already seen that 'high' meant something different in music to the ancient Greeks.

The dominance of conditioning over any innate emotional signposts in music is apparent in some cross-cultural studies. In one, Europeans and South East Asians were played a selection of extracts from Western classical, Indian classical and New Age music and asked to choose adjectives to describe the moods represented. Their responses suggested that cultural tradition exerted far stronger an influence than did any inherent qualities of the music.

These learnt associations can sometimes be positively misleading.* Westerners are apt to hear a mournful quality in some Javanese music that in fact expresses great happiness, simply because it uses a scale (*pélog*) and mode (called *barang*) that has an interval rather close to the diatonic minor third. Sadness in this music tends instead to be conveyed in pieces that use the *sléndro* scale, when some of the instrumentalists add 'foreign' notes outside the scale to create clashes of pitches that produce an anguished quality. All this is hard for Westerners to intuit. Nonetheless, some emotional qualities do seem to leap the cultural divide, since higher pitches tend to connote happiness in Java just as they do in the West, while slower tempos are used to convey sadness or calm.

Not what you expected

We have already seen many examples of how we develop expectations about what music will do, and how composers and musicians manipulate those to create lively and stimulating patterns. But this is not simply a matter of grabbing your attention or catching you out. Expectation and the violation or postponement of its realization has become the central pillar of modern cognitive studies of emotion in music – indeed, almost to the extent that it can seem taken for granted that this is the sole affective mechanism at play. With anticipation comes tension – have we guessed right or not? – and that in turn carries an emotional charge.

Paul Hindemith believed that expectation was central to the aesthetic effect of music, but only insofar as we derive pleasure from having our expectations met: 'the more closely the external musical impression approaches a perfect coincidence with [one's] mental expectation

* An extreme example of the sort of misreading that is possible in cross-cultural musical experience is given by music critic David Hajdu: 'In 1971 Ravi Shankar, the Indian virtuoso, performed at New York's Madison Square Garden. After hearing a few minutes of Shankar's ensemble, the audience of some 20,000 roared in approval. "Thank you", Shankar replied. "If you appreciate the tuning so much, I hope you will enjoy the playing more."'

of the composition, the greater will be [one's] aesthetic satisfaction.' Thus, he said, appreciating music entails 'the essential possibility of foreseeing and anticipating the musical structure as it arises in the performance, or at least, if the composition is utterly new to the recipient, his being given a chance to conjecture with a high degree of probability its presumable course'. If a listener can't do this, says Hindemith, 'music goes astray, disappears in chaos'. But he was vague about how those expectations were formed or what they corresponded to, suggesting only that they become more precise the greater one's extent of musical knowledge. This reliance on having well-established templates from past experience made Hindemith conservative about what music can contain (though you'd not think so to listen to him): 'the building material cannot be removed very far away from certain structural, harmonic-tonal, and melodic prototypes'. As we've seen, this seems seriously to underestimate even naïve listeners' ability to form new prototypes on the spot in the face of the unfamiliar.

But Hindemith's prescription is an odd one. His assumption that only perfect coincidence between expectations (however those might arise) and what we hear will achieve the ideal aesthetic experience seems to ensure that music will forever fall short of its potential, since who could ever guess, except for the most trivial of nursery rhymes, exactly how a piece of music will go? It appears to guarantee not that we will be transported by music, but that we'll always be left mildly frustrated and disappointed by it.

Just six or seven years after Hindemith made these comments, the American music theorist and philosopher Leonard Meyer presented a much more plausible account of how expectation produces aesthetic affect in music. The title of Meyer's 1956 book, *Emotion and Meaning in Music*, was itself somewhat revolutionary. Until that time, few people had dared to broach the subject of music's emotional effect, or even to link the two words at all. It was perfectly acceptable to discuss the aesthetics of, let's say, Bach's elegant formal structures, Palestrina's pure harmony or Beethoven's mastery of symphonic form. But to confess that one might be moved by any of this was to invite accusations of vulgarity and dilettantism. Eduard Hanslick had poured scorn on what he saw as the morass of cheap emotionalism in the responses of unschooled audiences, and the vogue in classical music in the 1950s matched the austere zeitgeist, whether with Stravinsky's spare neoclassicism or the experiments in serialism conducted by the followers of Schoenberg. Even today, the conventions of classical music would have us suppress all outward signs of emotion: one must sit in silence at a performance, listening respectfully and with as little body movement as possible. Emotions are to be observed only in their proper place: at the end of the event (where they are supposed to be celebratory), or in rote, prescribed and theatrical fashion on the last night of the Proms. Audiences at classical concerts routinely say that they experience deep emotion during the performance, and yet they display almost none.

Jazz too had become decidedly 'cool' in the 1950s, with the virtuosic displays of bebop supplanting the jaunty swagger of swing. Of course, mainstream popular Western music was steeped in emotional gesture, from Elvis' anguished 'Heartbreak Hotel' to the schmaltz of Perry Como and Pat Boone – but no serious musicologist was going to concern himself with that!

Moreover, methods of analysing music became steadily more technical during the twentieth century. While one couldn't overlook its emotive effects, there was a sense that, either these weren't terribly important, as Hanslick had argued, or that one couldn't even begin to explore them until there was a solid framework for understanding the formal structure of music. As this structure seemed to get more complex the more one delved, it began to seem as though the issue of emotion would have to be shelved indefinitely.

Meyer disagreed with that, and he opened a door that has since been pushed ever wider. A composer as well as a philosopher, Meyer studied with Aaron Copland and corresponded with Schoenberg. He offered a framework for talking about emotion in music that no longer seemed contingent on mere subjectivity and supposition. Fear of the latter also accounted for prior resistance to the question: with science in the ascendant and all academic disciplines seeming increasingly to aspire to the envied precision of physics, who was going to risk looking like a bungling, shoddy amateur by trying to speak about something as ill-defined as emotion? Hindemith betrayed that sense of inferiority: 'To the scientist our method – or, in his eyes, nonmethod – of looking at everything without ever fundamentally comprehending it must seem utterly amateurish.'

What Meyer realized is that the emotional effect of music doesn't come from having our expectations met, but from having them more or less thwarted. We don't simply want to feel happy or satisfied by music because it has gone the way we expected; indeed, that would be more likely to leave us only mildly affected by it, if not in fact positively bored. No, we like to listen to music because it sounds exciting, vigorous, poignant, beautiful, ennobled, sexy, and too many other things to list. And it achieves these things, Meyer argued, not in spite of but because of the mismatch between our expectations and the reality.

Meyer drew on the work of philosophers such as the American John Dewey, who supposed that emotions are born of conflict and frustration. In this view, 'nervous energy' is excited when a stimulus provokes a desire for action or resolution. If that resolution is inhibited by some obstacle or conflicting tendency, we experience emotion.

This sounds rather abstract and mechanical, but the point is not hard to grasp. A man reaches into his pocket for a cigarette, says Meyer (this is 1956, remember). But he has none. Then he discovers there are none in the house, and sees that it is late and the shops are closed. And this steady frustration of his desire creates first restlessness, then irritation and finally anger. But (Meyer could have added to complete the picture) just then his friend knocks on the door and turns out to have a packet of cigarettes in his coat. Anger and anxiety give rise to joy and relief. Meyer framed this image in rather grandiose terms, saying that 'In music the state of suspense involves an awareness of the powerlessness of man in the face of the unknown.'

This is precisely how he imagined a musical cadence working. Cadences – sequences of chords that end a music phrase – may bring about a feeling of partial or total closure by carrying the harmony on to a relatively stable chord. Western listeners have learnt to expect, even to long for, the final cadence of a piece of music to move to the tonic. The tonic here is that longed-for cigarette (let's be a little more contemporary and call it a glass of cool white wine, if that too is not now a forbidden pleasure). It's only when there is some frustration of that anticipated satisfaction that the emotions are truly engaged. If the man has only to remove a bottle from the fridge, there is no real tension and so no emotional pay-off when it is released. How much more he savours it when he was unsure that he'd get it – when the fridge was empty and the pubs closed. Similarly, in Meyer's view an authentic cadence tossed off at the end of a phrase is unexceptional, even mildly dissatisfying in its predictability. But when it is delayed – perhaps by first moving through a 'deceptive' cadence such as V-IV (see p. 183), or by sustaining the harmony on the V chord – we experience tension, because we sense what is coming but don't know when it will come. When it does, our satisfaction is greater. Even the slightest deviation from the most predictable outcome – a slowing down or rallentando of the kind that musicians commonly apply, and composers often specify, at the final cadence – is enough to create an emotional effect by toying with our expectations, creating hints of uncertainty about the outcome.

Why uncertainty is good

According to Meyer, music is full of moments like this in which we are given clues about what will come next but can't be sure that we've guessed right. Composers manipulate those expectations with more or less calculated gestures, pulling on the strings of our emotions. There are many ways of doing that, and we'll see some of the popular ones shortly.

But Meyer didn't really explain why we should *care* if our expectations were violated. He argued that we like to feel in control and not powerless to anticipate the future: 'emotions are felt to be pleasant, even exhilarating, when associated with confidence in the power to envisage and choose; they are experienced as unpleasant when intense uncertainty precludes the sense of control that comes from the ability to envisage with confidence'. And it's true that if we want a glass of wine and don't know if we'll get it, our feelings get awakened. But why should it make any difference to us if the final chord of a piano sonata comes on the regular beat or two seconds later? Why should we be bothered whether the cadence resolves itself at all? Even if we expect it to do so, we surely don't imagine that anything terrible is going to befall us if it doesn't.

But David Huron argues that a justification for Meyer's model can be found in the instincts implanted by evolutionary biology. The ability to develop expectations, he says, is a fundamental principle of survival, which can be seen to operate not only in humans but in other animals and even amoebae. Our prospects are improved if we can anticipate accurately what our environment is going to throw at us. If we hear a fearsome roar, we expect it to be produced by a large and possibly dangerous animal, not by an insect.* Early humans learnt to spot clues about the forthcoming weather, which could be crucial for hunting and agriculture. And as sportspeople demonstrate, we are remarkably good at predicting the trajectories of moving objects, whether to intercept or to avoid them.

This implies that we will have acquired mechanisms to reward good predictions and punish wrong ones. Just as the evolutionary benefits of sex, as a means of propagating genes, are supported by the fact that it is pleasurable, so the adaptive value of a skill at making predictions means that success stimulates good feelings and failures make us feel bad. The reward for guessing right ensures that we'll attempt it.

This idea receives support from brain-imaging studies conducted by Anne Blood and Robert Zatorre of McGill University in Montreal. They used a technique called positron emission tomography (better known as PET scanning) to monitor changes in blood flow in the brains of people listening to music that they had each chosen for its ability to induce 'intensely pleasant emotional responses', including chills. Like functional MRI, this method reveals which parts of the brain become active. Blood and Zatorre found that the areas that 'lit up' were ones known to be involved in emotion and, specifically, in euphoric 'reward' responses – these have been seen to become activated also by sex, good food and addictive drugs. In other words, when music makes us feel good, it may do so by arousing the neural circuitry involved in adaptive responses to stimuli such as sex and food. As dance DJ Lee Haslam claimed with his 2002 hit, music really is the drug.

But such evolved reward systems are rather indiscriminate. It is far cheaper in terms of mental resources to evolve a generalized

*Insects don't roar, of course, but they can make loud noises – think of cicadas. Yet we and other animals have learnt how to make predictions about the likely body size of an animal from the noise it makes, based on the range of acoustic frequencies the sound contains. This is an explicit example of expectations formed from auditory clues.

prediction-reward link in our minds – so that every right guess delivers a dose of 'feel-good' neurotransmitters, say – than to create a system that kicks in only when it really matters, for life-threatening situations.* After all, we won't always know until it is potentially too late what is life-threatening and what isn't. So it doesn't matter that there is really nothing at stake if our musical expectations are violated; our universal reward/punishment system will be activated regardless. Constantly erring on the side of caution, it responds even to the most unthreatening provocation. And that's our gain: as Huron puts it, 'Nature's tendency to overreact provides a golden opportunity for musicians. Composers can fashion passages that manage to provoke remarkably strong emotions using the most innocuous stimuli imaginable.'

Huron discriminates between instinctive and conscious responses. An unanticipated event may trigger a negative reaction in the moment, but this takes on a positive complexion when, on reflection, we realize it is a pleasant surprise. The 'gut response' happens very fast and travels along primitive neural pathways, from the thalamus (the relay unit for sensory information) directly to the amygdala, the principal seat of the emotions. The appraisal response comes later, and travels via the 'reasoning' sensory cortex, which can moderate the response of the amygdala – in effect, it sends out the message 'Relax – it's good news'.

You might think that if, by jumping to conclusions in music (or in life), we are constantly getting things wrong, we might eventually learn that it's better to wait and let events unfold rather than try to anticipate them. But evolution couldn't afford that luxury. Our ancestors needed to act fast on the basis of minimal data. Rather run a hundred times from the footfall of a deer than take the single risk that it could be a lion.

This could again seem to imply that listening to music ought to be an unpleasant experience, since we are bound always to be getting things wrong. But it's more complicated than that, Huron argues. For

* As Huron points out, one striking aspect of this process is that there is a reward for correctly predicting even an unwelcome outcome. This, perhaps, is why we sometimes encourage ourselves to expect the worst – if it happens, at least we have the slight compensation of having anticipated it.

one thing, we do get an awful lot right. When we listen to Vivaldi or Haydn, our expectation is that tonality will be strongly respected: the music won't be constantly veering away from the diatonic scales. Moreover, we can rightly expect that the relative occurrence frequencies of notes will follow more or less the tonal hierarchy discussed in Chapter 4, and likewise we have a reliable mental model for anticipating the frequencies of different melodic interval steps. In fact, this is to put things somewhat the wrong way around. Because we are instinctive prediction-makers, constantly seeking schemes to anticipate our environment, we assemble these mental probability distributions precisely so that we can make good predictions. So by definition they are bound to be right, as long as the music we hear is sufficiently similar to that we've heard before. As a result, we can by nature be expected to be somewhat conservative in our tastes, because we like to anticipate correctly. Since there aren't any common atonal nursery rhymes or mainstream pop songs, most people in the West develop strong tonal expectations, and so it stands to reason that music from the late nineteenth century onwards that challenges or discards tonality is often perceived as unpleasant to the ear: we are paying the price of our lost predictive power. This doesn't mean that such music is doomed to remain 'difficult', however, because our 'expectation templates' are constantly updated by experience. Most enthusiasts of Western contemporary classical music have traced a gradual path, say from Grieg to Shostakovich to Stravinsky to Berio, along which their evolving expectations are never challenged so much as to provoke outright revolt.*

So yes, we get a lot right – but we must nonetheless still often anticipate wrongly. This, however, is not necessarily a problem. In fact, it can be turned to advantage. Huron argues that the pay-off for correct anticipation is greater when it is less certain, just as the man who craved cigarettes is happier when he gets them after spending anxious moments unsure that he'll do so, rather than if he simply reached into his pocket and pulled out a packet. This, as I implied above, is

* Some people claim to be baffled by Bartók. But when I first heard his music in my late teens, I felt immediately at home. I suspect that his use of fast, stridently articulated modal ostinati (repeated melodic patterns) had a lot in common with the patterns I had become familiar with in rock music.

the crux of Meyer's theory. Delaying a cadence, or creating uncertainty about any other learnt musical pattern, boosts the eventual reward (assuming that it comes at all). And the hiatus sharpens our attention, obliging us to concentrate and search for clues about what the music will do, rather than riding on a wave of tepid, blithe predictability. The music, in short, becomes *more interesting* when it is *less predictable*. And when the final tonic chord arrives, what would have been a mildly pleasant sense of completion becomes a flood of delight, even awe.

A recognition of the emotive tension that a cadence can offer seems to have emerged around the late fifteenth century. Before this, cadences simply appeared in the required places as a formulaic way of marking musical boundaries. But the Renaissance composer Josquin des Prez exhibited a growing awareness of the pregnant affective potential of the cadence. He signalled their approach with protracted passages on the dominant (V) chord, suggesting to the listener that a resolving tonic is on the way and so heightening their expectation. But they might be kept waiting for some time. And even when the tonic arrived, Josquin tended to draw it out and thus delay the real conclusion, letting the melodic voices oscillate to and fro before finally coming to rest. His contemporaries clearly sensed what he was up to, and some of them accused him of failing to 'properly curb the violent impulses of his imagination', as though there was something indecorous in this orchestration of emotion.

During the nineteenth century, composers became ever bolder in toying with these expectations around cadences. Compositions of this time often build towards an anticipated climactic cadence with a ratcheting of volume and a slowing of pace, only to break off into a brief, airy gesture before the big finish. You can hear this in the first movement of Grieg's Piano Concerto, for example, and when such a moment occurs at the end of the first movement of Bartók's Second Piano Concerto, the reining in of our expectation is dizzying, the music seeming suddenly to soar into the stratosphere before crashing back to earth. Manuel de Falla pulled off a different trick in his *Fire Dance*, ending with the tonic chord thumped out repeatedly in a shifting rhythm like a jumping gramophone record until you wonder if the piece will ever truly end. Wagner was a master of cadential delay tactics, making music that seemed to take forever to reach a tantalisingly promised cadence. Sometimes, as in the Prelude to Act Three of *Parsifal*, it never did.

Meyer suggested that experience in listening to music provides us with 'schemas' against which to evaluate a piece of music as it unfolds. 'The delight of intelligent mental play and the excitement of its complement, affective experience,' he wrote, 'are significantly dependent on the deviation of a particular musical event from the archetype or schema of which it is an instance.' The truly great composers know precisely how to create and control such deviations. While some sprinkle their scores with exquisite little detours and puzzles, others will risk an implication that the music is about to fall apart into chaos, only to pull back from the brink of destruction and open up vistas of beauty, grandeur and awe. The French writer François Raguenet saw this contrast in the French and Italian styles of composition in the early eighteenth century. While the French will merely 'flatter, tickle, and court the ear', the 'more hardy' Italian musician is bolder – at times, his audiences 'will immediately conclude that the whole concert is degenerating into a dreadful dissonance; and betraying 'em by that means into a concern for the music, which seems to be on the brink of ruin, he immediately reconciles 'em by such regular cadences that everyone is surprised to see harmony rising again, in a manner, out of discord itself and owing its greatest beauties to those irregularities which seemed to threaten it with destruction'. Had he been around in the 1950s, one imagines he might have said much the same about John Coltrane.

How to keep them guessing

In looking for ways to toy with our expectations, composers and musicians are spoilt for choice. Just about every facet of music that we have considered so far is open to exploitation in this way. Here are some of them.

Rhythm. 'Certain purposeful violations of the beat are often exceptionally beautiful,' wrote C. P. E. Bach. We saw in Chapter 7 how eagerly we try to find regularities in the rhythms of music – and so we are highly sensitive to their disruption.

One of the oldest tricks for doing that is syncopation, which involves

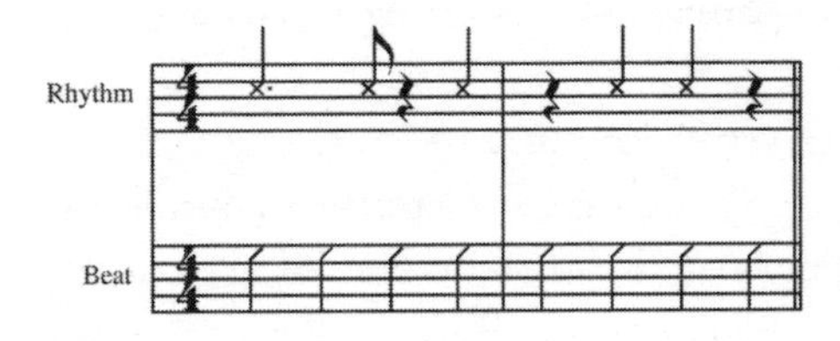

Figure 10.1 The syncopated 'Bo Diddley' rhythm. The symbols here denote unpitched percussion.

shifting an emphasis off the beat. Normally it is displaced to just before a strong beat. One of the classic examples from popular music is the 'Bo Diddley' rhythm (Figure 10.1), named after the rhythm & blues guitarist who made it famous. This pattern is used in songs ranging from 'Not Fade Away' by the Rolling Stones to 'Magic Bus' by the Who. But although syncopation is most often associated with the funky grooves of jazz (especially swing) and R&B, it was widely used by classical musicians since at least the eighteenth century. The uncluttered octaves at the start of Mozart's Symphony No. 25 in G minor, K183 (familiar from the dramatic start of the movie *Amadeus*) are made urgent by syncopation (Figure 10.2*a*), while the 'Ode to Joy' in Beethoven's Ninth contains perhaps the best known syncopated note in the classical repertoire (Figure 10.2*b*). The emotional effect of this rhythmic hiccup, with its early entry of the theme, is very clear: many

Figure 10.2 Syncopation in Mozart's Symphony No. 25 in G minor, K183 (*a*) and in the 'Ode to Joy' from Beethoven's Ninth Symphony (*b*).

people say it induces a thrill. (There must be something wrong with me – I just find it irritating.)

It's quite easy to show that syncopation confounds our expectations of beat: people asked to tap a beat over syncopated patterns commonly end up resetting the beat so that its pulse coincides with the syncopated notes. In other words we shift the beat to eliminate the syncopation: we will construe a rhythm as unsyncopated if we can. That might seem at odds with the widespread use of syncopation, seeming to imply that we can simply perform this mental frame shift to get rid of it. But in general syncopation occurs against a backdrop of a clearly defined beat, which serves to prevent such resetting. Indeed, we need to ensure that this *doesn't* happen if the syncopation is going to provide a vehicle for tension and emotion.

Another classic trick is to disguise one rhythmic structure as another – something that we encountered in Chapter 7. Beethoven does this in his Piano Sonata No. 10, Op. 14 No. 2, in which the third movement seems to begin in a duple time – pa-pa-*pah* pa-pa-*pah* – but is revealed only in the fourth bar to be in triple (3/8) time (Figure 10.3). We're left with a moment's confusion, which is quickly followed by pleasure at seeing that we have been tricked but have wised up to the fact. While most composers before the nineteenth century would only maintain this deception or confusion for a short time before confessing to it, later it became acceptable to sustain the ambiguity throughout. Alexander Scriabin, for example, often used melodic or rhythmic figures that violated the metre throughout the piece (Figure 10.4). If the rhythm never settles down, won't we just be deterred from listening? Not necessarily. Not only are there are plenty of other organizing features in the music (such as adherence to tonality), but the repeated figures set up an internal rhythmic norm of their own. There is ambiguity, but consistently so.

We don't necessarily need regularity to develop rhythmic expectations. A common percussion pattern in the music of Japan and Tibet involves a steady acceleration of pulses. This, it seems, mimics the speeding up of impacts of a bouncing object, and as such it is familiar from everyday experience.

Pop and rock music tend to recycle a rather small bag of rhythmic tricks to devastating effect. Their almost universal 4/4 beat provides a robust framework of expectation within which even small violations

Figure 10.3 The metre is disguised by the rhythmic structure at the start of the third movement of Beethoven's Piano Sonata No. 10, Op. 14 No. 2, forcing listeners to reassess their interpretation in the fourth bar.

can generate an arousing jolt. The Beatles were one of the first groups to systematically pepper their songs with the odd bar or two of different time signatures: you'll find them during 'Here Comes the Sun' and 'All You Need is Love' (Figure 10.5*a*). Burt Bacharach was another master of the irregular bar, inserting them in carefully judged doses to urge the song onward – there's an example in 'Say a Little Prayer' (Figure 10.5*b*). In the power riffs of rock, the effect of a sudden missing or added beat is exhilarating. Led Zeppelin's 'The Ocean' feels as though it draws a breath before leashing out again (Figure 10.6), while 'Sick Again' repeatedly throws around extra beats with a brash swagger. These jolts have to be earned in the way the music is structured – simply dropping a beat from a bland harmonic progression, as in Blondie's 'Heart of Glass', comes across as mannered, while the

Figure 10.4 The melodic figures in Scriabin's Prelude No. 1, Op. 11 defy the metre imposed by the bar lines, being mustered instead into groups of five quavers. (These are actually played as quintuplets – five in the space of four.)

Figure 10.5 Extra or omitted beats in the Beatles' 'All You Need is Love' (*a*) and in Burt Bacharach's 'Say a Little Prayer' (*b*).

tendency of progressive rock to stack up gratuitously mutable time signatures becomes merely tiresome.

Musicians repeatedly use little signals to increase our confidence in our expectations and thus maximize the pleasure of their verification. This is why cadences are generally 'prepared' in classical music: their approach is clearly signalled. In rock music, drum fills and rolls serve this purpose in a rhythmic context as universally recognized harbingers of change. Rock musicians commonly break out of an instrumental or improvised sequence with a frenzied burst on the snare drum, indicating to the listener that they should anticipate the onset of a different segment. The disruption of the steady drum pattern here acts not to confuse us but rather, to heighten our pleasure when the expected change arrives.

Cadences and harmonic sequences. A related rhythmic reinforcement of expectation is the use of a so-called anticipation tone in classical

Figure 10.6 Switching the time signature in the riff of Led Zeppelin's 'The Ocean'.

cadences. This is a note that sits between the two final chords of a cadence and which generally duplicates the final note of the cadence (Figure 10.7*a*). It sounds a little complex, but there are endless familiar examples: again, the 'Ode to Joy' obliges, both in its open and authentic cadences (Figure 10.7*b*). Here the anticipation tones give a little bounce to the end of each phrase, but their main function may be to modulate the expectation of the impending cadence. If we are anticipating a perfect (V-I) cadence, we think that the V chord will be followed by I. But we don't know for sure that this will happen, nor exactly *when* it might happen – conceivably the final chord could be delayed until the first beat of the next bar, for example. So the expectation, while fairly strong, is not as strong as it *could* be. If it can be enhanced, then (according to David Huron's theory) the positive pay-off for a correct anticipation is even greater.

This is what the anticipation tone does. Happening a half-beat before the final chord, it significantly increases the likelihood that this chord will indeed come on the third beat – it would sound very strange if this quaver remained orphaned until the start of the next bar. The same would be true of a whole-beat (crochet) anticipation tone (Figure 10.7*c*), except that its greater separation in time from the final tonic chord means that it offers slightly less of a clear signal about when that chord will sound. And when the anticipation tone in an authentic cadence is a tonic note, that also reduces the uncertainty about whether the cadence will indeed be closed and not deceptive. In other words,

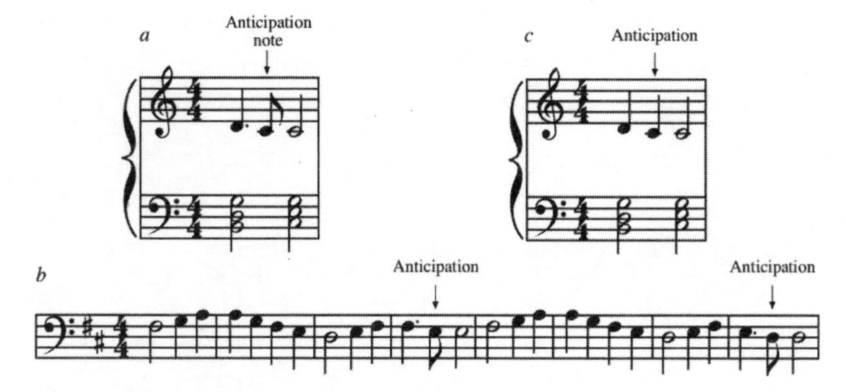

Figure 10.7 The anticipation notes of a cadence (*a*), and an example in the 'Ode to Joy' (*b*). An anticipation note of longer duration doesn't send out the same clear signal about when closure will occur (*c*).

the anticipation tone boosts both our rhythmic and our harmonic expectations.

Another common trick for focusing harmonic expectation is the suspension. Here the movement from one chord to another is carried out in stages, as though one of the notes of the first chord has got 'snagged' while the others move. Figure 10.8*a* shows an example in which a tonic chord in F moves to the dominant (C) . As the first chord is the tonic, there is no inherent sense of a need to move away from it. Following it with a simple dominant chord is fine but involves little expectation. But if the C chord contains a 'snagged' F, then something is wrong – the movement is incomplete. What's more, the 'suspended' C chord is then mildly dissonant. So we experience a strong expectation that the chord will resolve by a step down from F to E, as it does on the final beat of the bar here. Not only do we feel *what* is coming (an E), but we also have a fair idea of *when*. This is a very common pattern in the voice-leading practices of Baroque music. A similar figure, involving a suspension on the sixth rather than the fourth of the dominant chord, is familiar from the first line of the hymn 'Abide With Me' (Figure 10.8*b*). Suspended chords have an intrinsically open, unfinished quality to them: they sound as though they are on the way to somewhere else. This dynamic quality has been used frequently in rock music, for example in the breathless introduction to the Who's 'Pinball Wizard' or in the surging opening of Led Zeppelin's 'The Song Remains the Same'.

Slowing down. I have already mentioned that slowing down (rallentando) will increase tension at the end of a phrase, because this elongation of the rhythmic pulse introduces a little uncertainty about *when* the expected notes will arrive. In principle this should work anywhere in a composition, but in practice rallentandi almost always occur at

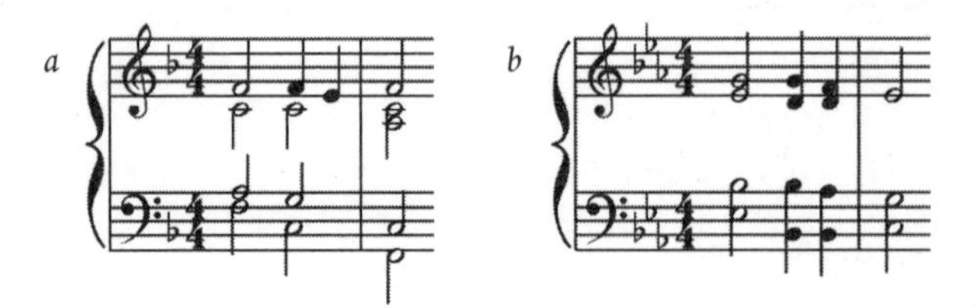

Figure 10.8 Suspension creates strong harmonic expectations (*a*), as in the hymn 'Abide With Me' (*b*).

boundaries between one section and another, or at the end of the piece. That's because it is in these places that music often becomes most formulaic, making use of stock patterns such as cadences. So it is here that we have the strongest expectations about what is coming next, which means that the emotional effect of a slowing of pace is greatest.

This is in fact such standard practice that Western classical listeners have come to *expect* a final rallentando, at least for most Classical and Romantic pieces. Baroque music is less overtly manipulative: the slowing here typically happens over a much shorter space of time, a kind of elegant final breath rather than an agonized or dreamy crawl to the finish.* And although the slowdown draws out the tension, it can't be arbitrary, for that would change a heightened expectation into mere confusion. Generally the slowing increases smoothly and is greatest for the final note (where expectation is usually highest). In fact, it seems that musicians instinctively give their rallentandi a familiar shape: music psychologists Ulf Kronman and Johan Sundberg have shown that the increasing delays between beats fit the same mathematical formula as the slowing of a rolling ball coming to rest.

The rallentando is in a sense just one formulaic instance of the elasticity of musical time. Western musicians constantly make small deviations from strict, metronomic time, speeding up or slowing down in order to make their playing expressive. This style of playing, called rubato, became especially favoured during the Romantic era, sometimes to the point where it tipped over into a vulgar simulacrum of 'feeling' rather than a genuine expression of spontaneous emotion. It's a delicate balance; according to the great pianist Maurizio Pollini, 'Rubato must emerge spontaneously from the music, it can't be calculated but must be totally free. It's not even something you can teach: each performer must feel it on the basis of his or her own sensitivity.' Rubato tends to become instinctive in classical musicians, but it's not a universal means of expression: African musicians observe metrical

* When a slowing of the beat is more or less immediate rather than gradual, it is called a ritardando. This tends to be used to distinguish phrases with different dynamical tone – impetuosity followed by sudden hesitation, say – rather than the dynamic modulation of a single phrase.

regularity with far greater precision, to the point that Westerners seem to them to be very sloppy timekeepers even when they are trying to keep strict tempo.

Endings. David Huron argues that the fade-out of popular music serves as an analogue of the rallentando that delays the resolution indefinitely. This might be expected to merely leave us frustrated, denied of the final closing chord. But because the fade-out has a slow, gradual trajectory, we can anticipate its vanishing point rather as we can anticipate the resting place of a rolling ball, and so we have a kind of substitute expectation that is fulfilled, leaving us with a mixed emotion that might be described as contented yearning.

Classical musicians have toyed with analogues of the fade-out, letting their music taper away to stillness rather than culminating in the Big Chord of Romantic symphonies or the resonant, exhaling ultra-bass of Chopin's piano pieces. Charles Ives' *Central Park in the Dark* fades back into the night, while Aaron Copland's *Quiet City* finishes on a gently pulsed unison C: a concluding tonic, but outside any predictable schema.

Perhaps because the endings of rock songs are often so formulaic, performers sometimes seem determined to undermine these strong expectations to occasionally bewildering effect. There is the unexpected cadence (the Beatles' 'From Me To You', which concludes not on the tonic chord but on the relative minor, vi), the formless frenzy (Jimi Hendrix's 'Voodoo Chile'), the abrupt 'pulled plug' (the Beatles' 'I Want You (She's So Heavy)'), the bizarre non-sequitur (ZZ Top's 'Ko Ko Blue'), the rant (PJ Harvey's 'Rid of Me') and the insouciant instant segue into the next song (Led Zeppelin's 'Heartbreaker' / 'Living Loving Maid'). The latter is taken to the extreme of generating a single, continuous stream of music in concept albums, pioneered by the Beatles' *Sergeant Pepper's Lonely Hearts Club Band* and later Pink Floyd's *Dark Side of the Moon*. Wagner could be considered the inventor of the unexpected segue. While we saw earlier how he liked to delay cadences, he sometimes simply abandoned an incomplete cadence and moved on to another musical section. It might be tempting to suppose that Wagner was here jettisoning the whole notion of cadences, but in fact the tension in his music *relies* on them – it is because he could take for granted that his audiences understood the language of cadences that he could subvert it in this way for emotional effect. As Huron points out, cadences are one of the 'big buttons' that classical composers could push in the late nineteenth century to thwart expectation. Audiences had strong expectations attached to them – just as they did to tonality and metre, which Schoenberg and Stravinsky exploited.

Tonal ambiguity and chromaticism. Chromaticism in Western tonal music – the use of 'accidentals' outside the scale or mode of the piece – has long been viewed as a vehicle for emotion, but the basis for this is sometimes misunderstood. It has been said that chromaticism in some Renaissance music 'always represents the extraordinary', being used to denote concepts such as crying, mourning and general spiritual pain. That may have been true in a symbolic way, but it has led to a belief that chromaticism is intrinsically anguished. As a result, some have claimed that, because atonal music is full of accidentals, it can access only a narrow emotional palette associated with unease and discomfort.

But there is nothing inherently 'painful' in deviations from a diatonic scale. We do typically experience chromatic notes as more 'tense', but that's not because they are non-diatonic; it is because we're used to hearing the diatonic notes as the norm. The tension comes from the deviation from our statistical expectation. If someone was brought up hearing only atonal music (an unlikely scenario that some might consider borderline abusive), it seems unlikely that they'd consider it tense and uneasy. Indeed, in some late Romantic classical music chromaticism became so common that its function of creating a sense of unease was undermined – it no longer sounded 'odd'.

And there's no reason why the emotion associated with chromaticism has to have a negative valence. In some classical music, chromaticism probably suggests anguish or mystery simply because that is the association we have learnt, just as we recognize the connotation in the 'jarring', out-of-scale melodies of today's horror and mystery movie soundtracks. Bebop jazz solos show that chromatic notes can be perfectly consistent with expressions of joy and exuberance.

Chromatic notes have a piquancy not just because they are statistically unlikely but also because they themselves create strong expectations. As we saw in Chapter 4, chromatic notes are considered unstable and have strong statistical links to their nearest scale notes:

a sharpened fourth note, for example, is nearly always resolved to the fourth or fifth. The powerful anticipation that this is where a chromatic note is 'headed' offers a correspondingly large 'reward' when it is confirmed. The pay-off is increased by delay – if a chromatic note occurs only in passing, all it does is to inject a little spice, if indeed it is noticed at all.

It's tempting to imagine that chromaticism is a relatively modern innovation, developing steadily from the nearly chromatic-free clarity of Palestrina to the almost atonal extremes of Wagner. Deryck Cooke suggests that its emergence in the late nineteenth century reflects the anxieties of modernism. But in fact chromatic notes were common in much medieval music, being employed in madrigals for expressive purposes sometimes almost to the point of manneristic distortion.

Some modernist composers have used chromaticism more for rhetorical than for emotional effect. Music theorist Deborah Rifkin suggests that when, for example, Prokofiev injects chromatic oddities into his tonal music, he intends these to be read as 'metaphors for modernity' – rather as if to say, 'Look, now I can do *this*.' Thus, whereas Chopin or Brahms used chromaticism to loosen the tonal constraints and open new expressive avenues, and will carefully prepare the ground for it, Prokofiev throws in 'wrong notes' without warning, precisely because they will be jarring and shocking – cracks in the tonal fabric, or 'agents of dislocation' as Rifkin calls them.

A favourite device of Prokofiev's, for example, would be to shift a melody note a semitone out of key. An example is the theme from *Peter and the Wolf* (Figure 10.9), where the 'expected' C♮ is flattened to a B♮. Here the effect is not so much jarring as pungent: we don't even

Figure 10.9 Chromatic aberrations are common in Prokofiev's music, for example in the theme of *Peter and the Wolf*. Here the 'out of key' B naturals are indicated with arrows.

necessarily notice that the note is 'wrong', but just that the melody is a bit crooked. Notice that Prokofiev hits this note twice, as though to say 'Yes, that really is the one I meant' – he takes care to make sure we don't subconsciously write off the odd note as a mistake. Sometimes he introduces chromaticism even more alarmingly, for example at a cadence where the expectations of the tonal tradition are particularly strong. It's crucial here that Prokofiev's music stays basically tonal: only then do we retain expectations that enable sudden chromaticism to do its spiky business. And if Rifkin is right to say that these gestures have more of a philosophical impulse – the 'frustration of conventional expectations that is an "emblem of modernist aesthetics"' – nevertheless it is inevitable that the deviations acquire an emotional charge.

As well as creating implications for the melody, chromatic notes send out signals about harmony. We've seen how modulation to a new key involves the introduction of notes that fall outside the old key but are diatonic in the new one. So a chromatic note raises the possibility that such a key change is under way. In the Classical era these modulations were usually effected rapidly and signposted clearly. Even the widespread use of the diminished chord – a series of minor-third intervals, which has no clear tonal centre – to enact modulations involved little ambiguity of purpose, since this transition was itself so formulaic (Figure 10.10*a*). But from the Romantic era, modulation became much more drawn out and progressed through heavily chromatic passages that obscured all sense of tonal destination. Chopin was a master at this, leading the music through strange, out-of-key harmonies with no end in sight. In his Waltz in A♭, Op. 64 No. 3, Chopin makes a transition from the minor to the major signature key. This is a relatively simple thing to do: as we saw in Chapter 4, the major and minor are closely stationed in harmonic space. But Chopin seems to make extraordinarily heavy weather of it (Figure 10.10*b*) – not because he is a bad navigator, but because that's where much of his genius lay. To switch from minor to major would be unremarkable; but after the convoluted little detour here, the major key emerges with fresh sweetness, like coming through a thicket into open fields.

In contrast to the thwarting of expectations with delay or surprise, ambiguity undermines them. We might be left thinking not 'oh, when

Figure 10.10 (*a*) Modulation via the diminished seventh chord was common in the Classical period, as in this example from Mozart's Rondo in D major (K485). Here arpeggios on the diminished seventh effect a modulation from D major to E major. (*b*) Chopin's modulation from A♭ minor to major in his Waltz Op. 64 No. 3 takes a detour in harmonic space to create ambiguity about the destination.

is that cadence coming?' or 'oh, the rhythm's changed', but rather, 'what on earth is happening now?' When the confusion resolves itself, our pleasure comes not so much from the delight of an expectation confirmed as from the relief of being able once again to discern pattern, structure and direction. We've already seen examples of how crossed rhythmic patterns may create uncertainty about the metre. But for classical composers before the twentieth century, it was harmony and tonality rather than rhythm that provided the main vehicle for ambiguity.

Tonal ambiguity does not even require chromatic notes. Consider the first four solemn notes of Bach's first Fugue in C♯ minor from Book I of *The Well-Tempered Clavier* (Figure 10.11): all are in the diatonic minor scale, but they are clustered within too narrow a tonal space – a major-third interval – to permit a secure guess at the tonal centre. We are gripped at once by Bach's masked intent.*

The emotional power of chromaticism isn't limited to Western music. Something akin to the potency of 'out-of-scale' notes can be

*There are other things going on in this four-note sequence to confuse our bearings. In particular, it contains the very unusual interval step from the seventh to the minor-third degree of the scale. Since we're not at all used to hearing this, we're puzzled about which key to fit it into.

Figure 10.11 The opening of the Fugue in C♯ minor from Bach's *Well-Tempered Clavier*, Book I.

found in Chinese music, where the *pien* tones that supplemented the pentatonic scale (p. 100) were not permitted in palace or temple music because, rather than performing music's intended function of soothing the passions, they were deemed to fill the soul with sensual lust.

Intra-opus. The predictability on which expectations depend can be generated *within* a composition rather than relying on schemas we have learnt previously. Indeed, this is really just one way of stating the obvious fact that musicians construct compositions from repeated themes and motives. These self-contained schemas in a musical piece are sometimes given the ugly term 'intra-opus' ('within the work'). They often work in opposition to our generic expectations, creating interesting opportunities for playing one against the other.

For example, the insistent rhythmic pattern of Holst's 'Mars' from *The Planets* drives home a regularity that is in itself rather unorthodox in Western tonal music, being based on a 5/4 rhythm containing both quavers and triplet patterns (Figure 10.12). If 'Mars' were to suddenly break into waltz time, we'd be surprised by it, even though waltz time is far more familiar generally than is quintuple time.

There is a nice little interplay of internal and external expectations in Dvořák's Ninth ('New World') Symphony, the fourth movement of which contains the motif shown in Figure 10.13. Here the rising line of the second bar E-F♯-G might ordinarily be expected to rise up to A, according to the gestalt principle of good continuation. But it drops back to F♯ in the third bar. So the listener thinks 'Aha, so that's the new schema for this little tune.' But when the figure repeats in bar six, it *does* go up to A – something that would have been unremarkable if we'd got it first time round, but which creates a little moment of surprise now.

Little is surely the operative word here – this extract is rather

Figure 10.12 The insistent 5/4 pattern of Holst's 'Mars' sets up its own norm.

Figure 10.13 Subtle creation and violation of expectations in Dvořák's Ninth Symphony.

mundane both melodically and rhythmically. We barely notice anything amiss in either case; certainly, we're not flooded with emotion. But that's the point about Meyer's theory: it is not necessarily a description of how music transports us to ecstatic heights (though it may well have something to say about that), but rather about the minute, moment-to-moment workings of any phrase, most of which are content simply to tweak our unconscious expectations in gentle ways to maintain the feeling that the music is speaking to us. More often than not, this is all music strives to achieve.

Form, style and genre. If someone tells you that you are about to hear a Mozart piano sonata or a track by Metallica, the slightest acquaintance with those artists prepares you for what is in store. Even a rudimentary knowledge of the conventions of styles and genres equips us with a host of expectations, and thereby sets up avenues for emotional direction. Metallica fans might like Mozart (who knows?), but they aren't likely to respond to it well at a Metallica concert. This works for particular songs too. Popular performers often take liberties with their best-loved material, not just to stave off boredom but to inject some freshness into it – in live performance, David Bowie sometimes reinvents his most famous numbers so drastically that only the lyrics identify them.

In blues music, the harmonic progression from the tonic to the subdominant chord (I-IV) is almost an inevitability – which is why, when Howlin' Wolf just strolls along indefinitely on the tonic, we can experience it not as monotonous but filled with thrilling tension. The classic rock progression I-♭VII-♭VI-V would sound odd in a Beethoven symphony. In short, each genre establishes its own schemas which it can then use to manipulate tension. Fred Lerdahl suggests that large-scale schemas such as the forms of classical music – sonata, rondo and so forth – are needed in order that we can develop any real expectations at all over these extended scales, for otherwise the possible ways the music could go are just too immense to allow any foresight. The better we know these schemas, the stronger is their potential to influence our expectations and responses. But you don't need to have any depth of musical knowledge to quickly sense the repeating structure of the rondo (typically with themes ordered in a pattern such as ABACADA), as in Mozart's Rondo alla Turca from his Piano Sonata No. 11 – and then to take delight in the returning theme. Likewise, the typical symphony structure, with its brisk first movement, slow second movement, a scherzo or minuet in the third and then the final allegro or rondo, quickly establishes itself in the classical listener's mind, so that a scherzo in the second movement would be something of a shock.

A neat example of the undermining of genre conventions is found in the Sensational Alex Harvey Band's 'Who Murdered Love', where a countdown ('Five–Four–Three–Two–One') leads not, as the rock fan will expect, to an explosive sonic climax but to an incongruous little motif on strings. The result is that the anticipated catharsis turns into laughter (itself a familiar emotion for the Harvey fan). This knowing manipulation of genre expectation is a frequent ruse for generating amusement in music, which earlier we saw Mozart utilize in his *A Musical Joke*. An altogether different inversion of informed expectation is found at the start of Stravinsky's *The Rite of Spring*, which began most unusually with an instrumental solo, and even more unusually with solo *bassoon* – and then triply so with a bassoon playing at the high extreme of its range.

Bringing music to life

What is it that makes someone prefer Alfred Brendel's rendition of Beethoven's piano sonatas to that by Vladimir Horowitz? Why will one person insist on Andras Schiff's recording of the 'Goldberg' Variations, and another on Glenn Gould's? One thing is sure – it's got nothing to do with Bach or Beethoven. All players have the same score, but the expressive quality of the performances can and usually does vary immensely.

It's one thing to explain how particular arrangements of notes, harmonies, rhythms and timbres engage and elicit emotions. We can argue there that the relevant factors are all laid out before us, literally in black and white. But a precise and faithful performance of the music can strike us as emotionally barren. That's why the online electronic renditions of the musical examples included in this book are hardly likely to set you swooning, I fear (but at the same time, their neutrality does have the virtue of focusing the attention on the point being illustrated, without the distraction of flashy dynamics). Music can be packed with emotional devices that will by and large misfire unless they are properly communicated by the performer. No amount of technical proficiency will guarantee that you will touch your listeners. And conversely, music that is dull and lifeless, even banal, on the page might be rescued by an expressive performance. But how?

Most musicians, musicologists and psychologists will agree that the key to expression is, to put it bluntly, *deviance*. As the early pioneer of music psychology Carl Seashore put it:

> The unlimited resources for vocal and instrumental expression lie in artistic deviation from the pure, the true, the exact, the perfect, the rigid, the even, and the precise. This deviation from the exact is, on the whole, the medium for the creation of the beautiful – for the conveying of emotion.

Yet the implications of this have sometimes been missed. Western musicology has now mostly outgrown its tendency to patronize folk and traditional music, but there are countless past examples of academics commenting on the poor pitch or rhythm of untrained performers, unaware that they are hearing a style of tremendous emotional nuance. For what folk musicians have long appreciated is that some of the most powerful tools of musical affect are those available only in performance, which cannot be notated even if one wanted to.

Béla Bartók, with his profound and humble interest in the traditional music of his native Hungary and its environs, was one of the first to recognize this. He noticed that peasants would often sing notes 'off key' according to standard Western tuning, but that they would do so systematically and reproducibly. This, he said, 'must not be considered faulty, off-pitch singing', like that done accidentally by urban amateurs – it was 'self-assured, self-conscious [and] decided'. Leonard Meyer understood this: 'the intonation of a traditional folk singer', he wrote, 'is not necessarily worse than [that of professional musicians], but it is often more flexible. A tone is often taken somewhat higher or lower than might be expected, for expressive and ornamental reasons.' Percy Grainger made a close study of styles of folk singing and realized that performers would often repeat more or less exactly the irregularities of pitch and rhythm that they injected into a song – these were not just random errors caused by poor technique, but intentional musical interjections (see Figure 7.15, in which Grainger annotates several ambiguous pitches). Grainger was motivated to invent what he called 'free music' in which pitches change not in steps but in smooth glides. He experimented with primitive mechanical and electronic devices that would provide the control over pitch that human singers found hard to achieve. This, however, was perhaps to miss the point of the deviations of intonation in folk singing – for as Meyer points out, they are used for expression, and are expressive precisely because they deviate from a norm.

This principle is found well beyond Western music. Meyer quotes from a collection of Javanese poetry called the *Tjentin*, in which it is said of a performance on the stringed rebab that 'Every now and then there was a little deviation from the correct pitch so as to enhance the charm of the music.' Similarly, blues and jazz musicians have long known to exploit the ambiguous pitch of the blue notes – especially the minor third and seventh – for expressive effect. Andre Hodeir comments that 'When an expressive intention shows up in the blues, whether instrumental or vocal, nine times out of ten it is concentrated on a blue note.' The ambiguity of pitch is often accentuated further by reaching the note in a portamento, a continuous glide (usually up) through pitch space. The importance of being able to manipulate intonation in this way no doubt accounts for the centrality of the guitar to the blues, and the saxophone to jazz.

Pitch bending is one of the modes of *ornamentation* that performers use to add expression to the notes they are 'given'. An expressive performer uses subtle changes of timing, loudness, melody and phrasing to bring out emotional qualities in music. The possibilities for such deviation are endless, but only a sensitive performer (which is not necessarily the same as a trained one) knows which of them to choose at a given moment, and how far to take them. With the wrong choices, a performance becomes mechanical, 'just a collection of notes', or conversely, affected and kitsch. We notice quickly enough when playing is larded with cheap emotion, expressed perhaps as overemphatic dynamics or rubato. Even ostentatious virtuosity can be seen as vulgar (the Chinese pianist Lang Lang suffers on this account). And yet studies show that players who are deemed truly 'world class' *do* give more exaggerated performances, on average, than other competent musicians. It seems we can be persuaded that excess represents profundity, when it is the right *kind* of excess – whatever that might be!

Ornamentation – the addition (often improvised) of ephemeral notes to a melody – is ubiquitous in music. According to C. P. E. Bach, 'Probably no one has ever doubted the necessity of embellishments', while an Indian treatise on the performing arts called the *Natya Shastra*, dated between 200 BC and AD 200, says that 'A melody without ornament is like a night without a moon, a river without water, a vine without flowers, or a woman without jewels.' Western classical music has grown abnormal in its relative rigidity, for many musical traditions, whether 'folk' or 'classical', offer a great deal more scope for injecting spontaneous invention into a performance, for which the 'composition' is little more than a skeletal framework. It is literally meaningless to say that 'All of Me' or 'Danny Boy' are 'expressive' songs: almost all the emotional content they might acquire is determined by the performer. What ethnomusicologist George Herzog wrote about Slavic music in 1951 applies to just about any traditional styles, whether that of a Tipperary fiddler, a Cuban dance trio or a Californian rock band: the appeal of the music 'may well be due to the contrast between the essential simplicity of its basic materials and the pulsing quality of life achieved through an abundance of expressive devices, including the ornamentation'. This is why, for example, analysis of rock music from transcriptions misses much of the point. Indeed, such transcriptions are themselves often meaningless: a piano score of an album by the Clash carries about as much information about its subject as a newspaper report of a football match.

As well as microtonal variations of pitch, ornaments may include such embellishments as grace notes (acciaccaturas), trills and vibratos, glides and turns (Figure 10.14). Put (and written) this way, they sound terribly formal, but that's a consequence of both Western musical notation and practice, which insists on categorizing and documenting spontaneity. The sustained, wavering trill is as much a feature of Jimi Hendrix's guitar playing as it is of the keyboard compositions of Couperin (who derived it from the same source, the hammering on and off of a stringed instrument, albeit in his case a lute rather than a Fender Stratocaster). All of these fleeting additional notes create an ambiguity of pitch – the trill in particular has a suspenseful quality as it teeters between upward and downward pitch motions. Even fixed-pitch instruments can have pitch ambiguity wrung out of them, as in the sweeping glissandi of jazz pianists or Hammond organists. An undulating pitch, or vibrato, is commonly experienced as emotive: Seashore claimed that vocal vibrato implies a state of high arousal, perhaps because of the resemblance to the quavering pitch of an excited or distressed speaking voice.

Western classical music once shared in the 'folk' practice of improvised embellishment. The great musicians of the Baroque and Classical periods made up astonishing cadenzas on the spot, only to see these gradually ossified in the score. Already by the time of the Bach family, ornamentation was being written down explicitly with codified symbols. Performers began to observe rather strict rules for where the embellishments might go and how they should sound. This is not wholly a habit of classical music; Irish folk, for example, has rather prescriptive formulas for inserting ornaments. In fact, it is fair to say that every genre tends to acquire its own repertoire of ornamentations and permissible micro-manipulations in performance; even the

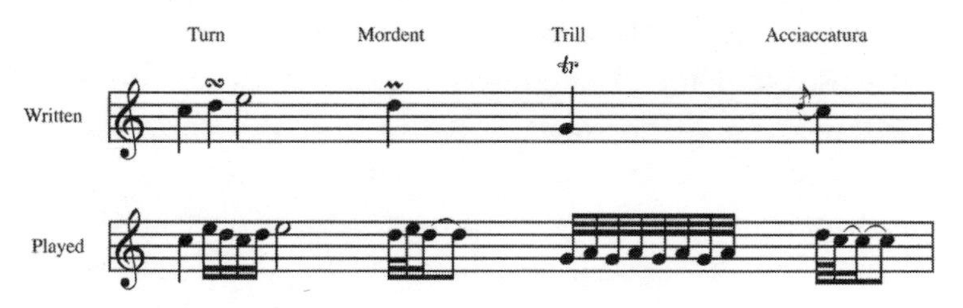

Figure 10.14 Common ornaments used in Western classical music: the turn, mordent, trill and acciaccatura.

'spontaneous' screech of electronic feedback is now something that rock guitarists seek to produce at key moments.

The systematization of expression in performance was also evident in the use of dynamic notation and expressive terms by Western composers. Since the Classical era, they made use of volume indicators (*ff*, *mp* and so on) and instructions to slow down or accelerate. And they would specify the overall mood of the piece at the outset: *giocoso* (merry), *grazioso* (graciously), *dolce* (sweetly) and the rest. This was not simply a codification of previously implicit intentions, but an indication of a changing attitude to expression: J. S. Bach did not necessarily think of performance in terms of these everyday emotional characters at all. In the late nineteenth century, Erik Satie reacted against such overprescription of affect with wicked parody, lacing his scores with injunctions such as 'Without your finger blushing', 'Dry as a cuckoo', or 'It's wonderful!'

How far can performers truly shape the emotional quality and message of a piece of music? Sure, they can presumably obscure the composer's intentions with a bad performance – but can they change them entirely? No doubt even the most skilled musicians would struggle to convey calm in Bartók's *Allegro Barbaro*, or frivolous joy in Beethoven's 'Moonlight' Sonata, but not all music is so prescriptive of emotional mood. In popular music, it's quite common for songs to be used as frameworks on which many different emotions can be hung, through interpretations that do transformative things to mood while keeping the original song recognizable: Joe Cocker's anguished 'With a Little Help From My Friends', say, or Jimi Hendrix's urgent 'All Along the Watchtower'.

A part of this question is whether ornaments and embellishments can convey emotion in a reliable way. In Baroque music, a standardized vocabulary of ornaments allowed for emotional signposting: the German music theorist Johann Mattheson stated in 1739 that these should be used to point 'toward the emotion, and illuminate the sense or meaning of the performance'. That's fine if you know the code, but what if you don't?

To test whether ornaments could offer definite emotional connotations, music psychologists Renee Timmers and Richard Ashley asked a violinist and a flautist to play a Handel sonata (HWV360) using standard Baroque ornaments in ways that expressed mild and intense happiness, love, anger and sadness. An audience, whose members were musically trained but not Baroque specialists, was asked to judge which of these emotions was being represented.* The performers played along to a pre-recorded accompaniment, which prevented them from using tempo variations to reinforce the emotions (although they could still exploit such things as dynamics and timbre).

The listeners' success in spotting the intended emotion was rather haphazard. They often did better than chance, but typically only fifty per cent or so guessed right. And intentions of love and sadness were often confused, which is perhaps understandable but nonetheless a trifle disturbing. What's more, there was no consensus between the performers about *how* to convey the emotions – to express 'happiness', say, the violinist emphasized trills while the flautist used turns. So not only are audiences rather limited in how much emotional content they can glean from these ornaments, but also the musicians have no agreed link between the two things.

Does that undermine the notion that ornaments and nuances of performance are a primary vehicle of emotionality? Not at all. It might mean that Baroque ornaments are no longer particularly effective at communicating to a modern audience, or at least not on their own. Or – and this is my suspicion – it might mean that, at least for Baroque sonatas, we're not actually looking out for messages of 'happiness', 'anger' and so forth when we listen to music. Rather, we respond to a sense that *something is being expressed*. One person's poignancy is another's serenity. As Hanslick said, when we ask people what is being expressed by a theme of Mozart or a Haydn symphony, 'One will say "Love". He may be right. Another thinks it is "longing". Perhaps so. A third feels it to be "religious fervour". Who can contradict him?' We are not passive recipients of musical emotion, but construct our own interpretations. All we ask is that the music provides us with the materials.

*That's not an ideal test: the listeners thus already had clues about what they were 'supposed' to hear. But as we've seen, it is hard to do any experiments at all without such constraints.

Mapping tension

Meyer might be regarded as saying that satisfying, 'good' music must find a path midway between the expected and the unexpected. In effect, it gives the listener puzzles to solve, and makes sure that they look solvable. These ideas are intuitively plausible, and Meyer's theory has conditioned much of what is now believed about musical affect. But is it right?

Well, there seems ample reason to suspect that he was on to something. For example, musical 'hotspots' that listeners report as inducing emotional responses often seem to make a particular play of expectations. John Sloboda has sought for specific musical events that trigger physiological signs of emotion, such as crying, shivering and increased heart rate. This isn't necessarily the best way to probe the less overt kinds of emotional response that are probably more representative of most musical experiences, but it has the advantage that such responses are memorable and often unambiguous – either you cried or you didn't. Sloboda asked participants in his study to identify precisely the point in a particular piece of music where they had these reactions, and found that people were surprisingly good at doing this: a third of the group could attribute it to a particular theme or smaller musical unit. And many of these involved the kind of manipulation of expectation that Meyer would have forecast, such as an acceleration towards a cadence or a delay in reaching it, the emergence of a new harmony, sudden changes of dynamics or texture, and rhythmic syncopation. One very common trigger for both tears and shivers was the so-called melodic appoggiatura: a kind of delay introduced to a relatively stable note by suspending it on a less stable one (Figure 10.15*a*). This has a near-universal current of pathos in Western music – it features, to offer one of innumerable examples in popular music, in Leonard Cohen's 'Bird On a Wire'. Albinoni's *Adagio for Strings* is a piece that moves many people to tears, and its opening melodic line contains three appoggiaturas in the first seven notes (Figure 10.15*b*). It's important here that the basic structure of this phrase is extremely simple, a descending minor-key scale. This transparency of form permits strong expectations to be created – namely, that notes on strong beats will be stable – which the appoggiaturas then fleetingly violate.

A common trigger for 'shivers' in Sloboda's study was the so-called

Figure 10.15 The melodic appoggiatura (*a*) is a common trigger for 'chills'. Albinoni's *Adagio for Strings* contains three in the first seven notes, marked with arrows here (*b*).

enharmonic change, in which the melody notes fit the key already established but the harmonic backing shifts so that these notes take on a new harmonic function. In Burt Bacharach's and Hal David's 'Walk On By', the key changes from basically A minor to F, via B♭, between the verse and chorus, the pivot being an A melody note on 'private' and 'tears' in the first and second verses – this functions as a tonic note in A, but as a major seventh against B♭, an unexpected and poignant change (Figure 10.16). Or in the chorus of Louis Armstrong's 'What a Wonderful World' ('And I think to myself . . .'), the melody line stays on the tonic note, but the harmonic backing drops down a major third (in the key of G, the appropriate chord is E♭), and we get a sense of the melody flattening out, suspended on a ledge before returning to the tonic. This kind of thing fulfils one expectation (the melody remains in the diatonic scale) while violating another (we suddenly have to re-evaluate the note's function within that scale). Such a simultaneous confirmation and frustration seems peculiarly potent as an emotional trigger.*

*Another signifier of mood change in rock and pop music is a change of mode. As I mentioned earlier, rock melodies are often modal, and a switch of mode may be used, rather like a change between major and minor, to emphasize a different emotional quality in the lyrics. In the Beatles' 'A Hard Day's Night', the verses use the minor-like Dorian and Mixolydian modes as John Lennon sings about his hardships, but then move to the major-like Ionian for the brighter sentiments of the bridge ('When I'm home . . .'), sung by Paul McCartney because Lennon allegedly couldn't reach the notes.

Figure 10.16 An enharmonic change in Burt Bacharach's 'Walk On By'. The A on the first syllable of 'private' (marked with an arrow) moves away from being the tonic of the A minor tonality in which the verse began, and, via a major seventh in B♭, towards the major third of the chorus in F major. The key point is that we can feel this change in its 'quality' here.

This looks all very supportive of Meyer. But to get beyond these qualitative explorations, we need to make his ideas a little more scientific. They have been recast in terms of quantifiable cycles of tension and release, in particular by Meyer's student Eugene Narmour. Narmour's model is more explicitly formalist: here the effect of music stems from formal relationships between notes. It draws on ideas we've encountered earlier, such as the differences in stability and status of notes in the diatonic scales and the gestalt implications of melodies. The expectations that arise from these mental schemas set up a constant flux of tension and release as a melody threads its way through pitch space. The composer artfully weaves that succession into a pattern of satisfying and stimulating contour. Narmour's description of this process is rather complicated, but musicologist Glenn Schellenberg suggests that it boils down to two issues: we expect successive pitches to be close rather than far apart, and we expect a reversal of pitch steps after large jumps. (As I indicated earlier, the latter may itself be just a statistical effect, since there are generally more available notes to choose from in the reverse direction after a large jump – see p. 113). Narmour asserts that we also listen out for 'closure' – the completion of a phrase or idea, as signified for example by an authentic cadence or a return to the tonic note on a stressed beat, or by rhythmic clues.

Narmour's criteria for melodic expectancies have received some qualified support from psychological tests conducted by music psychologists Lola Cuddy and Carole Lunney in Canada. Narmour argued that two successive melody notes – a melodic interval – create an implication about what the next interval (and thus the next note) will be. He postulated five gestalt-based principles that guide this anticipation: for example, whether the direction of melodic contour will stay the same or reverse, and whether the next interval will be large or small. Cuddy and Lunney tested a group of volunteers, some with musical training and some without, to see if these expectancies are truly evident. Their experiments were similar in principle to those of Carol Krumhansl for establishing a tonal hierarchy (p. 102): the experimenter supplies a stimulus (here a melodic interval) and then asks the listener to rate how well various possible continuations 'fit' with that context. Cuddy and Lunney found that four of Narmour's five predictive criteria (one in modified form) seemed to fit the results. There was no sign of his expectation for 'closure' – but perhaps that was because the test subjects were given to believe that this was just the start of a melody, so they weren't looking for how well it 'closed'. But the results also showed that the expectations bear a strong imprint of tonality, which Narmour did not include here: the listeners expect the continuation to remain in key too.

Narmour imagines that we apply similar predictive schema to harmony, rhythm, timbre and so on. And he proposes that we have our eye on the bigger picture too. We evaluate the degrees of similarity and difference in the musical patterns we hear, and use that to divide the stream of sounds into related units. This can work in hierarchical fashion: for example, we recognize the phrase endings in the lines of a song, but can also hear when a verse (a higher-order pattern) has concluded. In this way, we build a hypothesis about the structure of the music and the likely ways it will continue, at several levels of 'magnification'.

At each tier of the hierarchy, these expectations create tension and release. So evidently *there are lots of possibilities* for directing and frustrating our anticipation – and they don't necessarily coincide. All of these clamouring voices may all be doing different things, reaching closure or repeating patterns at different places, and so what we get is a complex, overlapping set of 'streams' competing for our attention. In just four bars of a Mozart sonata there is a dizzying interplay of affective forces, offering justification for the claim of the Danish physician Hans Christian Oerstedt that the lifetimes of several mathematicians might not suffice 'to calculate all the beauties in *one* symphony by Mozart'. This, however, may be precisely why the music

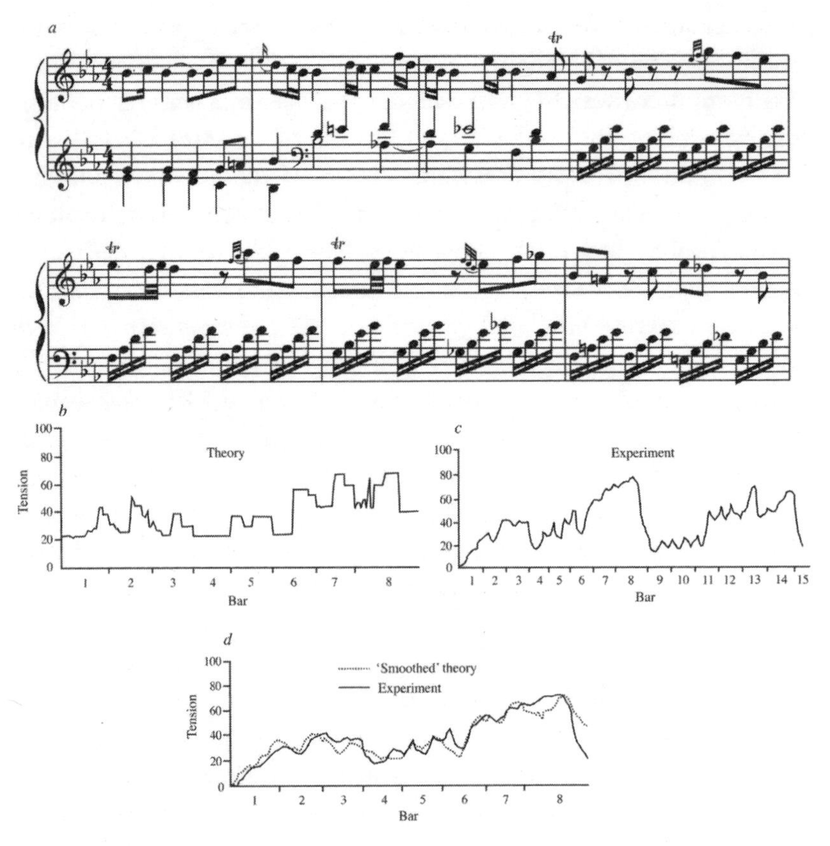

Figure 10.17 The 'tension profiles' of the first movement of Mozart's Piano Sonata in E♭ Major, K282 (*a*) has been studied in detail. The note-by-note theoretical version is shown in (*b*), as predicted by a theory of Fred Lerdahl (explained further in Chapter 12), and the judgements of tension by listeners equipped with 'tension sliders' is shown in (*c*). The two match fairly well, especially when the theoretical profile is modified on the assumption that listeners' responses slightly lag behind the acoustic stimulus and that they mentally average the perceived tension over short blocks of time – this makes the theoretical curve smoother (*d*).

engages us: the complexity is not too great to overwhelm us, nor too simple to bore us.

By stipulating approximate measures of how much tension accumulates from 'violations' – for example, how relatively unstable different notes and chords are within a tonal context – it becomes possible to plot the theoretical ups and downs of tension in the various musical dimensions of melody, harmony and so forth. Then a musical score can be assigned a kind of topographic tension profile – a mapping of the emotional roller-coaster ride, where peaks are moments of high tension and dips offer release (Figure 10.17). Some researchers have studied whether these maps possess any validity by giving listeners a 'tension slider' to push up or down in real time while they listen. The subjective tension profiles obtained this way are not only broadly consistent between different listeners, suggesting that there is something objectively real about the idea of tension and release, but also roughly follow those predicted from an analysis of the score according to Meyer's criteria for tension (Figure 10.17*d*).

But a host of questions remain. Is the tension listeners experience primarily melodic, harmonic, rhythmic or something else? Is it mostly determined 'in the moment', in the relationships between one note and those shortly preceding it, or does tension stack up in a hierarchical manner that takes account of longer-term structure such as phrases? Different 'tension' models say different things about these issues, and it's not clear that measurements of real responses will ever be precise and consistent enough to distinguish between them. For example, when they asked listeners (musicians and non-musicians) to rate the tension that developed during two musical pieces (a Chopin prelude and a composition created for the experiments), music psychologists Emmanuel Bigand and Richard Parncutt found that people seemed to base their judgements more on 'local' factors such as harmonic cadences than on the global, hierarchical stacking up of tension predicted in a model devised by Ray Jackendoff and Fred Lerdahl (which I discuss in some detail in Chapter 12). Jackendoff's and Lerdahl's model *did* fit the tension ratings quite well, but only because it does a good job of capturing the local as well as the global picture. Besides these findings are open to question because the experiments involved truncating the music at various points and asking, in effect, 'So how tense is it now?' But no one listens to music that way. Some studies with 'tension sliders' that involve continuous listening do imply that there might be hierarchical effects – that we carry over a sense of what has gone some time before to sense the 'deep structure' of tension and release. But the truth is that we're just not sure how much or what kind of information listeners use to make these assessments, or if everyone does it the same way.

Indeed, some studies have suggested that our emotional responses to music are so complex and idiosyncratic that attempts to explain them systematically may prove futile. In one of these, participants listened to five extracts of preselected music ranging in style from Baroque to jazz funk, and were told to press a button when they 'felt something' (the guidance was intentionally vague). Many of them pressed buttons at the same points in the music, suggesting that there was some characteristic intrinsic to the music that provoked emotion. And when some participants were played two of the same extracts a year later (when all but one of them had forgotten what they'd heard before), they gave much the same responses. Moreover, some of these musical triggers could be identified: buttons were pushed (literally) during peaks of crescendos, the entry of a singing voice, descending harmonic cycles of fifths, sudden rhythmic changes, and enharmonic changes. All of this offered some rough-and-ready support for Meyer's ideas.

But what about the listeners' perceptions of why they were moved, on which they were quizzed afterwards? Here there was almost no consistency in the responses, even though some of the participants had the musical training that would enable them to identify the events they had heard. Can we say that these people simply didn't 'know' why they were moved? Or that the answers they gave were 'wrong'?

Lowering expectations

Meyer himself was sceptical about formal, quantitative schemes for characterizing the emotional structure of music, saying that 'What is profound about the experience of a listener is not the "deep structure" of a piece of music, but the power of the rich interaction of sound and silence to engage our minds and bodies, to give rise to feelings and to evoke associations.' I suspect he is right: much of this analysis takes place at a level of theory far beyond that possessed by the vast majority of listeners, and which seems irrelevant to the way the music is experienced. Part of the challenge is to figure out how concepts such as tonal hierarchies, trajectories in harmonic space or enharmonic changes are mentally represented and experienced by listeners who haven't the slightest notion of such abstract constructs – how, in other words, these really function (or not) as platforms for emotionally laden expectations.

But there are also difficult unresolved questions about the whole picture of musical emotion as a product of anticipation. It is easy to enumerate the various ways in which expectation and violation might create emotion; the problem is that, once you apply these to real music, you find these features everywhere. Almost nothing in music can happen without violating some plausible expectation, because the possibilities are too great. As John Sloboda points out, one could argue that the repetition of the first two (tonic) notes in 'Twinkle Twinkle Little Star' create an anticipated third repetition, and the violation of this expectation should therefore produce 'emotion'. Or conversely, one might say that because repetition of notes is less common than a small-step change, each repetition intensifies an expectation of imminent change. You can argue it any way you will. Sloboda has compiled a list of ten musical features that seem to reliably provoke 'thrills' in listeners, but he admits that if all one needed was to insert one of these into music, 'there would hardly be a measure of music which did not contain one'. This makes expectation models of emotion self-fulfilling: no matter where in the music you feel moved, you're likely to find a stimulus that wasn't 'expected'.

It's not even obvious what an expectation really is. Let's say, for example, that we hear a C followed by a D. We might anticipate that the next will be E. But is this because we literally expect an E (that's unlikely, unless we have absolute pitch), or because we expect the next note to be a whole tone above D, or to be the third note of the major scale, or just to follow the same general upward contour? There are several ways in which even this simple expectation can be mentally represented. And some of these may conflict. For example, if we have decided that this little fragment of melody is in C major, then C is tonally more stable than E and so might be expected to command a greater expectation. And if the third note turns out to be E♭, this confounds our major-scale expectation but conforms both with a more general expectation of melodic continuation – an upward trend in small pitch steps – and with the 'expectation template' that we might also be expected to have for minor keys. Perhaps, as some neuroscientists have suggested, we maintain a battery of different types and representations of expectations, and are constantly shifting between them depending on which seems at any moment to be the most successful – a notion dubbed 'neural Darwinism'. This mental juggling act puts a new complexion on what looks at face value to be a relatively simple idea. Fred Lerdahl argues that Meyer's model of frustrated expectations offers too simple and linear a picture of this process. Rather, there is an entire multidimensional 'field' in which different features of the music – melodic, rhythmic, harmonic – set up degrees of attraction in particular directions. And the shape of the field itself depends on what has already happened. Then what counts, he says, is

> the entire interaction between the attractional field and the unfolding of events. The field pulls events in certain directions, whether they go there or not, and at every point the events cause the field to evolve. Out of this swirl of force and motion arises effect.

One objection frequently raised to Meyer's theory is that it looks as though it can only work once, or at least that the effect of expectations should decline very quickly as we grow familiar with a piece of music. But it seems that in fact our appreciation of music in relation to familiarity has, in graphical terms, an inverted-U shape: we may like it rather little on first hearing, grow to like it more on relistening, and then eventually become bored by it. So not only does the 'surprise' not disappear right away, but its effect seems to be enhanced by repetition. One possible explanation, suggested by Ray Jackendoff, is that some aspects of music processing are subconscious, pre-empting the role of memory: surprise is invoked by basic cognitive processes no matter how many times we hear the piece, just as pleasure is evoked by a delicious meal even on the tenth serving. Another possibility is that the ways in which violations and resolutions interact as a piece of music progresses, especially if it is relatively complex, are so numerous that we can't consciously assimilate them all: we are repeatedly 'tricked' into having a response because we can never properly grasp everything that is happening. Although that seems possible, it has a corollary that challenges the whole 'expectations' model. In some dense works, particularly those of the modern era, there are so many potential directions for violations to take that it's not clear how we might develop meaningful expectations about them at all. Beyond a rough sense of shifting tonal centres, it makes no sense to me to believe that I anticipate what is coming when confronted with the thick, multifaceted slabs of sonic matter that make up, say, Arthur Honegger's First Symphony.

Another deeply troubling problem with Meyer's model is that it hardly seems sufficient to explain the immense variety and subtlety of musical emotion in terms of 'bad' feelings of wrong expectations and 'good' feelings of validated ones. David Huron offers a plausible argument for how a variety of feelings such as amusement, frisson and awe might be produced by different kinds of violation and resolution, but even that seems much too limited – as Huron admits, it's hard to see how expectations could play a role in the initiation of emotional states such as sadness. As John Sloboda and Patrik Juslin say, 'The interplay of tension, release, surprise, and confirmation does not amount yet to a full-blown emotion. It is better characterized as "proto-emotion", because it has a strong tendency to grow into emotions through the addition of further mental content.' But it's not clear what this further content might be, or where it comes from.

So although explaining the emotional, affective qualities of music in terms of the creation and violation of expectations has much to be said for it, it falls pitifully short of telling us why we enjoy music. Unfortunately, however, it has tended to be the only game in town, and so some accounts of music cognition just trot out this explanation and leave it at that. It seems to me that there are also, for example, emotional aspects of musical timbre and texture: we 'feel' very differently about Stravinsky's economical scoring of strings than about the full-blooded way Tchaikovsky uses them. Our reaction to a trumpet is not the same as that to a flute, say, even if they play the same notes. Voicings carry subtle affective qualities. Take the final reprise of the theme in the third movement of Beethoven's Piano Sonata No. 17, Op. 31 No. 2 (Figure 10.18). Does the added high A exert its effect (and there certainly is one) because it wasn't there on the earlier occasions and therefore subverts expectation? Perhaps we have by then heard the theme often enough for that to play a role, but I suspect that primarily we are responding simply to a local thickening of the harmonic texture.

Theories of art are prone to becoming prescriptions, whereby art that fails to deliver what the theory demands is deemed inadequate. Heinrich Schenker belittled music that did not fit his formal scheme

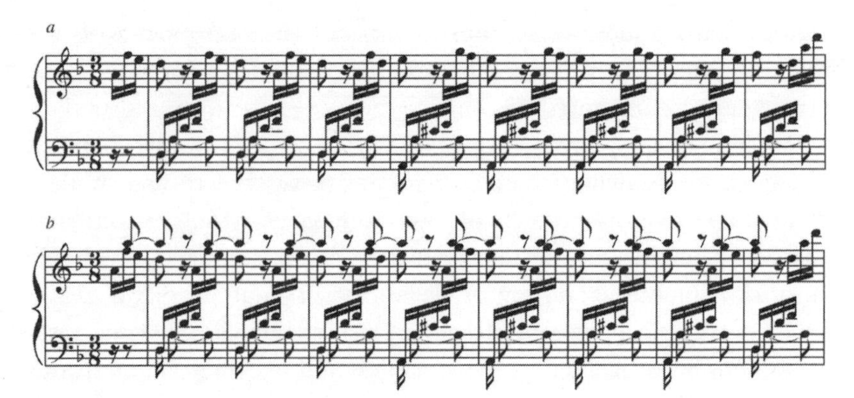

Figure 10.18 The subtle change of voicing in the initial (*a*) and reprised (*b*) theme of Beethoven's Piano Sonata No. 17, Op. 31 No. 2.

of musicological analysis (described on p. 366), and in similar fashion Meyer criticized minimalist and popular music for failing to establish sufficient expectations about future events. I don't think Meyer is right about that even in his own terms; but even if it was so, that doesn't mean these forms can't provide a rich musical experience, so long as they provide other means (timbral, say, or lyrical) of stimulating interest and emotion. Meyer complains that we have lost the ability to tolerate the long-term uncertainties about the future that are offered in a Beethoven symphony, saying of 'bing bang' popular music that its 'aggressive force implies no future goal, but rather calls for compliance *now*'. Yet not only do I suspect that many, perhaps most, enthusiasts of Beethoven or Mozart have tended to derive their enjoyment from fairly localized, 'superficial' aspects of the musical structure, but also it is hard to see why Meyer's complaints wouldn't apply equally to the popular songs of the Renaissance.

What's missing

These are some of the reasons why I believe we need a wider general framework for constructing a picture of musical emotion – one that reflects more precisely and explicitly the true nature of what we feel when we take pleasure in music. I want to end this chapter by looking at where we might find such a thing.

Not everyone agrees that musical emotion can be given a cognitive explanation, partly I suspect because this appears (and only appears) to reduce music to formulas. Philosopher Roger Scruton insists that 'we should resist cognitive theories of expression; for, however sophisticated, they miss what is really important – which is the reordering of the sympathies that we acquire through our response to art'. But Scruton has nothing to put in its place: he holds strong views about the kinds of emotional responses we *ought* to have, but can say nothing about how music evokes any response at all. When Scruton says 'the great triumphs of music . . . involve this synthesis, whereby a musical structure, moving according to its own logic, compels our feelings to move along with it, and so leads us to rehearse a feeling at which we would not otherwise arrive', it is hard to disagree. But compels us how? And in which direction?

And while it is easy to concur again with Scruton's injunction that our response to music should be sincere and sympathetic, he can't explain why we can have an emotional response to patently insincere and manipulative music, often against our better judgement. 'Many people are moved by music that is not moving at all, but merely sentimental,' he frowns. I know that's true, because I am. I suspect you are too. And I don't think it is a bad thing, provided that you recognize it as such. To an aesthete like Scruton, such a response is uninteresting; from the perspective of music cognition, it is critical. What is it about bad, sentimental music that moves us? Probably much of that is due to the artful deployment of clichés – but why are they clichés, and why do they work even if we know that they are hackneyed? Besides, since it seems absurd to suppose that there are clear boundaries to be drawn between the sentimental and the profound (Fauré is one composer who offers both), we can't simply rule this out as a question that is irrelevant to 'real' music. It's possible that understanding it might even guide our aesthetic judgements.

Where most current cognitive analysis seems to fall short most egregiously is in presenting musical emotion as some kind of oscillation between positive and negative emotional responses. Obviously, the positive must largely win out, because we enjoy music so much (that's to say, we enjoy the music that we enjoy!). But I wonder whether a considerable part of that positive response comes not from pleasure at having expectations (eventually) met, but from the sheer pleasure of mental arousal, which itself carries a basically neutral valence. In other words, the music itself produces the aural equivalent of fireworks, and we delight in this sense of activity. No one, I think, would argue that fireworks are enjoyed because of the tension between our expectation of darkness and its temporary violation by explosives. There is an aesthetic beauty in the way each sparkling trail combines with the others to produce a transient pattern, and in the rich contrasts of glowing colours against velvety black. I'm not sure, to be honest, quite *how* a psychologist might explain the pleasure in all of this (although Freudians would probably see some degree of mimesis); the fact remains that it seems to be intrinsic to the dynamics, and not linked to any kind of expectation and reward. There is probably an optimal level for this sort of stimulation: fireworks filling the entire canopy of night would be overwhelming and confusing.

This position has been argued most cogently by philosopher Peter Kivy. He points out that some of the finest music can't meaningfully be described in terms of 'garden-variety' emotions such as happiness, sadness, anger or 'lovingness'. Think, he says, of the sacred music of the Middle Ages or the Renaissance, of Josquin des Prez and Thomas Tallis. This is beautiful and magnificent to a degree that can easily move us to tears – and yet the music is in a real sense *not expressive*, certainly not in terms of attempting (as Mozart's music might) to elicit a particular emotional state. Significantly, this kind of music, and the intricate counterpoint of Bach, is rarely included in psychological testing of musical emotion, precisely because it is so hard to say what 'emotion' is being represented.

Yet this is not, Kivy reassures us, because the music is invoking some abstract 'aesthetic emotion' without a name. We know what this feeling is: it is 'excitement', or 'exhilaration', or 'wonder', or 'enthusiasm'. I would suggest that it is, more precisely, 'the excitement that you feel when listening to music' – that we can best understand it with explicit reference to its cause or object, just as we can describe love as the feeling we get in relation to our loved ones. This doesn't necessarily mean that the fundamental feeling is music-specific, but it does suggest that it is not well captured by reference to the 'garden-variety' emotions that are the focus of a great deal of music psychology.

Kivy argues that this view solves the puzzle of what is going on when we call music 'sad' or angry', without actually feeling sad or angry in the way we do in everyday life. That 'sadness' seems to be more like a contemplation of sadness. Kivy suggests what we are actually experiencing is the feeling of musical excitement or joy *at* the 'sad' connotations of the music. Similarly, we can be exhilarated by the 'funereal melancholy' of the slow movement of Beethoven's Seventh, or moved at how 'magnificently fearful or angry' music can be – whereas the actual experience of fear, anger or sadness would not have this positive valence at all. Philosopher Stephen Davies offers a somewhat similar view: that the emotions invoked by music, devoid of any object or cause to which we might attach them, give us an opportunity to reflect on these states free from any need to 'do' anything about them. Sadness that is 'about' something may well leave us wishing its causes were otherwise. But music can offer a kind of image of sadness which invokes no compunction to 'put it right' – because there is obviously nothing 'wrong'.

One might object that this is a kind of juggling with mental states that have never been clearly characterized and that people do not articulate. But to me it comes closer to the truth than do psychological studies in which listeners are required to tick boxes marked 'happy', 'sad' or whatever. We seem to broadly agree on how that task should be conducted for musical stimuli; but it does not say much about what we actually feel.

One could also object that Kivy's notion of exhilaration or excitement sounds suspiciously close to the idea that music is a kind of sonic cheesecake, a mere massaging of the auditory sense. But this essentially hedonistic quality is not like the satisfying stimulus of sugar and fat, or the sensual nature of a warm bath (at least, it is not *just* that). It holds within it something of the joy of being alive and in community with others. It is partly a kind of wonder at realizing what other minds are capable of creating.

11

Capriccioso
Going In and Out of Style

What are musical styles?
Is music about notes, or patterns, or textures?

Throughout the world we are listening to a greater diversity of music than ever before. We can wander into a big music store in Tromsø or Tokyo and pick up music for virginal by William Byrd, Tibetan Buddhist chants, and Scandinavian death metal. We may find entire genres of music we never knew existed. But who now goes to music stores anyway? The musical universe is online, available at the click of a mouse.

This richness in musical styles is accompanied by a remarkable capacity to tell them apart. It's true that many people over the age of thirty may struggle to hear a difference between hip hop, gangsta rap, reggaeton, snap music and crunk; but we can sometimes make surprisingly discerning judgements without knowledge of a genre at all. Well-informed listeners can often identify specific composers or bands from compositions that they have never heard before, sometimes after only a few bars.

How people develop their musical stylistic preferences is surely bound up with experience and environment, including musical 'training' (not necessarily formal) and self-image. We may not understand much about how that happens, but we can make some good guesses. Musically sophisticated listeners become adept at spotting patterns and structures, and are likely to seek satisfaction in the complexity that derives from discrepancies in these schemas: they enjoy the kinds of 'puzzles' we explored in the previous chapter. Naïve listeners, meanwhile, prefer simplicity through repetition. In other words, some want Messiaen and Ornette Coleman, others Dolly Parton. Or rather, perhaps there may be times when we want Ornette, and times when Dolly hits the mark. Sophisticated appreciation of one musical genre doesn't always transfer to others – some rock enthusiasts never progress beyond Beethoven, and some great classical musicians have excruciating popular tastes. The interesting question, however, is not how one sort of music should be graded against another in a kind of talent show or intelligence test, but how each type of music does what it does.

That is the real mystery of what we call musical style or genre. Even if all the notions we have encountered so far about, say, the use of tonality, the principles of melodic cohesion and harmonic progression, and the violation of expectation, are true descriptions of how we perceive and respond to music, they do not begin to explain why music is not (within constraints set by scales and intonation) a homogeneous affair. How can someone who responds to these things love bebop and loathe Berlioz, or love Parisian café music but have no time for Beijing opera? I suspect you may know what I mean when I confess that there are certain genres of music I feel I should like but simply cannot, no matter how I try.

It may not be quite true to say that there's no accounting for taste; but I suspect that it is seldom very interesting to try to do so. It seems unremarkable that we will tend to find easier on the ear the music that we've heard since childhood, rather than that from a wholly different culture, or that we often find our musical tastes begin to ossify in early adulthood. But what is it that enables people to distinguish Beethoven from Mozart, or Irish fiddle music from Scottish? What gives composers and musicians their own distinctive soundscape?

Elegant variation

Some sonic signatures are obvious and well understood: Thelonious Monk's percussive tone clusters, the Beach Boys' lush harmonies, Debussy's whole-tone scales and parallel movements of seventh and ninth chords. But I want to focus here on how some of the considerations that I have introduced earlier – the achievement of particular cognitive effects through the manipulation of pitch, rhythm, harmony and timbre – are expressed in the context of stylistic variation. In short, I consider the question of style from the point of view of how composers make choices from the available musical options.

Take melody. There are about 60 million ways of combining just four notes on the piano keyboard (even before we allow for variations in rhythm), so in one sense the space of all possible music is astronomical. But cognitive principles, whether explicitly acknowledged or (more often) empirically intuited in composition, narrow down the range of acceptable note permutations considerably. Observance of tonality cuts almost in half the number of notes likely to be used, and the laws of pitch steps place tight constraints on where a melody might go next. This might be imagined to frustrate composition, but in fact all artists know that rules and limitations are essential: too much choice is paralysing. That is why, when Schoenberg abandoned tonality, he felt the urgent need for a new system of constraint, and entered one of his most productive periods when he found it in twelve-note serialism.

Such limitations on the choice of notes and melodic figures mean that inevitably some composers have ended up using very similar tunes and harmonizations. Take the simple figure in Figure 11.1*a*, and look at what Mozart, Beethoven and Dvořák made of it (Figures 11.1*b*-*d*). There's no reason to think that these composers were lifting material from one another, although there was no law against that – composers commonly 'quoted' from one another's works, sometimes as an act of homage and sometimes for satirical effect. (Such borrowings are less permissible in the competitive, commercial world of pop music,

Figure 11.1 A simple melodic figure (*a*) as used by Mozart (*b*), Beethoven (*c*) and Dvořák (*d*).

as George Harrison famously discovered when he was deemed by the US Federal Court to have 'unintentionally copied' the melody of the Chiffons' hit 'He's So Fine' in his 1969 song 'My Sweet Lord'.) The rules of tonal composition throughout most of the eighteenth and nineteenth centuries were so tight that it was fairly inevitable that composers would sometimes hit on the same ideas. One can hear many little echoes throughout the works of Western composers, and certain harmonic progressions became hackneyed even in the hands of the greatest composers. As the example above illustrates, these similarities have little significance in themselves; the interest lies in how the composer has, through choices of phrasing, rhythm and harmony, imposed his own style on material that is at root so basic that anyone could have come up with it.

There's a warning here for cognitive musicology, which sometimes appears to regard composition as the art of finding nice or 'effective' sequences of notes. Of course it is much more than that. You don't even need to be a good melodist to write great music. Bach composed some wonderful tunes, but he was also happy to make use of the most bland, inexpressive snippets of melodic material, which he assembled via fugal counterpoint into structures of staggering beauty.

Moreover, one can take the same melody and adjust the harmonic backing, completely altering the mood and implication. Duke Ellington's version of 'Summertime' seems to place it not in the Deep South but in the even deeper south: in Latin America. Even a variation of loudness or tempo can turn the music on its head. 'Always remember', said Aaron Copland,

> that a theme is, after all, only a succession of notes. Merely by changing the dynamics . . . one can transform the emotional feeling of the very same succession of notes. By a change of harmony, a new poignancy may be given the theme; or by a different rhythmic treatment the same notes may result in a war dance instead of a lullaby.

In Western classical music the fingerprint of personal style seems to be so deeply impressed even at the smallest scales that it becomes feasible to distinguish one composer from another even from just a tiny snippet of a work. Art experts have long been familiar with this idea. A common method of assessing whether an unsigned work attributed to a particular painter is genuine is to look at what they call 'minor encoding habits': how the artist has rendered small details such as hands or stones, which, often being painted almost without thought, may be more distinctive than larger-scale features such as composition or lighting. This method has also been used to attribute authorship of unsigned literary texts. In 1964, communications researcher William Paisley at Stanford University in California considered whether this technique would work for composers too. He picked several themes written by different composers, and showed that just the first four notes encoded distinctive melodic habits of each individual, so that, for example, a theme by Beethoven could, using the habits discerned from the test cases, be reliably distinguished from ones by Mozart or Haydn.

Impressed by such a simple measure of style, psychologist Dean Keith Simonton decided to investigate what it might reveal about the readiness of composers to depart from conventional melodic tropes and display originality. He deduced what is 'conventional' by studying the statistics of successive pitch steps in the first six notes of over 15,000 themes composed by 479 classical composers working before the mid-twentieth century. In other words, he calculated the probability profile of 'two-note transitions' – the likelihood that the first six notes of a theme will contain an interval of, say, tonic to second note, or tonic to fifth, or third to fourth. Not surprisingly, given that most of the composers worked solidly in the tonal tradition, pitch steps between the tonic and fifth notes (including tonic-tonic and fifth-fifth) were the most common, while transitions to notes outside the major or minor scales were very rare.

One might question the validity of any average that stretches over such a range of times and styles, but it's not unreasonable to think that this procedure nonetheless reveals something about the norms of the Western tonal tradition. Simonton then measured the 'originality' of a theme in terms of the extent to which its two-note transition probability profile departs from the average. For example, the final movement of Beethoven's 'Waldstein' Sonata (No. 21, Op. 53) begins with a rather conventional theme (Figure 11.2*a*), in which all of the two-tone transitions are common ones. But the initial theme of Liszt's *Faust Symphony* (Figure 11.2*b*) contains some very unusual

Figure 11.2 Originality? The opening of the final movement of Beethoven's 'Waldstein' Sonata No. 21, Op. 53 (*a*) uses conventional two-tone transitions, whereas the theme of Liszt's *Faust Symphony* (*b*) uses rather unusual ones.

pitch steps. Simonton could work out, from the average transition profile, just *how* unusual such a six-note theme was, and assign it a numerical value – a measure, he said, of originality. Of course, this simple method ignores all the originality (or banality) that might be located in rhythm, harmony and orchestration. What's more, it shouldn't be confused with any measure of musical quality – a 'conventional' theme can still be very beautiful, while an 'original' one, by this measure, might sound cacophonous.

Nonetheless, the originality measure tells us something about the readiness of composers to innovate melodically at different times and in different contexts. For example, instrumental themes tend to be more 'original' than vocal ones, possibly because instruments can accommodate unusual intervals more readily than the voice – or maybe because composers tend to keep vocal lines more transparent so that they can communicate the words clearly. There is also more 'originality' in chamber works than orchestral ones, perhaps because composers regard smaller-scale works as better suited to experimentation. And the originality scores for the first and last movements of many-movement compositions such as symphonies are higher than those for 'inner' movements – might the composer feel a stronger need to grab the listener's attention with novelty at the outset, and to wake them up again at the end?

It's also suggestive to see how 'originality' changes over time. It increases fairly steadily since the Renaissance, but not smoothly – there are ups and downs, with a trough coming during the Classical period of Mozart and Haydn. One might argue that this music was more formulaic: the rules of tonal harmony having been established in the early eighteenth century, we saw earlier that composers of this period tended to make use of standardized themes and melodic structures with well-defined emotive connotations. But again, this doesn't by any means imply that the music of that time was less engaging or inventive; it means merely that experimentation with melody was not such an important consideration. It's perhaps telling too that the 'originality' measure for individual composers typically increases during their lifetime, peaking at around the age of fifty-six (Beethoven's age when he died). Today we tend to imagine that artists are at their most creative in their youth. But in earlier times music was, like painting, a somewhat conservative tradition in which originality was not especially encouraged early in one's career – compare, for example, the early and late works of J. M. W. Turner, or think of the contrast between Beethoven's early and late string quartets.

It's curious to note that the increase in this crude measure of 'originality' doesn't seem to be accompanied by an increase in the use of chromatic notes, at least for J. S. Bach, Haydn, Mozart, Beethoven and Brahms. In other words, Beethoven seems to have used much the same notes in more original ways in his themes as he got older. (It's still an open question whether Beethoven's *harmonies* became more chromatic and dissonant over his career.)

Some themes are so memorable that they can't be recycled: they become inseparably associated with a particular work. No one can reuse the *dah-dah-dah-dum* motif that opens Beethoven's Fifth Symphony (unless they wanted explicitly to invoke Beethoven). The recurrent themes and motifs of a classical composition were, until around the mid-twentieth century, considered the main focus of music analysis. (Many later-twentieth century works dispensed with themes altogether, as we'll see.) But there was nothing particularly systematic about the way this was done – the rules for determining what was a 'theme' could seem rather arbitrary. David Huron has argued that a theme characteristic of a particular work should not only recur frequently, in some basically recognizable form, within that work, but should also be distinct from patterns found in other works – a common pattern such as a rising scale is not really a theme at all. Such structures can be identified by analysing a piece of music for its 'information content', according to definite mathematical principles. A theme offers no or little 'new information' when it recurs in a work, but its first appearance does supply information relative to what we have 'learnt' from other works. This is really just a way of formalizing the intuitive idea that a composer strives to deliver something memorable, like the 'hook' of a popular song.

But traditional musicological analysis can get too hung up on themes. For one thing, as Charles Rosen points out, 'A classical composer did not always need themes of any particular harmonic or melodic energy for a dramatic work . . . the drama is in the structure.' Great music does not stem from finding a fresh tune and then running it through a sequence of permutations and transformations, but from assembling the component parts into a coherent whole. There is no prescription for doing this; music that moves us seems to have a necessity about the way it unfolds, but it is not something that can be derived from a formula. That is as true of a Mozart sonata as it is of a Johnny Hodges solo. You need to be able to hear imagination and creativity at work.

Music by numbers

If, as it seems, statistical analysis can identify at least some of the basic elements of a composer's style, that would seem to raise the possibility of systematic pastiching: of creating music according to a set of rules that 'sounds like' Bach or Debussy. Of course, we can do this without explicit rules, precisely because the human mind is such an extraordinary pattern-recognition device. I am confident that I can sit down and write 'Debussy-like' music which, although musically vacuous, would be recognized as such to any informed listener. Playing a piece of music 'in the style of' a different genre is a favourite parlour game of musicians.

But the notion of formalizing the 'rules' governing a compositional style may have more profound implications. Let's say, for example, that one could write down a set of rules for composing music in the style of Bach, and that if these were enacted by a computer program, it could generate music that even well-informed listeners couldn't differentiate from real Bach. Does this mean that we will have discovered 'how Bach composed'? Of course it doesn't: natural and social scientists are well aware that, just because a computer algorithm can produce something that looks like a phenomenon in the natural world, that doesn't imply that the generative rules capture those that apply in reality. All the same, they *could* be Bach's own rules, even if he didn't employ them consciously.

The real impact of a program like this, however, is that it would seem to undermine our feeling of what making music is about, namely that it is a creative, even numinous, human activity that can't be automated. If you were to say that it seems utterly implausible that any computer program could come up with a composition to rival the great (or even the minor) works of Bach for profundity and beauty, I would agree with you. Might it not be feasible, however, that such an algorithm could at least generate short segments of music sufficiently 'Bach-like' to persuade us that this is indeed one of the more routine, formulaic passages that, in all honesty, can be found in his works if you look hard enough? And if so, can it really be meaningful to suggest that everything of true musical worth lies in the gap between *that* and the 'Brandenburg' Concertos? On the contrary, a feat of that kind must surely have already got a considerable step along the way.

Well, as you might have guessed, this is no mere thought experiment. In the 1980s, a computer scientist named Kemal Ebcioglu created a program called CHORAL that 'harmonizes chorales in the style of Bach'. Clearly this is quite a different challenge from that of 'writing' a piece of Bach-like music from scratch – the basic melody was taken from genuine Bach chorales, which made it possible to compare what CHORAL did against what Bach did (Figure 11.3). The

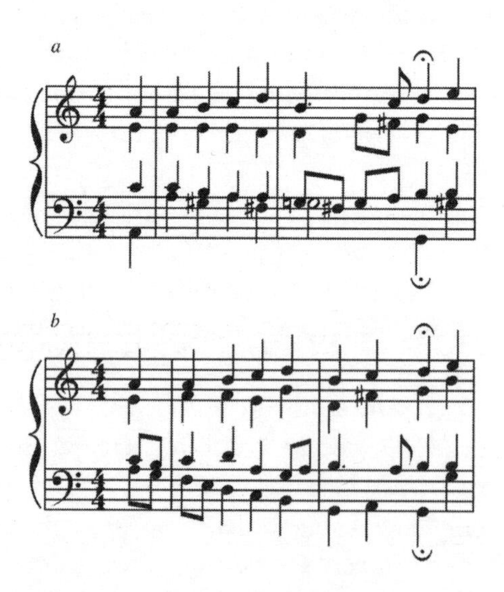

Figure 11.3 Bach's Chorale No. 128 (*a*), and as harmonized from the melody alone by the computer program CHORAL (*b*).

program used purely 'music theory' principles, such as the rules of voice-leading, whereas Bach was likely also to have considered such things as how the harmonization should reflect the words or how easy to play the parts were. Yet the results were reasonably convincing, at least superficially – they weren't, at any rate, obviously nonsensical. However, because Ebcioglu was more concerned to show that a somewhat abstract product of the human mind could be plausibly mimicked by an algorithm, rather than to assess the musical quality of the product, sadly he didn't conduct any tests of how knowledgeable listeners would judge CHORAL's efforts.

In the 1970s, Swedish researchers Johan Sundberg and Bjorn Lindblom created a very different style-mimicking musical algorithm. They were trying to uncover a set of 'grammatical principles' behind the Swedish nursery rhymes composed by Alice Tegnér in the late nineteenth and early twentieth centuries. These tunes are fairly simple, even stereotypical, and Sundberg and Lindblom wondered whether they possess a so-called generative grammar, akin to those proposed in linguistics (see Chapter 12), that could be computerized so as to churn out 'Tegnér-like' songs ad nauseam. It's hard to judge the results (Figure 11.4) unless you're familiar with Tegnér's *oeuvre*, but the idea that such rudimentary, highly tonal tunes have stylistic features that are sufficiently narrowly defined to be automated seems inherently plausible.

If the thought of computer-generated music makes you shudder,

Figure 11.4 A nursery rhyme by Alice Tegnér (*a*), and 'Tegnér-like' tunes generated by a computer algorithm (*b*).

remember that experiments in the 'automation' of composition were well known in the eighteenth century, when some composers dabbled in so-called *Musikalisches Würfelspiel* or 'musical dice games'. Here rolls of the dice specified the order in which pre-composed musical fragments would be assembled. One of these games was published in 1792 by Mozart's publisher Nikolaus Simrock in Berlin, and although the author is unknown, the piece has been attributed to Mozart himself. A well-authenticated manuscript by Mozart, denoted K516f, is believed to be some kind of dice game, although the instructions are not specified. It consists of many two-bar melodies designated by small or capital letters, along with an example of their combination assembled by Mozart. Hadyn too created his *Philharmonic Joke* (1790) this way. Both Mozart and Haydn were famously playful, and it's not clear that these pieces were intended as anything more than amusing diversions. But a random, algorithmic approach to composition has been pursued more soberly by some modern composers; most notably, Iannis Xenakis used computers to create stochastic music. And Schoenberg's serial method was a pseudo-mathematical device that later adherents expanded to leave ever less scope for intervention by the composer.

Algorithms based on randomness are, however, only likely to produce something that most people would regard as musical if the ingredients are highly constrained from the outset. The fragments jumbled about by Mozart and Haydn, being tightly bound to a given key and anodyne in themselves, were pretty much guaranteed to produce something easy on the ear yet also somewhat bland. Ebcioglu's algorithm, on the other hand, was kept within 'respectable' bounds by the fact that the principles of tonal melody and harmony were built into the rules. Baroque music was in any case rather rule-based, which is why auto-Bach seems a much more viable proposition than auto-Stravinsky or even auto-Beethoven.

Computer-generated music today typically takes a different approach. Rather than relying on a priori, 'bottom-up' rules about how each note should be chosen, it works at a more global 'top-down' level by inferring rules and guidelines from real musical examples. Research into artificial intelligence over the past several decades has been characterized by the use of adaptive 'neural' networks that can generalize from experience, learning to recognize typical patterns in the stimuli rather than relying on exact matches to some predetermined criteria. This is probably closer to the way the human mind works, although it's not at all clear that there are any precise parallels. One of the best-known attempts to use adaptive learning to create music was that of computer scientist John 'Al' Biles of the Rochester Institute of Technology, whose algorithm GenJam learns to improvise jazz. Biles, a trumpet player, performs live alongside GenJam under the name the Al Biles Virtual Quintet, saying that this makes his scheme perhaps 'the only evolutionary computation system that is a "working musician"'. Biles is modest about the results, saying that 'After sufficient training, GenJam's playing can be characterized as competent with some nice moments.'* 'Training' here is the procedure by which GenJam's learning algorithm finds pleasing variants of the music it seeks to emulate. This is a kind of Darwinian process in which a listener assesses GenJam's efforts and provides a 'thumbs up' or 'thumbs down' to help identify 'good' results that are mutated and used as the basis for the next generation of attempts. Biles says that GenJam's improvisation is usually tolerable after about ten such generations, but that the first few generations can be 'quite numbing for the mentor'.

Well, everyone has to learn, and we all sound pretty grim to begin with. But musicality is not simply a matter of trying out new licks until your mentor is satisfied – most composers have only had themselves as a critic before they put their works before an audience. Can computerized musical systems be autonomous, evaluating themselves? Some researchers are trying to create 'synthetic artists' that are also critics, capable of comparing their attempts against some kind of automated aesthetic criteria. One of these systems, devised by cognitive scientists Lee Spector and Adam Alpern of Hampshire College in the mid-1990s, is called GenBebop, and tries to create improvised solos in the style of Charlie Parker (there's no harm in aiming high). Like GenJam, it is a 'genetic' algorithm that relies on training, in this case by 'listening' to Bird himself; but the algorithm also has its own internal critic. After twenty-one generations, the program was able to create four-bar improvisations that satisfied this critic. But, said Spector and Alpern, 'it does not please *us* particularly well'. They were philosophical about it: 'Nobody said it would be easy to raise an artist.'

* You can judge for yourself at www.it.rit.edu/~jab/GenJam.html

A big part of the problem seems to be finding 'sieving' criteria that do not end up producing something trite. Much of the skill in creating music involves knowing when to *break* the rules yet without producing sheer cacophony. Finding an auto-critic who is both 'open-minded' and discerning is a huge challenge. But isn't that just like life?

Among other things, studies like this serve as a reminder that we still understand rather little about how performers develop their improvisational styles in the first place. That neglect perhaps reflects the Eurocentric nature of music-cognition research, since a great deal of music in the world is improvised, and only Western classical music of the eighteenth to the early twentieth centuries tends to be prescribed down to the last note. Arguably the most profound musicality resides in spontaneity. Good improvisation is not merely a stringing together of stock phrases and licks, but an intelligent and sensitive exploration of the musical surface on which the composition sits. Even superb improvisers struggle to explain how they do it; some would doubtless prefer to let the process remain mysterious.

New ways of listening

Musical styles become codified with depressing inevitability, and as a result virtually every innovation is opposed – Bob Dylan was in good company when his use of the electric guitar provoked such hostility in 1965. Minor 'riots' were a common, almost habitual response to new compositions in the early twentieth century, and the works of Schoenberg and Berg got worse receptions (especially in ultra-conservative Vienna) than that of the notorious Paris premiere of *The Rite of Spring* in 1913. Some of this resistance is understandable, though: it takes time to retrain the 'ear' to new sounds, and until we do they might seem a dreadful din.

Cultures that use music ritualistically are likely to be among the most musically reactionary, as borne out by the restrictions on musical experimentation in the medieval and Counter-Reformation Catholic Church. It is easy to scoff at such narrow-mindedness, but we should remember that some conservatism results from a determination to take music seriously, rather than seeing it as hedonistic frippery. If, like the Greeks and the medieval churchmen, you believe that music has moral implications, and that the 'wrong' music can lead to the wrong morals, it is natural that you will resist change.

Something of the same attitude can be detected today in diatribes against popular music: Roger Scruton equates what he regards as 'the decline of musical taste' with 'a decline in morals'. Punk, which to its critics was a morass of nihilistic anomie, at its best contained more integrity and even morality than the bloated, numbing excesses of the 'easy listening' that sugar-coated the hypocrisies of the post-war decades. And proposing that musical taste has declined since one's youth is never a wise move, given that this has been more or less universally asserted since at least the Enlightenment, and probably back to the Counter-Reformation. Scruton's paragons of pop musicality, the Beatles, were of course once the corrupters of youth.

But Scruton's case for musical decline is interesting because it is not simply couched in vaguely moralistic terms but arises out of an analysis of the music. As he rightly points out, much pop and rock music ignores genuine rhythm for the sake of relentless beat, and relinquishes any harmonic movement in favour of repetitive block chords strummed on guitar. Some 'melodies' wander sluggishly between two or three notes. If one analyses the music using the same tools and criteria that are applied to Mozart, a lot of rock music has rather little to offer.*

Yet as rock musicologist Allan Moore has demonstrated, these are often *not* the right tools at all, any more than it makes sense to look for cadential structures, harmonic progressions and rhythmic invention in the music of Ligeti or Stockhausen. As Scruton recognizes, allusion is a central aspect of any rich musical style, and that is nowhere more true than in rock music, which juggles genres and idioms (consider David Bowie or Talking Heads, for example) and is engaged in a constant dialogue with its past. And as I argued in Chapter 6, discussing rock music without reference to timbre and texture is like discussing African music without reference to rhythm.

This is equally true of a lot of recent and contemporary classical

*It doesn't help that Scruton, apparently considering rock music much of a muchness, seems unconcerned where he takes his examples from, and draws mostly on second-tier fare – rather like assuming that classical music is equally well represented by Czerny or Beethoven.

music: we are simply not going to get anything from it if we insist on forcing it through the mesh of nineteenth-century romanticism that now constrains so much of the public perception of the classical music world. We need to find new *listening strategies*. That, perhaps, makes it sound like awfully hard work, but it's really just a matter of being open to what the music offers us as organizing principles. What Judith and Alton Becker say about the music of non-Western cultures applies equally to music outside the Western tonal tradition: 'it has become increasingly clear that the coherence systems of other musics may have nothing to do with either tonality or thematic development . . . What is different in a different musical system may be perceived, then, as noise. Or it may not be perceived at all. Or it may be perceived as a "bad" or "simple-minded" variant of my own system.' (The last is certainly true of many traditional musicological critiques of rock and jazz.)

For example, contemporary art music is often less about sequences of notes or beats than about sculpting sounds. We hear them *as* sounds: as discrete, almost material entities occupying space. Such music does not have the quasi-narrative flow of Mozart or Beethoven, but exists in a kind of timeless present, or what music theorist Jonathan Kramer calls 'vertical time', in which events are stacked up rather than sequential. Modernists such as Webern, Stravinsky and Messiaen began exploring this mode of composition, and it reaches its most extreme from in the ultra-minimalist works of La Monte Young and his sometime mentor Stockhausen. Young's *Composition 1960 ♯7* consists of the instruction to sustain a perfect fifth 'for a long time', while Stockhausen's *Stimmung* (1968) is basically a chord that slowly mutates as it is sung by six vocalists for about seventy-five minutes. One might argue that, in their pseudo-random, disjointed and erratic barrage of notes, Boulez's *Structures I* and *II* attain the same 'vertical' character: they might be seen as the random molecular motions that, seen from a sufficient distance, attain the average uniformity of a gas. It is no wonder these works find no space for cadences, for in vertical time there is no real beginning or ending. Now, I don't suggest that finding the right listening strategy will unlock untold riches in these works: *Stimmung* has a rich, meditative quality, but the listener who starts to yawn or wriggle can reasonably conclude that her body is telling her something about the cognitive content of this music. The point is, however, that not all music requires the same kind of listening.

The most successful of experimental contemporary compositions are those that provide genuine structures for cognition to work on. (Boulez's 'structures' above are not cognitive ones, but theoretical.) We might, for example, sense an organized interlocking and overlapping of timbral layers to create novel, shifting textures. The texture of sound became a central compositional element in the 1960s in the music of Xenakis and Stockhausen, Luciano Berio and Krzysztof Penderecki (see p. 238). György Ligeti created a remarkably diverse set of new musical textures which need to be heard as such – as almost visual objects, some monolithic and others ethereal, rather than as movements in tonal space. He recreates sunbeams streaming through coloured glass in *Lontano* (1967), while *Melodien* (1971) bubbles with countless quasi-melodic fragments. This is music that doesn't flow as streams but is carved from blocks of architectural sound, an idea again prefigured in the works of Stravinsky. The music fills up and saturates sound space just as Jackson Pollock's drips filled and saturated his canvas. Early serialism has this quality too: Anton Webern once wrote that he felt a piece was over when all the chromatic space had been used.

It is no shame to admit that this is often not easy music to listen to. Sometimes a single listening is all we really want. Its emotional palette can be rather constrained – although joy can be found there, and wonder too. Sometimes the only thing that really stands in our way is our own preconceptions.

When style becomes cliché

One of the welcome corollaries of Meyer's theory of musical emotion is that it punctures musical purism. If emotion stems from deviation from norms, then music *can never have a fixed form* if it is to avoid sterility. Deviations that prove effective at stimulating an audience are likely to become widely used – in which case they end up no longer as deviations at all, but merely clichés. All string players today rely heavily on vibrato for the post-Baroque repertoire. This wavering of pitch would have once created emotion from its slight ambiguity, but it is now simply what string playing sounds like. History shows that, when one route for deviation is cut off – by ecclesiastical decree, say

– others arise. A rigid insistence on invariant 'authenticity' such as that demanded in 1960s folk music by Pete Seeger and Ewan MacColl, may be motivated by an understandable resistance to vulgarization, but is a profoundly unmusical position.

On the other hand, Meyer saw decadence in music in which 'traditional modes of deviation are exaggerated to extremes and where these deviations are, so to speak, pursued for their own sake'. The traditional artist, he said, 'is one who understands the relationship of norms to deviants and who works within this relationship'. And the true innovator, by implication, is one who understands that relationship but knows how far to step outside it.

Eduard Hanslick remarked on the way innovation can become tired mannerism, leading to a high turnover of forms in classical music:

> Modulations, cadences, intervals and harmonious progressions become so hackneyed within fifty, nay, thirty years, that a truly original composer cannot well employ them any longer, and is thus compelled to think of a new musical phraseology.

That was precisely Arnold Schoenberg's rationale for inventing the twelve-tone method of atonal composition a decade or so later. He argued that the entire system of tonality had become repetitive, flabby and sentimental, and needed replacing. Schoenberg felt that his serial technique, with its prescriptive tone rows, offered an alternative to diatonic tradition that 'unifies all elements so that their succession and relation are logically comprehensible, and which is articulated as our mental capacity requires, namely so that the parts unfold clearly and characteristically in related significance and function'. We've seen that serialism, especially in its extreme forms, failed to do precisely this, and that whatever coherence remains in its *oeuvre* is there in spite of, rather than because of, its handling of pitch.

But was Schoenberg right in the first place to say that tonality was dead on its feet? We need only consider that he made this claim when Bartók, Stravinsky, Shostakovich, Hindemith and Honegger were at the height of their powers to see how groundless it was. However, to some of Schoenberg's supporters, notably the sociologist and musicologist Theodor Adorno, this wasn't a solely musical issue. Adorno held that the edifice of tonality was a product of a complacent and selfish bourgeois capitalism, representing all the forces of nineteenth-century conservatism that modernism would replace in an act of liberation. Not only did this conveniently ignore that fact that this same bourgeois audience was grumbling at the newfangled strangeness of Stravinsky's extraordinary new *tonal* language, but it also failed to anticipate that serialism would fast become a much more elitist and rigid style than tonalism ever was (see Box: *Who cares if you listen?*).

Besides, Schoenberg and Adorno never really explained where the supposed banality of tonality lay. While it's true that one can detect a certain reliance on easy affect and overused structures in the music of, let's say, Sibelius and Vaughan Williams, the complaint was also that classical conventions had been debased in popular music. The serialists poured special scorn on the diminished seventh chord, whose 'shabbiness and exhaustion', said Adorno, is evident 'even to the most insensitive ear'. This chord, we saw earlier, had been used as a pivot for modulation at least since Mozart's day. But Roger Scruton points out how absurd it is to say that a particular *chord* can be banal, rather than the way it is used. 'What would remain of the art of painting', he asks, 'if individual shades could simply be deleted from the painter's palette by those who use them tastelessly?' Moreover, he suspects that Schoenberg had it in for the diminished seventh precisely because its ambiguous nature, devoid of any clear tonal centre, opens up for the tonal musician the liberation from the tonic that Schoenberg wished to claim for himself.

The most important lesson to be drawn from serial atonalism, however, is that musical styles cannot be invented by fiat or by a set of rules. Certainly, they can be the work of individuals – pioneers from Monteverdi to Louis Armstrong created genuinely new styles of composition or performance. But they do so by making a new synthesis of old approaches, not by rejecting them all and starting from scratch. Indeed, even Schoenberg didn't draw a line under everything that went before, since he used traditional classical forms to frame his atonal works. His protégés sometimes went further: Berg resurrects tonality in his serialist Violin Concerto with a tone row split into two tonal segments, in G minor and B/F♯ major. For the ear needs a bridge in order to assimilate new ideas: it can adjust to novelty, but only if there is something familiar to help organize it.

Who cares if you listen?

I suspect most people would agree that music is a kind of communication. (I look in the next two chapters at what and how it can communicate.) Whether the composers feel that everything they wish to say is expressed in the formal relationships between musical events, or whether they have some programmatic intention, or whether they want the listeners to dance, or be amused, or impressed by skill and virtuosity, or be battered by a wall of sound, the implication is that the composer or musician acknowledges an audience that can potentially be engaged and in a musical sense, spoken to.

Of course, plenty of music happens without an audience in the usual sense – at rehearsals, for example. Some players never develop the confidence or feel the need to perform in public. Even in such cases, I suspect musicians play as though an audience existed.

The notion, then, of a composer who painstakingly creates a work, perhaps of immense intricacy, while being utterly indifferent to the existence of a listener could seem odd, even perverse. Some composers assert that they pay no heed to what they think an audience might enjoy or demand – Harrison Birtwistle is, by his own admission, one such – and no doubt that sort of freedom is important for creativity to thrive. But no one questions that Birtwistle writes 'for' an audience.

Yet the image of a composer who eschews listeners was presented several decades ago as a desirable norm. In 1958 the American composer and music theorist Milton Babbitt wrote an article in the music magazine *High Fidelity* that still provokes furious debate today. His title said it all: 'Who Cares If You Listen?'

Babbitt was responding to widespread accusations, during this period of extreme serialism, that classical music was incomprehensible to, and disliked by, the great majority of the 'music-loving' public. He argued that not only should modern composers be unconcerned at this animosity, but they should welcome it. If the public had no interest in their new music, he said, then the composer should stop worrying, stop compromising and indulging in exhibitionism to attract an audience, and simply get on with his (hers was not in the picture) craft – or perhaps, he might have been tempted to say, his science.

Babbitt admitted that the use of pitch, rhythm and other musical parameters in the atonalism then in vogue 'makes ever heavier demands upon the training of the listener's perceptual capacities'. This, however, was no acknowledgement of the difficulties with which a listener was confronted, but a prelude to angry criticism of the inadequate resources with which they were typically equipped. Because they are so bad at remembering precise values of pitch, register, dynamics, duration and timbre, these audiences ended up 'falsifying' the intentions of the composer.

And to make matters worse, they then blame the composer for the unsatisfying experience. But why on earth should they *expect* to understand this advanced music, Babbitt asked, any more than they would advanced mathematics? Mathematicians and physicists can go about their recondite business without being accused of decadence and of shirking their social responsibility; why not musicians?

And so, he concluded,

> I dare suggest that the composer would do himself and his music an immediate and eventual service by total, resolute, and voluntary withdrawal from this public world to one of private performance and electronic media, with its very real possibility of complete elimination of the public and social aspects of musical composition.

We can understand some of the reasons why Babbitt was so irate and contemptuous. The lament that 'modern music' is 'hard' often comes from a complacent audience which seems to want only a stodgy diet of romanticism in which a touch of Stravinsky and Prokofiev is tolerated to show breadth of taste. Many people today who claim that 'modern music' is incomprehensible seem not to have listened to any since Babbitt's era, and are unaware of the resurgence of tonality (as one strand among many). And Babbitt is rightly impatient with the idea that music should be undemanding. What's more, he was prescient in suggesting that electronic media could free musicians and composers from the normative and homogenizing demands of commercial success – all sorts of odd sonic experiments thrive on the web, and our musical world is the richer for it. Babbitt is right that the marketplace is not the best arbiter of what is valuable in music.

Yet in ignoring public taste entirely, musicians and composers from Babbitt's milieu became increasingly dogmatic and narrow about what music *is*. (At one point Boulez refused to speak to non-serialist composers, and was even condescending to Stravinsky.) Composers cannot expect to do whatever they like and have it recognized as music. I know that will sound terribly proscriptive – and certainly the old question 'is it art?' quickly becomes a sterile semantic debate. But for the reasons described in this book, we can be rather more precise as far as music is concerned. Music can mean many things, but it doesn't seem tenable to allow it to mean *anything*. Our minds use particular cognitive tools to organize sound into music. If we eliminate all possible modes of organization from the acoustic signal, we are left with just sound.

This is what some of the later experiments in serialism threatened to do. If, for example, we remove any means to develop expectations, we remove much of music's affective power. If the formal relations between musical elements cannot be perceived, but only demonstrated on the score, there is no audible coherence. If something of this sort were done in literature, would we accept it as such? I am not talking about *Finnegans Wake* (though one might reasonably baulk at the suggestion that this should become the only acceptable type of novel). I mean that if a writer were to put down letters at random from the first to the final page, could he or she claim to have written a work of literature? (Of course he/she could *claim* it, but should we be expected to agree?) If there was an academic community devoted to producing such works, should we be expected to respect and support it? Could we regard it as in any sense a worthwhile enterprise?

I should say that even the atonalism of Boulez and his school is not quite that extreme. I am happy to regard it as music. I am, in fact, happy to listen to it from time to time, and let it exert its strange, dislocated effect. But I'm not convinced that, when I find transient form in little flurries of notes, they stem from the composer's skill, sensitivity or intention. They seem instead to be piggybacking on the musical resources I bring to bear from outside the ultra-serialist world.

Babbitt is often disparaged for his denial of the musician's alleged duty to communicate, but he seems not to have been asked to account for what he actually proposed to do with music. How, one should demand, will the hermetic sounds that you seem to envisage be turned into music? How will we be able to assess and respond to them? If your colleagues are going to congratulate one another for doing clever things with combinations of notes, regardless of whether these create perceptible sound structures, how will you avoid being trapped in a glass-bead game? What, musically, will you do?

Babbitt posed a serious question, but ducked any serious answer. Those answers are highly contingent on understanding music cognition. Some of the 'new' music of Babbitt's era made excessive demands on the listener not because of the listener's laziness and lack of knowledge or aesthetic sense, but because of the composer's indifference both to the natural contours of musical space and to the mechanisms of the listener's mind. As Fred Lerdahl says, 'the best music utilizes the full potential of our cognitive resources'. This doesn't mean that we should go on composing like Mozart – indeed, that would be doing precisely the opposite – but it means that composition that disregards our neural and cognitive limits is likely to end up unloved.

How do you know what you like?

There's no accounting for taste. Or is there? Music psychologists Patrik Juslin and Daniel Västfjäll suggest that our preferences are strongly emotion-driven, and as such, are often more dependent on context – on 'where we were at' – than on the music itself. They suspect that our feelings about music in everyday life are based largely on unconscious associations, governed by the emotional centres of the subcortical 'primitive' brain in the amygdala and the cerebellum. We may also have conscious recall of these associations – the 'Darling, they're playing our tune' effect, which resides in other parts of the brain. Most adults have a particularly strong 'reminiscence bump' for the period between the ages of about fifteen and twenty-five, and it's well known that we tend to prefer, know more about, and have stronger emotional responses to the music we heard during this period of our lives than to that which we hear earlier or later.

It's little surprise that our tastes are highly contingent on non-musical factors. But occasionally, 'blind tasting' experiments have revealed just how arbitrary, not to say prejudiced, those judgements are. Classical, rock, folk and jazz enthusiasts all have their snobberies, inverted or otherwise, which depend much more on peer pressure than on objective musical assessments. In 1973, sociologist Göran Nylöf at Uppsala University showed that the evaluations of jazz music heard by jazz fans was dependent on what they were told about the prestige or the race of the musicians (black musicians were deemed more 'authentic'). And sociologist Karl Weick and his co-workers tested the preconceptions of jazz musicians by presenting two orchestras with two unknown pieces for them to perform (plus a third as the control). They were led to believe that one piece was by a 'serious' jazz composer, the other by a basically commercial writer – but these designations were inverted between the two orchestras. The players were found to pay more attention to the 'serious' piece, making fewer mistakes and recalling it better later. That seems a disturbing message for new composers who have yet to build a reputation, as they struggle to secure for their pieces a decent airing.

But there does seem to be some more abstract musical basis to our preferences too. During the 1960s, psychologist Daniel Berlyne at the University of Toronto began to investigate whether aesthetic judgements can be ascribed to measurable features of the objects under scrutiny, and in particular to their 'information content'. One might guess that there is a trade-off: too little 'information', such as a musical composition that uses only two notes, leads to negative judgements because of monotony, whereas too much information, such as a piece full of microtones or chromatic notes chosen at random, turns people

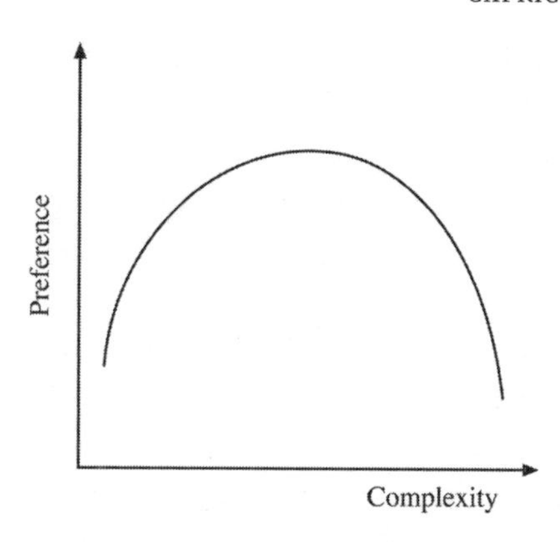

Figure 11.5 People generally prefer melodies with an optimal complexity – not too simple, not too complex.

off because it is incomprehensibly complex. That's essentially what Berlyne found in psychological tests: a graph of preferences plotted against complexity has the shape of an inverted U, with the highest preference being for objects that have a moderate degree of complexity (Figure 11.5).

Berlyne mapped out this response for visual perception. Its relevance for music was established in 1966 by the American psychologist Paul Vitz, subsequently better known for his work on links between psychology and religion. Vitz presented listeners with strings of tones with randomly chosen pitches, and found that their preferences followed the inverted-U shape as a function of complexity: as the sequences got increasingly random, the subjects at first liked them better but then less so. Interestingly, Vitz had failed to find this pattern in a set of similar experiments two years earlier, in which he took the pitches from the tones of the major scale. In that case, the preferences just kept rising with increasing complexity. That seems to have been because the subjects here already possessed an organizing scheme for what they heard: the tonal hierarchy that relates notes of the scale to one another. The implication is that the 'complexity' of a piece of music is partly a subjective quantity that depends on whether we possess a mental system for making sense of it.

These studies confirm our experience that more complex music requires persistence, but is potentially rewarding to those who stick with it. There is again a rather disheartening corollary for young classical composers who struggle to get their music performed: if their music is at all ambitious (and with new music it typically is), several performances may be required before listeners have sufficiently absorbed the style and schemas to appreciate it.

Where does the peak in preference-versus-complexity lie? For most people, it's not hard to anticipate that it probably sits closer to Benny Goodman than Cecil Taylor. And in popular music, hits are rarely 'sleepers' that grow on their audience after repeated listening: the song has to make an impact right away. A bias towards lower complexity has been revealed by music psychologists Tuomas Eerola and Adrian North for the songs of the Beatles. They developed ways to measure the complexity of melodies based on several of the theories described earlier of how musical expectations arise: for example, taking account of the tonal hierarchy of notes, pitch-step expectations of melody, and rhythmic patterns such as variability and syncopation. The researchers assigned to each tune a series of complexity measures that quantify how clearly the melody sets up and satisfies expectations in each of these dimensions. They then discovered which of the measures best reflected listeners' perceptions of melodic complexity by comparing them with ratings (from zero to ten) made by participants. This highlighted five key parameters in the model, which could be combined to give a single rating, calculated purely from the music itself, that reliably predicted how listeners would rank its complexity.

Eerola and North then used their measure to calculate the 'complexity' of all of the songs the Beatles composed – 182 in total. They found that the songs became steadily more complex over time between 1962 and 1970. Since this confirms what any Beatles fan would suspect (compare, say, 'All My Loving' (1963) and 'Mean Mr Mustard' (1969)), we can feel fairly confident that the researchers were measuring some real property of the music. How did this trend affect the preferences of the music-buying public? Judging popularity by the number of weeks singles spent in the UK hit chart, the increase in complexity systematically reduced success. It was the same story with the group's albums, rated by the mean complexity of their songs.

The implication – that even the simple melodies of the early Beatles are already at or beyond the preference peak of the public's inverted-U curve for complexity – is perhaps rather dismaying, suggesting as it does that many of us may have melodic tastes not so far advanced beyond the nursery-rhyme level (of course, the Beatles ingeniously exploited that in songs like 'Yellow Submarine' and 'Ob La Di, Ob La Da'). But unless you are thinking in purely commercial terms, it's debatable that public opinion should be held up as a meaningful measure of artistic success. After all, it is the later works of the Beatles that are universally regarded as offering the clearest expression of their genius.

Interestingly, a very similar relationship emerged from Dean Keith Simonton's work on 'originality' in the classical repertoire. Simonton's measure of originality – the degree of departure from convention in the first six notes of a theme or melody – can be regarded also as a measure of complexity: the higher the originality, the more complex and less predictable the melody is. (Simonton's measure is simpler than that of Eerola and North, taking account only of the statistics of pitch steps.) Simonton found that listeners showed higher arousal – they were more attentive – for music with greater 'originality', as you might expect. But what did they *prefer* – originality or predictability? To make some assessment of that for the vast list of compositions in Simonton's sample (and recall that the originality measure is still based just on the first six notes of the themes), he looked at how often works had been performed. This presumably indicates how well each work has been accepted into the classical repertoire. Of course, one can adduce all sorts of factors beside popularity which might influence the frequency of performance (the size of the orchestra, say), but these figures should give a rough guide. A graph of repertoire popularity against 'originality' again had the inverted-U shape (actually more like an inverted J, since it began at a higher point than it ended) (Figure 11.6).

Popularity is not, of course, a good guide to artistic value. But Simonton wondered whether his originality measure said anything about aesthetics and accessibility. Those are inevitably subjective matters, but he decided to rely on one man's judgement – not his own, but that of Richard S. Halsey, whose mammoth 1976 book *Classical Music Recordings for Home and Library* contained ratings of both the 'aesthetic worth' and the 'listening level' (the accessibility) of all the works listed. For Halsey, aesthetic value attached to works that will withstand repeated listening without palling, while 'accessible' works are of the type suitable for school and college teaching

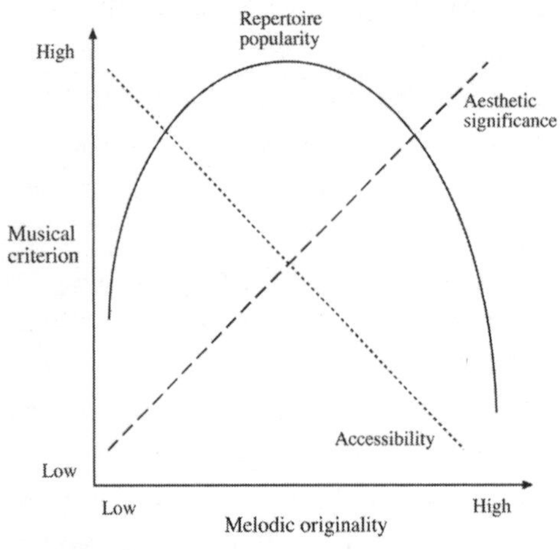

Figure 11.6 The repertoire popularity of classical music, as gauged from the number of performances they have had, has an optimum value of a measure of 'originality'. This may reflect a trade-off in popularity between accessibility and 'aesthetic worth'.

of music. Simonton found that Halsey's scores for aesthetic significance increased steadily as the originality rose, but that accessibility steadily declined. He figured that the high point of the popularity curve arrives when the increase in the music's interest, as measured by its aesthetic significance, is outweighed by its decreased accessibility. This is the point at which people start to deem the music too hard to be worth the effort of trying to appreciate it.

This sort of exercise in quantification of aesthetics will doubtless horrify some. And with some justification, given that it is based on melody alone, and only a mere morsel of that. (To judge from appearances, however, popular tastes in rhythm and harmonic progression are probably even more simplistic, as the 1970s disco boom attested.) But the popular-music industry is rarely very concerned either about the aesthetic high ground or the judgement of posterity. What it wants are hits. And studies like this seem to imply that there is *something* in a hit record that is measurable and predictable: some way of anticipating what will go down well with the music-buying masses. No one, as far as I know, has yet seriously considered whether a 'hit formula' might be extracted from measures of musical complexity. But they have certainly sought elsewhere for a prescription like that.*

*Not all such attempts are wholly serious. Neuroscientist David Sulzer attempted to write the most popular pop song ever by asking hundreds of people what they liked best: which instrumentation, what kind of lyrics, which tempos and so forth. The

(I suspect the answer they seek is as much sociological as musicological.)

While record companies want hits, most of us just want to find music we like. In the West, where there is now a potentially paralysing range of music to choose from, we tend to simplify the choices by breaking them down into genres and styles, so that we might proclaim confidently that 'I can't stand acid jazz', or 'I adore Renaissance motets'. We may become almost surreally attuned to the niceties of genre: music psychologists Robert Gjerdingen and David Perrott found that people could sometimes classify common genres of Western music (such as jazz, blues, rock, classical, country & western) after hearing just a *quarter of a second* of musical excerpts. That is too short a time for them to be able to use clues such as rhythm or metre – it would have to rely on more or less single tones, and so was presumably done on the basis of something like timbre. This would seem to support my contention earlier that timbre is a far more important aspect of music cognition than is often acknowledged. What's more, it rather undermines the usual way that musicologists think about distinctions of style, which they do on the basis of larger-scale structures such as melodic or harmonic patterns. Of course, this sort of rapid identification relies on considerable familiarity with the styles and genres in question; few Western listeners could distinguish so quickly (if at all) the Mbaquanga of South Africa from the Makossa of Cameroon or the Mapouka of the Ivory Coast. And naturally, we tend to classify and differentiate most highly the music we hear most. On the other hand, some studies suggest that we have a surprising ability to intuit stylistic differences without any prior experience of the music concerned. Music psychologists Isabelle Peretz and Simone Dalla Bella found that both Western and Chinese non-musician listeners rated excerpts of Western classical music as being more similar the closer they are historically, despite knowing little or nothing about the music itself. It seems that the listeners used cues based mostly on variations in tempo and rhythm to evaluate similarities (the duration of neighbouring notes in Western music steadily increases over time), on the basis of ad hoc 'rules' that they figured out for themselves.

result, which he called 'Most Wanted Song', is reportedly a sickly abomination, both musically and lyrically, proving Sulzer's point that this 'scientific' approach to musical consensus is absurd. Worryingly, however, 8,000 people bought the record all the same.

It's for good reason that many musicians deplore the pigeonholing of music. But genres are not just there for the convenience of music stores. It seems likely that they allow us to 'insulate' one type of music from another, so that the differing expectations we develop about the musical structures and norms of each don't interfere. This is why, for example, we can find dissonances shocking in Bach's music that we accept with equanimity in Bartók.

Yet the increasing diversity of new musical styles, and of technologies for accessing them, is beginning to throw into question some of the conventional ways in which genre and style are defined. As anyone who shops online will know, retailers and advertisers are eager to figure out what kind of products we like, so that they can offer us more of the 'same'. In music, some of these potential suggestions are obvious: other recordings by the same artist. But increasingly these selections are identified not by a crude division of genres but by a statistical sampling of what you and other customers have bought in the past: Amazon never fails to tell you what 'customers who bought this item also bought'. Data like this can point to clusters of 'taste' that may or may not reflect common genre distinctions.

Even when you find music you like, it can be hard to keep track of it. Part of the appeal of the random shuffle modes on MP3 players is that many people lose track of what is in their digital library: random searches can bring pleasant surprises.* But equally, this could deliver a rude shock if you're not in the right mood, a phenomenon dubbed 'iPod whiplash'. If you've just been lulled into a mellow mood by Miles Davis, you may not be ready for a blast from the Pixies. If, as Apple claims of the iPod, 'random is the new order', it has its drawbacks.

Choosing music by genre may do little to guard against that: if you hanker after a lilting Irish air, what if your machine alights on the Pogues? How do you tell it to find 'happy' Mozart rather than the *Requiem*? Conventional musical genres don't necessarily capture

*These algorithms are in general not truly random, for that tends to generate repetitions or closely related selections more often than you would expect. Special measures to avoid repetition are needed to persuade listeners that the selections are random – by making them less so, they sound more so.

the emotional qualities of a particular song or composition. That's one reason why some people make their own categories on their iPods, sorting music by its suitability to mood or scenario: one list for high-paced jogging and another for romantic dinners. But with personal digital music collections now easily stretching to more than ten thousand songs, that's not easy.

Several commercial products offer to help you. In the process, they are suggesting new ways to think about what 'style' really is, and what are the acoustic characteristics that we use to judge similarity in music. The Music Genome Project, launched by Californian company Pandora Media in 2000, is one of the most extensive of these systems, claiming to identify musical attributes or 'genes' that, in the marketing spiel, 'capture the unique and magical musical identity of a song – everything from melody, harmony and rhythm, to instrumentation, orchestration, arrangement, lyrics and vocal harmony'. A team of thirty musical analysts classify songs based on up to 400 different attributes, to create an online streaming music station that, for a subscription fee, supplies music from the region of the 'musical universe' corresponding to a particular song or artist specified by the user. Pandora's library contains more than 400,000 songs from 20,000 artists, and its recommendations are fine-tuned by user feedback.

Two rival programs under development are Playola, devised by a student at Columbia University in New York, and Sun Microsystems' Search Inside the Music. Playola looks for patterns in songs that allow them to be fitted into somewhat non-traditional genres. After listening to a song, the user moves sliders to adjust the genre preference: a little more 'singer-songwriter' and a little less 'college rock', say. Search Inside the Music is the brainchild of Paul Lamere, a software engineer previously engaged in speech synthesis who now works at 'music intelligence company' The Echo Nest in Somerville, Massachusetts. The system looks for 'acoustic similarity' between songs, such as comparable melodies, tempos, rhythm, timbre and instrumentation. A listener can then search for songs that 'sound like' ones they enjoy. These purely acoustic parameters can be combined with others' listening habits to refine the search. Lamere warns, however, that this sort of social feedback can create 'popularity bias', where hit songs get more heavily recommended than lesser-known ones, so that fashion drowns out novelty. A study in 2006 by social scientists at Columbia University in New York showed that feedback about other people's choices can severely skew the popularity ratings of pop songs relative to the case where everyone makes choices independently: runaway feedback may boost seemingly mediocre songs to unwarranted popularity, for example. When automated 'recommender systems' use only acoustic properties, these biases are removed.

Are the acoustic properties relevant to our subjective classifications and tastes the conventional ones of melody, rhythm and so forth? Some 'music retrieval' systems don't make that assumption. In one prototype, called AudioRadar and devised by computer scientist Otmar Hilliges and his colleagues at the University of Munich, music is classified along four different axes: fast/slow, rhythmic/melodic, calm/turbulent and rough/clean, the last two pairs being measures of the amount of rhythmic or melodic change within a song on long and short timescales. This gives every song a set of coordinates in four-dimensional space, and the distance between songs is deemed a measure of their similarity. To determine which music is selected from this space, the user specifies the preferred levels of each of the four attributes. The researchers have developed a way of projecting these results on to a two-dimensional display in which a 'reference song' appears at the centre and the other songs are placed in concentric circles surrounding it. Users can decide to move in a specific trajectory from the current song, for example to speed up the tempo without altering other characteristics. Or an MP3 player could be set to meander randomly in a particular 'mood space'.

There are still teething problems with this new approach to genre-busting. Hilliges admits, for example, that the system 'can't distinguish an uninspired rip-off from the original great song'. And the classification scheme struggles when faced with very different types of music: in AudioRadar space, classical music is very close to heavy metal, a result that would doubtless horrify some listeners.*

* Early versions of Search Inside the Music seemed to make the same confusion, apparently because (rather splendidly) its analysis of timbre registered harpsichords as sounding similar to distorted guitars.

Some might reasonably conclude that this sort of confusion damns the whole notion of music-retrieval systems, showing that their measures of musical quality are just too crude. Any system that groups 'classical music' (whatever that means) alongside heavy metal doesn't seem yet capable of telling us much about how we *hear* music. And automation seems sure, at present, to lack anything like our aesthetic sense: the difference between good, mediocre and poor judgement in Western classical music, and probably in any other genre too, is extremely fine. Even if we could find some way of quantifying the difference between, let's say, Franz Danzi (German composer, 1763–1826) and Ludwig van Beethoven, it seems unlikely that it could decide which is the better composer.

Yet the attempts of music information retrieval technologies to identify new classifications of and similarities in music are not worthless. They might, for example, enable a kind of audio indexing system, in which musical recordings could be scanned to find particular items (such as solos by specific jazz session musicians). They might help film composers find the mood music they need from vast libraries, or even identify tunes from a hummed recollection (such a search engine, called Musipedia, already exists).

What is rather more disturbing is that people might build their musical world through automated searches for tunes they already like. Of course, there is nothing new in that: plenty of art patrons commissioned the Old Masters with instructions along the lines of 'give me something like that Titian'. And like it or not, many people (at least in the West) do use their everyday music listening for mood regulation. The kind of system envisaged by Lamere and others, in which a user could define a particular 'mood profile' and then let the computer dictate the choice of music accordingly, sounds ideal in that case. It may be not only pointless but unfair to deplore this way of 'using' music.

If all we ask of music, however, is that it provide us with more of what we already know and like, then we are the poorer for it. Any technology that broadens our accessible universe of choice seems laudable at face value, but if the end result is that, on the contrary, it makes it easier for us to stay within our comfort zone and hear only the voices to which we are already sympathetic, we are in danger of lapsing into thoughtless passivity, our creativity and imagination and intellect fed on nothing but lukewarm, monotonous reassurances. And we risk becoming blind to what music can do.

12

Parlando

Why Music Talks to Us

Is music a language? Or is it closer to the non-verbal arts?

The musician, said French composer Albert Roussel, is 'alone in the world with his more or less unintelligible language'. It is a striking image – first, because it implies that music is indeed linguistic, and second because it suggests that no one can understand it. One pictures the composer or performer as a person in a foreign country, desperately trying to communicate and being greeted with blank stares.

Happily, music does not appear to be quite as unintelligible as Roussel implied, and we have seen some of the reasons why.* But is it a language?

The association of music and language is deep-rooted. We saw in Chapter 2 that some researchers believe these two psychoacoustic phenomena were, in evolutionary terms, once the same thing: that human communication began as a kind of 'musilanguage'. If music is still language-like today, however, it should possess syntax and grammar. Leonard Bernstein thought this was so, and in his 1976 book *The Unanswered Question* he proposed several linguistic characteristics of music.

This is not a question of whether music can have semantic meaning – whether it can 'say' anything. That is the contentious topic of the next chapter; but the comparison with language is not contingent on it. The parallels are drawn because music and language seem so alike at the *structural* level.

*Claude Lévi-Strauss perhaps chose his words more carefully. It is not, he said, that music is unintelligible, but that it is untranslatable. We understand, but cannot put that understanding into words. More on that notion later.

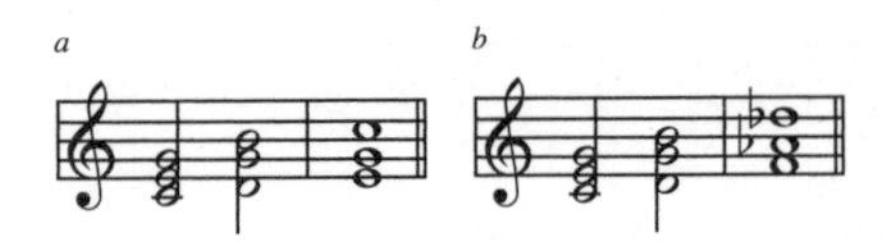

Figure 12.1 A simple harmonic sequence ending in an authentic cadence seems syntactically sound (*a*), whereas one that ends in an unexpected chord seems to violate the normal syntax of Western tonal music (*b*).

To see what I mean, think of the cadential structure I-V-I: let's say, a C chord followed by G and then C (Figure 12.1*a*). What does a C chord 'mean'? Nothing at all! But what does this sequence suggest, provided that you have been exposed to the conventions of Western music (that's to say, provided that you listen to the radio, sing songs, like a 'good tune'). I'd maintain that it says, 'Here's a little story: a beginning, a middle, an end.' What's the story about? It's not about anything. (Or as some formalists might say, it is a purely musical story.) But it has a definite trajectory.

What I'm really saying here is that this musical sequence has a *syntax*: a set of organizing rules that governs the relationships of its elements. What if the C and G were to be followed instead by a D♭ chord (Figure 12.1*b*)? What's the story now? I don't know, but it was an odd one. It didn't seem to end properly. It didn't make a lot of sense. In other words, it seemed to violate the syntax.

Now look at these two sentences:

I went to the shops.
I went to of beguiling.

Which would you match with the first sequence of chords, and which the second? It's obvious, isn't it? One obeys English syntax, we can parse it (that is, uncover its syntactic structure, or 'the way it fits together') without any conscious effort. The other leaves us floundering, bereft of syntactic satisfaction.

I'm aware that this might seem to be a trivial analogy, so let me say that it is not mine. It derives from the work of one of the most influential music theorists of the twentieth century, Heinrich Schenker. He proposed that musical phrases can be progressively simplified until we arrive at their 'deep structure', a basic musical idea called an *Ursatz*, of which only a few types exist. The I-V-I

Figure 12.2 Different flavours of the I-V-I sequence. All seem syntactically 'valid'.

progression is the most important. The dominant (G) chord sets up a rival to the tonic, containing notes (here D and B) that the tonic triad does not. Then the recurring tonic resolves the matter: the story is told, and we're back where we started. We go from stability to disturbance to stability, a narrative arc that can be found in countless stories. But the instability is itself controlled: the new notes remain in the diatonic scale of the tonic key, and so are not utterly alien or arbitrary.

Whether or not you accept Schenker's notion of *Ursätze* (we'll explore it shortly), the most universal and pertinent features of his theory are twofold:

1. Music unfolds according to rules.
2. These rules are *hierarchical*: they govern a nested structure of clauses and subclauses.

Both of these are characteristics of language. In a sense, 'I went to the shops' and 'I went to the cinema' are the same sentence: their syntax is identical. They both refer to essentially the same sort of event. Similarly, we can play the I-V-I sequence in various ways (Figure 12.2) – they have somewhat different flavours, but it makes a kind of sense to say that they 'mean' the same thing. We can construct more complex sentences:

I went to the shops, and there I bought a radio.

This two-clause structure can be heard in countless simple melodies, such as Bach's well-known Minuet in G (Figure 12.3). The first clause ends with a kind of pause, after which the second clause seems to qualify, respond to, or 'explain' it.

Again, these might seem like rather trivial, superficial analogies at first encounter. But there's more to it than that. Music, like language, can't be put together any old how, at least not without sounding confusing or 'wrong'. Imagine reversing the order of the two-phrase Bach melody: it doesn't sound terrible, but just peculiar. Similarly, reversing the order of the two clauses in the sentence above doesn't produce complete gibberish, but the sentence then clearly isn't right, because the syntax is warped.

In this chapter, I look at what musical syntactic rules are, where they come from, and why they help us process music. Among the questions they raise, one might ask whether we apply the same mental machinery to musical and linguistic syntax. If so, does this support the idea of an evolutionary link between them?

Composing in tongues

To some degree it's trivial to posit a link between speech and music. For most of the world's music is song, and so must be dictated, or at least influenced, by the metre and rhythm of the words. In ancient Greece poetry was all but inseparable from music: 'lyrics' were poems accompanied by the lyre.

But even non-vocal music seems to share speech-like patterns, and not just rhythmically: pitch contours in speech, called prosody, have long been thought to be echoed in music. The Czech composer Leoš Janáček was fond of notating the tonal rise and fall of spoken phrases that he heard, as if they were little melodies. He became convinced that prosody might hold the key to the emotional content of music. 'Whenever someone spoke to me', he said,

Figure 12.3 The initial four bars of Bach's Minuet in G major seems to consist of two phrases that are clearly related, with the second following naturally from the first.

> I may have not grasped the words, but I grasped the rise and fall of the notes. At once I knew what the person was like: I knew how he or she felt, whether he or she was lying, whether he or she was upset. Sounds, the intonation of human speech, indeed of every living being, have had for me the deepest truth.

Béla Bartók held the same suspicion that speech patterns could offer a model for the musical expression of emotions. Both Janáček and Bartók were avid collectors of folk tunes, which they saw as repositories of the true 'musical voice' of a nation. The Russian composer Mikhail Glinka claimed that 'Nations create music, composers only arrange it.' But might a nation's music actually embody its *language*?

The flag-waving revellers of the Last Night of the Proms at the Royal Albert Hall in London might assert that there is something objectively 'British' about the music of Elgar, whose 'Pomp and Circumstance' from the 'Enigma' Variations is an obligatory part of the event. They might claim to find similar qualities in the music of Vaughan Williams, Holst and Bax. By the same token, aren't Debussy, Ravel and Fauré characteristically French? You might think that these are merely learnt associations, but neuroscientist Aniruddh Patel and his co-workers have found that there's more to it than that. They have evidence that language patterns of rhythm and melody have shaped the music of English and French composers from the late nineteenth and early twentieth centuries.*

Patel and his colleague Joseph Daniele investigated a carefully selected sample of 318 musical themes by sixteen English and French composers, excluding those based on folk melodies, chorales, serenades and other forms that might have had some link to song (and thus possibly have linguistic attributes imposed on them). They looked for similarities with prosodic rhythm and melody in the respective languages.

There are some rather clear distinctions in the rhythms of English

*Their focus on this period was intentional, both because it was a time when many composers consciously strove to articulate a musical nationalism and because it would be hard to make a meaningful comparison with the imperfectly known language patterns of more distant periods.

and French. For example, English words tend to have stresses on the first or early syllables, while French words place them at the end – I might solipsistically offer as evidence the English 'Philip' and French 'Philippe'. Some linguists maintain that English and French in fact belong to distinct categories, called respectively stress-timed and syllable-timed languages. In English the timings of stresses are spaced roughly evenly (think of the nursery rhyme 'Jack and Jill') whereas in French it's the timing between the onset of each syllable, regardless of stresses, that is evenly spaced. German and Dutch are supposedly also stress-timed, while Italian and Spanish are syllable-timed.

One manifestation of this is the way that, in an English sentence, adjacent vowels tend to have rather different durations – long and then short, say – whereas in French the durations are more similar. Perhaps this is why French songs seem to have a rather regular pulse ('Frère Jacques'), while English songs have more of a skip ('In an English Country Garden'). These distinctions can be made precise using a measure fearsomely called the normalized pairwise variability index (nPVI), which measures the variation in duration between successive vowels in a spoken phrase. The average value of nPVI for British English is significantly higher than that for French.

Patel and Daniele applied the same measure to note durations in instrumental music. They found that each composer tends to favour a particular range of nPVI, and at a glance these values don't seem to show any nationalistic bias: the ranges for Debussy and Fauré, for example, both lie within that spanned by Arnold Bax. But when all of the composers from each country are averaged, there is a clear and significant difference: as with spoken language, the nPVI for English music is higher than that for French.

There are other arguments that language preconditions rhythmic musical structure. For example, children's songs in 6/8 time are common in the English language ('Boys and Girls Come Out to Play', 'Hickory Dickory Dock') but rare in Japanese, perhaps because these triple rhythms are rare in Japanese speech.

It seems plausible that rhythmic processing in music and language share mental resources. Some language dysfunctions have a genetic origin, being caused by mutations in a gene called FOXP2 that is linked to brain development. These deficits seem to extend to rhythm production and perception in music, while leaving pitch perception untouched. And there are several formal analogies between the patterns of stress used in speech and those in music, although the latter are much more regular. In poetry and rhymes particularly, we often speak the verse with a sense of the underlying metrical pulse, even to the point of putting in 'rests' like those in music. When you recite limericks, you'll generally include a silent 'beat' at the end of the first two lines (but not the third and fourth) to maintain this regularity:

To listen is good, in a way, [rest]
But it's better to join in the fray. [rest]
Musicians agree
They hear what they see;
The audience hears what they play. [rest]*

How about melodic parallels? There are some clear similarities between pitch in prosody and in music. For example, both tend to follow an arch-like contour, moving up at the start of phrases and down at the end. (Questions in English have a rising inflection, suggesting a phrase that expects a 'continuation'. This lack of closure is perhaps what Britons over thirty-five find so grating in the ubiquitous rising inflection of young people today.) And tonal languages, such as Chinese and some West African languages, make use of pitch contour to convey semantic meaning. (It's seldom remarked that more than half of the approximately 5,000 languages spoken today are tonal.) Musical training makes it easier to perceive and learn tonal languages, while we saw earlier that Chinese speakers are much more likely to have absolute pitch. Some interaction between pitch prosody in language and pitch use in music might therefore be anticipated. Patel and his colleagues John Iversen and Jason Rosenberg wondered whether this might be manifested in national differences in the pitch characteristics of music and speech. For example, spoken English seems to have more pitch variation

* True, this doesn't quite have the classic limerick form of 'There was a young lady from Prague . . .' But it seemed especially apt, coming from a little collection for musicians called *The Well-Tempered Limerick*, by Virginia C. Albedi.

than French, the prosody of which is more monotonous. Is that true of the respective nations' music? That suspicion is long-standing – Donald Jay Grout's 1960 history of Western music contains the suggestion that Elgar's melodic lines mimic the intonation patterns of British speech, while Wagner's operatic music was described in 1935 as 'an intensified version of the actual sounds of the German language'.

But proving this idea is not easy. For one thing, speech doesn't share the stepwise pitch quantization of Western classical music. The voice glides smoothly between one pitch and the next, and even sustained pitches conform to no scale. So it's not obvious how to compare musical with prosodic pitch. Here Patel and colleagues were inspired by the work of linguist Piet Mertens, who suggested that although voice pitch is constantly gliding within a sentence, listeners often perceive a discrete pitch for each syllable, largely defined by the pitch of the syllable's vowel. So a spoken sentence can be reduced to a series of steps between these 'vowel pitches', one for each syllable.

The researchers found that the changes in pitch from one vowel to the next tend to be more uniform in spoken French than in spoken English. That's to say, although the average vowel-pitch jump is the same in both English and French, the jumps are more variable in English. They found precisely the same distinction for the melodic intervals of their selections of French and English music. Thus, both rhythmically and melodically, the music of each nation reflects the prosody of its spoken language.

It's not yet clear if the same applies to other nationalities. Patel and Daniele found that even though German is a 'stress-timed' language with a relatively high value of nPVI, the rhythms of German and Austrian music have a relatively low nPVI during the Baroque and Classical eras. They suggest that this is because the music of these countries was strongly influenced by that of Italy – a country with a syllable-timed, low-nPVI language – over that period. The nPVI value of German and Austrian musical rhythms increased steadily since the seventeenth century as these countries literally found their own musical voice.

Ironically, Patel's findings seem to undermine Janáček's intuitions about the prosodic basis of folk music. Janáček thought that such 'speech-like' music would be universal – he did not count on the distinct imprint created by different languages. 'If I took a Czech, or an English, French or other folk song and did not know that it is a Czech, an English or a French song,' he claimed, 'I could not tell what belongs to whom. Folk song has one spirit, because it possesses the pure man, with God's own culture, and not the one grafted upon him. Therefore I think that if our art music can grow out of this folk source, all of us will embrace each other in these products of art music.' It sounded very inspirational, but in fact it seems that a nation's music is distinctly nationalistic.

The musicality of the spoken word is of course celebrated in just about all vocal music, but rarely has the musical potential of prosody been made as explicit as in Steve Reich's *Different Trains*. Here Reich drew his basic melodic material from fragments of interviews about the development of the US rail system. The effect is surprising in its emotional richness, and it's notable that *Different Trains* is considered one of Reich's most 'human' compositions.

How to string a sentence together

These studies offer intriguing glimpses of how language can inform a musical vocabulary. But they refer only to the 'surface' features of music, and do not imply that music shares the formal 'deep' structural attributes of language: that the two are put together in analogous ways. That is where the question of syntax enters.

In language, *syntax* refers to the rules of combining words into phrases and sentences, while *semantics* refers to the meaning these assemblies impart. Sentences can be syntactically correct but semantically void ('I saw a green smell'), and vice versa ('She have six tomato'). Syntactic rules in language don't imply that there is only one way to make a particular utterance – or, one might say, the fact that language possesses syntax doesn't stop you from saying the same thing in different ways. Indeed, that's what makes language such a wonderful thing: the syntactic rules provide a guide both for speaker and listener, while keeping available an immense variety of tellings. (You might like to think about the musical analogies throughout this discussion.)

It's not known how we acquire a grasp of linguistic syntax. One possibility is that we simply learn it by deduction during early language development. The American linguist Noam Chomsky has claimed that this is improbable: the rules are too abstract and subtle, and as young children we don't get enough exposure to their variety of use to be able to make reliable generalizations. Chomskyists propose that instead there is a 'universal grammar' hard-wired into our brains.

One of the most familiar grammatical orderings in simple English specifies that a sentence may have a subject and an object linked by a verb: 'I [subject] saw [verb] the tree [object].' The conventional rule is that the three elements are arranged in this order. Other languages may use different orderings: in Germanic languages, for example, the verb may come at the end. In some languages the order of words may be more fluid, since other indicators (such as word endings, as in Latin) specify the syntax.*

In some cases, altering the word order makes a sentence sound odd and harder to parse, but not impossible to decipher: 'I the tree saw' has a more or less well-defined meaning. But 'Mary saw John' is clearly not the same as 'John saw Mary': here the syntax suggests a distinction between subject and object which is inverted by transposition. In a musical phrase too, the order of notes matters: it conditions our interpretation of the phrase. In the key of C, say, a C falling to a B creates the expectation of more to follow, whereas the opposite sequence suggests finality.

Linguistic syntax displays a branching, hierarchical grammar. Consider this sentence:

The driver whom the policeman stopped got a caution.

At root this has the same [subject][verb][object] structure as before: the subject is 'the driver who the policeman stopped', the verb is 'got' and the object is 'a caution'. But the subject is now a rather more complex entity: he is, in effect, specified as the object of the embedded sentence *The policeman stopped the driver.* There is an additional noun-verb branch (*the policeman stopped*) (Figure 12.4). This complexity can continue:

* English syntax has a fair amount of fluidity too, particularly in times before its principles had crystallized. There is plenty of shuffled syntax in Shakespeare's plays, such as when Edgar implores Gloucester 'Sit you down Father; rest you.' We'd now say 'You sit down, Father; you rest.'

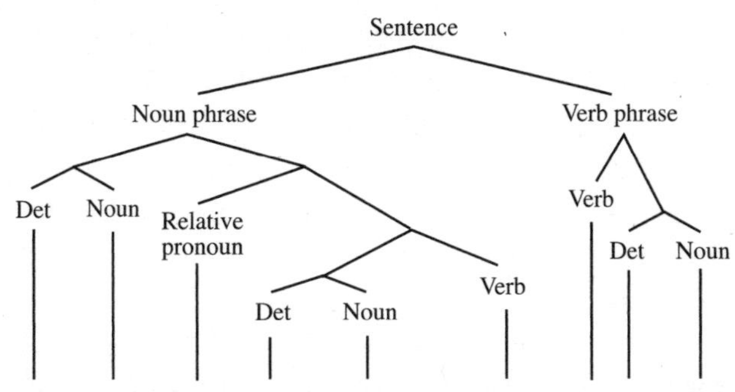

Figure 12.4 The hierarchical, embedded structure of linguistic grammar (here somewhat simplified). 'Det' indicates a 'determiner'.

The driver whom the policeman waiting at the corner stopped got a caution.

and so on. You'll see that merely observing syntax doesn't guarantee easy comprehensibility. But we can decode sentences like this because we know how the syntactic rules work. Not only do they show us how to create simple sentences, but they allow us to make complex sentences that are not merely linear strings of simple ones. We could say:

The policeman waited at the corner. He stopped the driver. The driver got a caution.

But most languages have syntax that permits a denser, hierarchical structure without introducing ambiguity.

Is music like this? It certainly seems that way. For one thing, just as language produces related sentences in which certain elements can be legitimately changed – one verb or noun substituted for another, say – so too can different notes or chords be interchanged at particular positions in the music. We might say:

I saw you.
I regarded you.
I inspected you.

and we might play the cadential structures in Figure 12.5 with various different chords in the second position, each of which has recognizably the same overall form but with quite different implications.

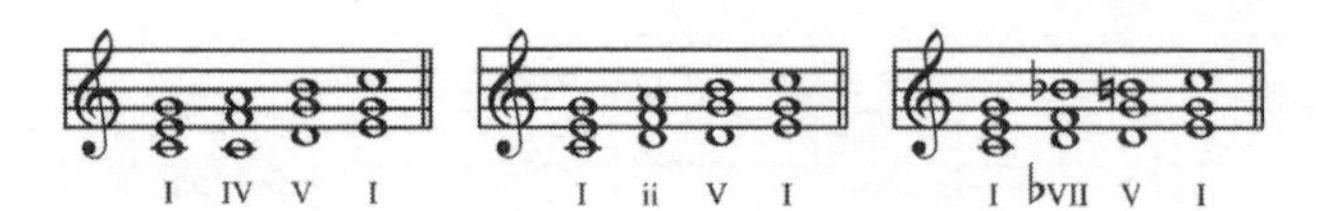

Figure 12.5 These variations of simple tonic-to-tonic progressions can be considered akin to sentences with the same grammatical structure but substituted verbs or nouns.

Not only are there basic music-syntactic structures like cadences, whose 'meaning' is governed by the order of notes or chords, but also pieces of music are not simply series of these very simple formulas strung together. We can create a very clear sense of musical subclauses and embeddings. Let's see how.

Heinrich Schenker was one of the first to consider the question of music's syntactic structure. His postulate that every musical sequence can be iteratively simplified, by considering some notes to be embellishments of others until we arrive at the deep structure of just a few notes and harmonic progressions, draws partly on the way tonal melodies tend to observe the tonal hierarchy. It is generally the more stable notes (such as the tonic and fifth) that shoulder the main load, anchoring the tune at key points. Chromatic, out-of-scale notes are passing tones, and many tunes remain easily recognizable if we omit them (Figure 12.6). Likewise, any notes of short duration that fall off the beat are unlikely to be too badly missed. In this way we can progressively strip a melody down to its bare skeletal form. At some point we may start to lose sight of the tune itself, because we lose the features that distinguish it from other, similar tunes. But even then we can probably still discern the outlines of the original tune if we know what it is.

Schenker formulated specific (although in fact not terribly rigorous) rules for 'reducing' music by gradually eliminating those notes considered to be embellishing and ornamental. Ultimately this leaves just the skeletal *Ursätze*, which Schenker claimed are mostly simple descending lines of scale tones: **3-2-1**, say, or **5-4-3-2-1**, ending in an authentic cadence.* Schenker wasn't really interested in whether we *perceive* these basic frameworks; he was looking for a way to study

* This is reminiscent of how the American writer Kurt Vonnegut once claimed, only partly in jest, that plots in fiction can all be represented by a few fundamental contours showing the ups and downs of the protagonists' fortunes.

Figure 12.6 The Schenkerian reduction of Variation 3 of J. S. Bach's *Aria variata* (BWV989). The original is shown in (*a*); (*b*) is the so-called 'rhythmic reduction', which removes 'superficial' sixteenth-notes, while (*c*) shows the harmonic and melodic skeleton. In (*c*) notes with stems are the key notes of the theme, while slurs indicate that the included notes are in some sense dependent on one another.

music formally, so that one might identify similarities and differences between compositions or parts of a single composition. The success or failure of a composition, according to Schenker, depended on how well the composer has grasped the dynamics of the underlying *Ursatz*. (His theory, developed in the early twentieth century, carried an unspoken agenda of 'proving' that the compositions of the (predominantly German) tonal masters of the Baroque and Classical periods were structurally superior to and more coherent than those of modernists such as Stravinsky and Schoenberg.)

The key problem with this approach is that it is *under-determined* – the rules for how to decompose a piece are somewhat arbitrary, and so there is no unique way of regressing from the musical 'surface' – the full score – to the 'deep structure'. Worse still, there is no way of going in the other direction. Even if the deep structure is a plausible skeleton, it is generally as interesting musically as stick drawings of the paintings of Titian would be visually. And there are no rules for putting flesh on those bones – even though Schenker believed that his methods reflect the subconscious processes that a composer uses to relate the surface to the deep structure, it seems more likely that the former provides the primary musical drive. No doubt classical tonal composers have some sense of the deep harmonic trajectories that their melodies seem to call for, but there's no real sense in which Schenkerian analysis offers a theory that 'explains' why the *Eroica* or the 'Jupiter' symphonies sound the way they do – a theory that, if you will, explains the observations from fundamental postulates. Besides, Schenker's approach only really works for compositions that follow the conventional harmonic rules of the eighteenth and nineteenth centuries. It is of little help in analysing much classical music before Bach or after Brahms, or popular forms such as rock music. (Schenker himself would have taken that as a reflection on the music, not the theory.)

All the same, there is surely some validity in Schenker's notion of a hierarchical unfolding of musical structure, and that is why Schenkerian analysis informs most of the systems used today to uncover the 'grammar' of music: these all tend to assume that musical phrases can be boiled down according to the differing status of their component notes. This is sometimes called pitch reduction. It can be regarded as a kind of 'coarse-graining', analogous to the way digital pictures are divided up into larger and larger pixels as the resolution worsens, with each pixel containing the key colour and brightness information for that area of the image.*

Take, for example, the first line of the Beatles' song 'Norwegian Wood', whose hierarchical structure has been analysed by music theorists Fred Lerdahl and Ray Jackendoff (Figure 12.7*a*). The most important notes, they say, come on the words 'I', 'girl', 'say' and 'me' (Figure 12.7*b*). This substructure is particularly obvious because the key notes fall on strong beats, and are the ones with the longest duration, and furthermore they are all relatively tonally stable, comprising the notes of the tonic triad in the song's key of E major: B, G♯, E, B. If you sing

* The analogy isn't perfect, because coarse-graining in digital imaging generally averages over the colour/brightness values in the pixels, whereas in the musical case we extract only the most salient pitch.

Figure 12.7 The syntactic structures of the first line of 'Norwegian Wood' (*a*). The main notes form a descending triad (*b*), which, when elaborated, make up a diatonic (modal) scale (*c*).

just these four notes, you get an immediate sense of the full phrase. These anchor notes not only make up an E major chord, but sound in descending order, as in an arpeggio. Thus the overall contour of the phrase is descending. In fact, if we include a few more notes of the melody we can make a descending scale (Figure 12.7*c*). (Note that this isn't a major scale but a modal one, because the D on 'she' is a natural, not a sharp. This is what gives the song its folky feel.)

Like Schenkerian analysis, the 'generative theory of tonal music' (GTTM) developed by Jackendoff and Lerdahl puts forward a formal procedure for carrying out these pitch reductions, so that the skeleton of any piece of music can be gradually revealed even when it isn't obvious from inspection how to break the phrase down.* The result is a tree structure in which the main branches end up on the most significant notes, while shorter side branches alight on the embellishments. Figure 12.8 shows how it looks for the phrase in 'Norwegian Wood'.

* And like Schenker's method, the GTTM has been criticized for the apparent arbitrariness of the rules. 'The reader of Lerdahl and Jackendoff will be struck by the number, variety and ad hoc character of their transformation rules,' complains Roger Scruton. He also points out that these pitch-reduction hierarchies can't explain why one phrase sounds better than another, even though their trees might be identical. It's a fair criticism in a way, but could equally be said of linguistic syntax – yet no one doubts the latter's existence on that account.

Figure 12.8 The hierarchical tree structure of pitches for 'Norwegian Wood' according to Lerdahl and Jackendoff's Generative Theory of Tonal Music.

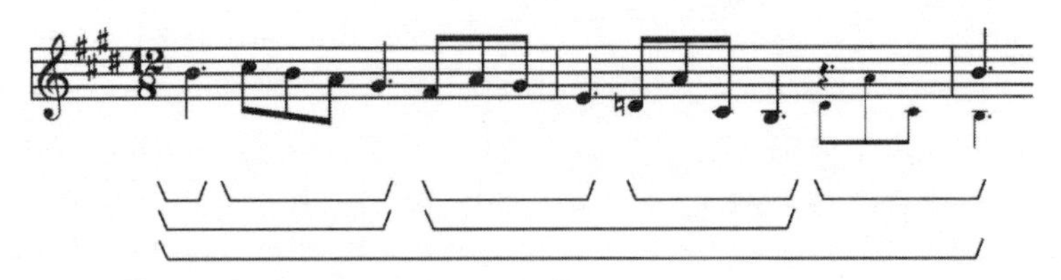

Figure 12.9 The rhythmic hierarchy for the phrase in 'Norwegian Wood'. Notice how this differs from the pitch hierarchy.

Notice that the two most fundamental notes here are the first and the last: the upper and lower B. So we can regard this phrase as an octave step downwards, embellished by intervening notes. Second, the 'significance' of the notes in this analysis doesn't necessarily equate with their perceived importance. For example, the A on 'once' is, in the GTTM tree, very much an incidental note. But it leaps out at us in the melody itself, because it is the only point in the phrase where there is a large jump in pitch – two, in fact, from D up to A and then back down to C♯. What this means is that the A racks up the tension we experience: the jump takes us by surprise, leaving us wondering where the melody is going. This *is* nevertheless reflected in the tree's shape, because the rules of the GTTM imply that the slope of a branch reflects the degree of tension or relaxation associated with the corresponding note (at least so far as pitch is concerned): the flatter the slope, the greater the tension. By this measure, the A on 'once' evokes the greatest increase in tension of all the notes in the phrase.*

*Recall from Chapter 4 that large pitch jumps rarely occur on their own in a melody: they would seem unmotivated and arbitrary, whereas a repeat of such jumps normalizes them. Is 'Norwegian Wood' an exception? No, because the phrase ends with an instrumental lick played on the sitar in which this jump is repeated (see Figure 12.7*a*).

But this is not all there is to the hierarchical structure of music. Rhythm is involved too: the important notes tend to be on strong beats. The pitch hierarchy tree links successive notes into small groups of three or so: [I once had] [a girl or] [should I say] [she once] [had me]. But when you read this (or rather, you should sing it), you'll notice that it doesn't easily fit with the rhythmic flow of the line, which is more like this: [I] [once had a girl] [or should I say] [she once had me]. The GTTM includes rules for identifying these groupings too, and again they cluster into several levels of hierarchy (Figure 12.9).

The richness of musical content in this song comes from the way that these patterns – the grouping by pitch, by rhythm, and the tension created by the tonal stability of pitches – overlap, interact and intersect. Leonard Meyer's model of emotion in music suggests a further, not entirely independent, set of salient factors too, based on expectations and their violations: chromaticism or changes in rhythm, say.

And that's just a single line in the song. The hierarchies continue over larger scales, for example in terms of how successive lines relate to one another: do they repeat, or change subtly, or launch on to totally different trajectories in contour or key? Eventually we start to encounter the large-scale elements of structure that are familiar at the colloquial level: verses and choruses, themes and developments, movements and the whole large-scale architectural frameworks of Western art music, such as the sonata and the symphony. These too are part of the hierarchy of structure, but they don't generally seem to have fundamental *cognitive* significance. Although some music theorists talk of how the experience of music may draw on the violation of expectation in these long-range musical forms, cognitive studies show that they are largely inaudible even to trained musicians when they hear pieces of unfamiliar music. Of course we can hear where movements begin and end, and we can (usually) distinguish an andante from a scherzo, but our memory for detailed structure does not seem to extend beyond a few minutes. So we should take with a pinch of salt all those programme notes about the clever things composers do with form: these manipulations may have provided a creative stimulus to the composer, but they don't have much to do with the way we perceive music unless we have studied the composition in advance.

Musical syntax and grammar are not arbitrary constructs but have a logic that helps to make the music comprehensible even to naïve listeners, who may rather quickly and subconsciously intuit the rules of organization. Most children acquire some grasp of musical grammar by the age of around six or seven without any specialized tuition.

Yet the linguistic parallels have limits. In language, syntactic structure tends to impose only one correct reading; creating truly ambiguous, yet grammatically correct, sentences takes some effort.* But in music there is often no unique way to parse a phrase, and we may try out several strategies, switching between them in a search for confirmation or contradiction. If that happened in literature (poetry may be different), we'd soon get tired of it; in music, it seems to be something we value. A parsing based on melody or harmony may or may not be supported by one based on rhythm. Interesting music often plays one against the other, forcing us to constantly revise and update our interpretations. Melodies that are wholly transparent in their syntactic structure may please us for a while, but ultimately we realize that we are listening to the equivalent of a nursery rhyme.

By the same token, we can state with some confidence that certain sentences are grammatically wrong (even if some rules have blurred boundaries), whereas it is debatable whether the same can be said for music – we can construct music that seems grammatically odd or unexpected, such as ending a cadence on the 'wrong' chord, but we can't rule out such things as impermissible. These apparent syntactic solecisms can potentially be redeemed – explained, you might say – by what comes after.

And just because schemes like the GTTM offer a formal way of dissecting or telescoping music into hierarchical patterns, it doesn't follow that *this* is the scheme the mind uses. Beyond a vague sense of rightness and wrongness, do we actually perceive syntactic structures at all? That is an open question. It is quite conceivable that many listeners hear in 'Norwegian Wood' only the melody's swaying

*Some linguists would debate this, saying that actually many sentences are rather ambiguous from a formal point of view. We are, however, very adept at untangling such ambiguities from the context. That's especially noticeable (to this English speaker) in Chinese.

rhythm and its descending contour. Certainly, I don't imagine that anyone sees Jackendoff's and Lerdahl's tree sprouting as the tune unfolds. Indeed, they *cannot* do, because the 'correct' tree can only be decided in retrospect. In principle, the question might be explored by testing whether subjects report the same rise and fall of tension as predicted by the theory. As I mentioned in Chapter 10, this sort of testing has been done in a few cases, but the results aren't conclusive because qualities such as musical tension are not easy to measure unambiguously. Fred Lerdahl suspects that people experience music both hierarchically and sequentially – they might intuit some of the recursive structure, but they are also carried along by the moment-to-moment trajectory of the melodies and harmonies. He thinks that it is partly a matter of experience: naïve listeners stay floating along 'closer to the surface' while more seasoned listeners hear the deeper levels of structure. In the latter case, Lerdahl argues, listeners can accumulate and carry forward tension embedded in recursive relationships, while the naïve listener's tension levels are dependent more on how each note relates to those that have just sounded – whether, for example, they resolve towards more tonally stable notes or move further away from them.

As the name implies, Lerdahl and Jackendoff developed their tree-mapping method specifically for tonal music. It's not clear that anything comparable can exist for atonal music, because the grammatical and ranking rules are no longer obvious. Indeed, they might not exist at all. There is no atonal equivalent of a grammatical form such as a cadence, and no reason why any particular chord should follow or precede another (beyond the constraints of the tone row, in serial composition). So when Jackendoff and Lerdahl apply their approach to atonal music, they find structures that are 'perceptually fragile' (that is, it's difficult to make out the start and end of phrases, at least based solely on pitch) and 'of limited hierarchical depth': they are all surface, with little recursive branching. Tests on music students and specialists conducted by musicologist Nicola Dibben seem to bear this out: the subjects were unable to identify accurately which of two pitch reductions of pieces of atonal piano music best matched the original pieces, indicating that they couldn't clearly perceive any hierarchical structure that might be present. Moreover, there is little sense of tension being manipulated according to well-defined relationships between the musical elements, as they are in tonal music – apart, perhaps, from the variation of sensory dissonance as a new 'tension dial'. What all this implies is that, although serialism has governing rules (and typically applies them inflexibly), they are not of the kind that permit a well-formed musical grammar. Syntactically speaking, this music is shallow.

That may be cognitively significant. One of the reasons why music commands our attention may be that its syntax and grammar fool us into hearing it as a kind of pseudo-language. This doesn't imply that music is pleasant gibberish; rather, our mechanisms for linguistic processing might be marshalled to give a perceptible logic to complex music. Our ability to develop musical grammar means that we are not doomed to remain at the level of nursery rhymes (which, I might add, already have a simple syntax). But music without a clear grammatical framework may struggle to amount to more than a linear series of notes and motifs, lacking in depth. Could this be why some of Schoenberg's most potent compositions are miniatures?

Shared resources

If music and language exhibit at least some structural similarities, do our brains treat them both the same way? This has been contentious. The earliest stages of auditory processing must surely process all stimuli the same way: the 'ear' can't decide whether it is hearing speech or music, and direct the signal accordingly to different parts of the brain. But this distinction is evidently made at some stage, since we don't generally get the two confused. So it seems possible in principle that high-level attributes such as syntax might be processed by different neural circuits. Yet if they invoke similar principles, why would the brain be so profligate with its resources?

For many years, a neurological connection between music and language was disputed because people could suffer brain damage that prevented them from processing words (aphasia) but not music (a condition known as amusia), and vice versa. Most famously, a stroke in 1959 left the Russian composer Vissarion Shebalin's linguistic abilities so impaired that he could not repeat three simple short sentences. Yet he continued composing, and his Fifth Symphony in 1962 was described by Shostakovich as 'a brilliant creative work, filled with highest emotions'. And then there are cases, sometimes awe-inspiring, of people with severe cognitive deficits in language use, such as pronounced autism, who display great musical virtuosity, not just replaying it like tape machines but improvising it with aplomb. Others might lose the ability to recognize melody while retaining a clear perception of prosody in speech. How could these things be possible, if language and music are processed along the same mental channels?

Well, it is clear that they are *not* in any universal sense: language is a relatively specialized brain function, whereas music processing is distributed throughout the brain. But might the syntactic elements of both share a common pathway? It's not clear whether cases of aphasia without amusia can tell us much about that, because they are extremely rare (only a handful of instances are described in the clinical literature), and tend to be documented for people with unusually high musical ability, such as Shebalin, who are not necessarily representative of the general population.

Modern neural imaging techniques now allow this question to be probed more directly. Aniruddh Patel and his co-workers have found that incongruous chords in a harmonic sequence may create the same characteristic signals of brain activity as violations of linguistic syntax. Such harmonic infelicities are rather like hearing a sentence that doesn't make sense: they elicit a 'Huh?' moment. But that doesn't in itself mean that the responses involve the same mental processes, less still that they are both related to syntax violation – they could be triggered by lack of semantic meaning in language, and by discordance in music, both of which are often the consequence of the respective syntactic violations without being essential to them.

Fortunately, however, the brain supplies telltale fingerprint signals specific to a *syntactic* 'Huh?' Measurements of electrical activity in the brain using electroencephalography, in which electrode sensors are attached to the cranium, have shown that violations of syntax in language elicit well-defined patterns of activity that are different from violations of semantics. Patel and colleagues looked for a kind of electrical pulse called P600, which rises to its peak intensity about 600 milliseconds after a syntactically incongruous word. A P600 can be detected, for example, when subjects hear the sentence 'The broker hoped to sell the stock was sent to jail'. This isn't totally incomprehensible, but something is obviously wrong: the rules of syntax insist that the word 'was' shouldn't follow 'stock' in this context. The peak of the P600 signal follows 600 milliseconds after subjects hear the onset of 'was'.*

Patel and colleagues constructed harmonic sequences of chords in a popular style, somewhat like little jingles. At a certain place in the sequence, they inserted chords that seemed more or less out of place. For example, a sequence in C major might suddenly include a chord of E♭ major, which is unexpected but moderately related to the tonic (and consonant in itself), or D♭ major, which is more distant from the tonic and so even more incongruous. In both cases, the rogue chords represent violations of normal harmonic syntax – which is quite a different matter from, say, having the pianist simply hit the keyboard with her fist at these points. The researchers found that both moderate and strong violations of musical syntax provoked a P600 response in their subjects, and the size of the response was larger as the harmonic incongruity increased.

These findings were supported by studies of Stefan Koelsch, a German neuroscientist then working in Leipzig, and his co-workers. They played subjects recordings of various cadential chord sequences, some that progress as normal authentic cadences and others that involve unconventional chords. They probed for different types of 'syntactic stress' brain signals: a so-called early right anterior negativity (ERAN) signal, which appears around 180 milliseconds after the stimulus, and a later bilateral frontal negativity (N5) about half a second after the stimulus.

The researchers used a set of five chords in a harmonic progression starting and ending with the tonic, and looked at the effect of inserting an odd chord, called a Neapolitan sixth, either in the third or the fifth position (Figure 12.10). A Neapolitan sixth is based on

* The P600 signal is not exactly a mental warning of syntactic error, but rather, a response to syntactic difficulties – it can be elicited by sentences that are syntactically correct but not syntactically transparent, requiring some effort to understand correctly.

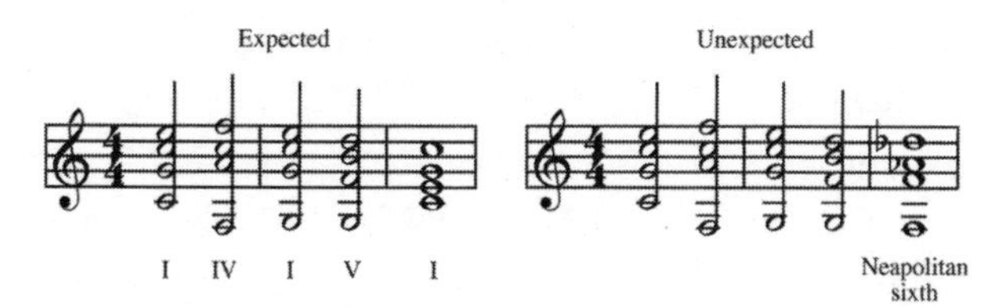

Figure 12.10 Examples of the harmonic progressions used by Stefan Koelsch and colleagues to test how the brain processes musical syntax. The expected sequence is a normal cadential structure, while the unexpected one ends on an unusual (but not wholly discordant) chord.

the minor subdominant chord – that's to say, if the tonic is C, the chord is based on F minor, but in place of the usual fifth (here C) it has a flattened sixth (D♭). That sounds strange, but not utterly dissonant – it can be found in classical music from the Baroque to the Romantic periods.* So as in Patel's studies, Koelsch and colleagues reasoned that these Neapolitan sixths will be experienced not as simply wrong but as a perplexing deviation from the normal harmonic grammar.

The researchers found that both ERAN and N5 signals were elicited by the Neapolitan sixths, both in musicians and in non-musicians familiar with Western tonal music. And this response was stronger when the unexpected chord appeared at the end of the sequence than when it was placed third, since in the former case the expectations created by conventional harmonic grammar – that the sequence was heading towards the tonic – were more intense. The researchers later showed that ERAN and N5 responses can also be produced by harmonic irregularities created *intentionally* by composers. They selected excerpts from sonatas by Beethoven, Haydn, Mozart and Schubert in which the composers had chosen somewhat unusual chords in place of the obvious ones. As well as proving that the findings aren't just an artefact of using 'weird laboratory music', these studies are a reminder that, as in language, musical syntax does not specify what an artist must do: it encodes a norm that can be more or less violated

*The Neopolitan sixth can also be considered to be a major chord on the minor second of the scale (here D♭), played in its first inversion (starting on F). It was used by Alessandro Scarlatti and other eighteenth-century Italian composers of the Neapolitan school, but was also favoured by Beethoven, who used it in the first movement of the 'Moonlight' Sonata.

for expressive purposes. If bad syntax were outlawed, poetry would be drab fare indeed.

In subsequent experiments Koelsch and colleagues were able to identify which parts of the brain the ERAN signals were coming from: the area in the left hemisphere called Broca's area, and the corresponding location in the right hemisphere. An analogous signal comes from the same source in language syntax processing. What's more, experiments in which syntactic irregularities of both words and music are presented simultaneously show that there is considerable overlap between the neural resources used in the two cases. This suggests rather strongly that the brain uses the same mechanism for interpreting linguistic and musical syntax. It does not prove that the two types of syntax are equivalent – it could be that this part of the brain has a somewhat more generalized function of interpreting the expected ordering of mental stimuli – but that's quite possible.

Aniruddh Patel believes that studies like this offer good reason to think that syntactic processing in music and in language do share resources in the brain. But this doesn't mean that linguistic and musical information are simply shovelled into the same mental box labelled 'Syntax', to be decoded according to identical rules. Rather, the basic *elements* of syntax – words and word categories for language, notes and chords for music – seem to be encoded in distinct locations in the brain. But the neural circuits that figure out how these 'syntactic representations' are combined and integrated may be shared between both music and language.

Leonard Meyer's violation-of-expectation theory of musical emotion implies that Koelsch's manipulations of harmonic syntax should arouse some emotional response as well as a 'does not compute' signal. That appears to be so: Koelsch and his co-workers found that the 'irregular' chord sequence ending in a Neapolitan sixth stimulates activity in the amygdala, one of the key emotion centres of the brain, and that listeners judged it less pleasant than the regular sequence ending in a tonic. (Notice that there's nothing intrinsically 'unpleasant' about a Neapolitan sixth, which is in itself a consonant chord; it is the syntactic implication, the oddness of the chord's position in the sequence, that creates the response.) Moreover, the harmonic syntax violations generate increases in the conductance of the listener's skin, a response commonly produced by changes in emotional state. And the magnitude of the N5 response to syntax irregularities increased when the excerpts were played with more expression (in this case, expressive dynamics were introduced or eliminated by tampering with digital recordings), suggesting that our registering of 'surprise' depends to some degree on the emotive features of the performance. These findings add some further support to Leonard Meyer's fertile theory.

It's curious that the ERAN response, showed no such sensitivity to expression. This suggests that we may have at least two pathways for processing musical syntax violations, one linked to emotion and the other purely cognitive. In speech, emotional cues seem to be processed independently of semantic and syntactic ones: some people with brain lesions can 'hear' the emotional connotations of speech without properly understanding the meaning (indeed, this may also sometimes happen in healthy individuals), and vice versa. But the musical studies show that there is no easy divide in music between the *logos* of syntax and the *eros* of expression.

Musical primes

I said that I would shelve the thorny matter of semantics until the next chapter, but there is one aspect of the question that seems so relevant to the interplay of music and language that it begs to be included here. Neuroscience has now provided evidence that music *can* possess intrinsic semantic content: it can carry a certain kind of meaning.

If two semantically related words or sentences are read or heard one after the other, the second is more easily processed by the brain, as though the first phrase has got it into the 'right mode of thinking'. Thus, we make sense of the word 'music' marginally more quickly if it is preceded by the sentence 'She sings a song', whereas that sentence confers no such advantage on the processing of the word 'sock'. This is called semantic priming, and is a well-established phenomenon in linguistics. It is again revealed by electroencephalography: a spike of electrical activity about 400 milliseconds after the second word is sounded is diagnostic of the brain assigning a meaning to the word. This spike, known as N400, is smaller when the word has been primed.

Might music be able to substitute for the priming sentence? Stefan

Koelsch and his colleagues wondered whether well-chosen passages in Beethoven's symphonies might prime for the word 'hero' but not, say, for 'flea'. They first selected what seemed to be good candidates for musical priming by asking listeners to assess the relatedness of several ten-second extracts of music to likely words, and then used the most robust pairings to see whether other subjects showed the N400 signature of priming.

They did. The links seemed surprisingly varied; for example, words such as 'narrowness' or 'wideness' could be primed by correspondingly sized interval steps; low tones could prime for 'basement' (but not 'bird'); and the German word for 'red', semantically related to 'fervent' (the subjects were German) was primed by a 'fervent' excerpt from Stravinsky. In about eighty to ninety per cent of cases, the subjects could rationalize afterwards why the priming link had been made.

The researchers concluded that music 'can prime representations of meaningful concepts, be they abstract or concrete, independent of the emotional content of those concepts'. Music, they decided, 'transfers considerably more semantic information than previously believed'.

Of course, it is one thing for a piece of music to set us thinking about heroics, and quite another for it to depict in an unambiguous manner, let us say, the triumph of heroic optimism over the threat of despair. Not only is it rather dispiriting to imagine composers feeding a kind of musical Morse code to our unconscious, but it is hard to credit that the option of such conceptual specificity exists. Are there, however, other ways in which musicians and composers can impart *meaning* to their works?

13
Serioso
The Meaning of Music

What are composers and musicians trying to say?
Can music in itself say anything at all?

What was Beethoven trying to tell us with his Third Symphony, now called the *Eroica*? It's obvious, surely. He said himself that it was a paean to an heroic icon, Napoleon Bonaparte. This has led some musicologists to ascribe highly prescriptive meanings to the piece, even down to Beethoven's choice of the 'heroic' key of E♭ major. True, the composer's disillusion was so great when his hero proclaimed himself emperor in 1804 that he is said to have scribbled out the initial title on the manuscript, 'Bonaparte', with such force that he scratched a hole. But the thing was written by then.

Yet this doesn't tell us what the music of the *Eroica* 'means' – it speaks only to Beethoven's inspiration for writing it. Aaron Copland makes the insightful suggestion that the greatness of a composition follows in inverse relation to our ability to say what it is about. 'It is easier to pin a meaning-word on a Tschaikovsky [*sic*] piece than on a Beethoven one,' he wrote – and 'that is why Beethoven is the greater composer'.

All this has left musicians themselves uncertain of what manner of art they are engaged in, and what, if anything, can be said with or about it. 'Is there meaning in music?' Copland asked. 'My answer to that would be "Yes". Can you state in so many words what the meaning is? My answer to that would be "No". Therein lies the difficulty.' For Gustav Mahler, that was the whole point: 'If a composer could say what he had to say in words he would not bother trying to say it in music.'

Felix Mendelssohn agreed, taking issue with those who contrasted music's vagueness of meaning with the precision one can achieve in literature:

> People usually complain that music is too many-sided in meanings; it is so ambiguous about what they should think when they hear it, whereas everyone understands words. For me it is exactly the reverse. And not only with whole speeches, but also with single words. They too seem so ambiguous, so vague, so subject to misunderstanding when compared with true music, which fills the soul with a thousand things better than words. The thoughts that are expressed to me by the music I love are not too indefinite to put into words, but on the contrary, too definite.

This was not simply the boast of a musician that his art could do more than that of the writer. For the French poet Paul Valéry held much the same view:

> Language is a common and practical element; it is thereby necessarily a coarse instrument, for everyone handles and appropriates it according to his needs and tends to deform it according to his personality . . . How fortunate is the musician! [Musical] elements are pure or are composed of pure – that is to say, recognizable – elements.

This ability to 'say precisely what is meant' seems to lie behind Walter Pater's famous claim that 'all art constantly aspires towards the condition of music'.

But others have denied that there really is anything to 'say' in the first place. 'If, as is nearly always the case, music appears to express something, this is only an illusion, and not a reality,' wrote Stravinsky. Jean-Jacques Rousseau, Immanuel Kant, Georg Wilhelm Friedrich Hegel and Hermann von Helmholtz all agreed that music has no 'subject' as such – it is not *about* anything, but just *is*. Or as Jack Westrup, professor of music at Oxford University in the mid-twentieth century, put it, 'Strictly speaking you cannot write about music; music expresses what it has to say in its own terms, and you cannot translate these into language any more than you can translate a picture.'

Many people are uncomfortable with this notion, which seems to (but certainly does not) imply that music is mere sound. And the denial of any musical meaning that can be articulated may become a restrictive dogma of itself. One of the motives behind Deryck Cooke's interesting but misguided attempt to decode the 'language' of musical meaning was that he perceived his contemporary critics in the late

1950s to have more or less declared it out of bounds to ask if a composer was trying to 'say' anything. Fearing that this led to arbitrary subjectivity, they seemed to have capitulated to Stravinsky's dictum that the analysis of music must limit itself to questions of form. This, Cooke complained, made music into a mere 'decorative art', rather than, as he saw it, an exploration of the human condition.

Most of what has been written and said about *semantic* meaning in music addresses itself to Western classical music. It is possible to defend that focus (although doubtless part of it stems from cultural bias), because this music has in many respects one of the most extensive, refined and sophisticated musical vocabularies and so crystallizes the question of communication with particular clarity. Moreover, Western art in general has served as a vehicle for exploring philosophical and existential ideas. It's not unique in that, but if Western classical music cannot find a way to express semantic meaning, we might reasonably anticipate that other traditions will struggle to do so too.

Yet that must not disguise the limitations of this particular window on musical meaning. Popular music tends to wear its message on its sleeve: the lyrics and the attitudes of the performers often leave little doubt about what it wants to say. And we have seen that in many non-Western societies music has a fairly well-defined social function that must be regarded as its true meaning: it validates ritual, accompanies dance, or encodes cultural beliefs. (It should go without saying that these aspects of musical meaning are clearly evident in Western music too.)

Besides, it depends on what you want to call 'meaning'. Some have argued that purely instrumental music conveys very specific narratives. Others think the meaning resides in emotional qualities. Copland held that 'Music expresses, at different moments, serenity or exuberance, regret or triumph, fury or delight. It expresses each of these moods, and many others, in a numberless variety of subtle shadings and differences. It may even express a state of meaning for which there exists no adequate word in any language.' He felt that this sort of 'meaning' wasn't fixed for a given composition – for 'music which always says the same thing to you will necessarily soon become dull music, but music whose meaning is slightly different with each hearing has a greater chance of remaining alive'. Yet the questions that Copland did not answer are, first, whether it is right to equate such moods with meaning; second, whether or not this should be considered music's primary goal or function; and third, what exactly 'expresses' means in this context. We'll look at all of those issues shortly.

The real question I want to explore in this chapter, then, is not whether music can have meaning – the obvious answer to that is 'yes' – but whether it can convey ideas in ways that are not strictly symbolic and determined by social consensus. Can the notes alone say anything to us, and to all of us?

The morality of melody

Aristotle believed that music was fundamentally moral, because it directly imitates the passions of the soul and can therefore induce them: 'when men hear imitations, even apart from the rhythms and tunes themselves, their feelings move in sympathy'. Listening to the wrong kind of music makes you a bad person, he insisted in his *Politics*, while good music cultivates the mind just as gymnastics cultivates the body. Boethius echoed this in the sixth century AD, saying that music 'has the power either to improve or to debase our character'. Thus, music may be a tool for shaping the character of the populace: 'Let me make the songs of a nation,' wrote Plato, 'and I care not who makes its laws.' This was why music was subject to strict regulation in the early constitutions of Athens and Sparta. (Remember, however, that in ancient Greece musical meaning was generally made explicit by the words that accompanied it. Plato disapproved of purely instrumental music precisely because 'when there are no words, it is very difficult to recognize the meaning of the harmony and rhythm, or to see that any worthy object is imitated by them'.)

St Augustine sanctioned the use of music in Christian worship when he claimed in his *De musica* that it may guide the soul towards noble and pious behaviour.* Martin Luther, while otherwise keen

* *De musica* is mostly not 'on music' at all: the first five of its six books are about metre in poetry. But Augustine had some profound insights into music, not least that it requires cultivation and active participation rather than just passive listening: as Hindemith put it, 'music, like humus in a garden soil, must be dug under in order to become fertile.'

to strip the Christian service of all superfluity, delighted in music, saying that:

> The riches of music are so excellent and so precious that words fail me whenever I attempt to discuss and describe them . . . In summa, next to the Word of God, the noble art of music is the greatest treasure in the world. It controls our thoughts, minds, hearts, and spirits.

Music is, he concluded, 'a discipline, and a mistress of order and good manners, she makes the people milder and gentler, more moral and more reasonable'. In the twentieth century Paul Hindemith shared this belief in the spiritual and moral values of music. And the censorious view that makes music a *causative* principle of behaviour is alive and well today in accusations that the 'wrong' music corrupts, although this position tends to draw more on xenophobia and the need for scapegoats than on classical scholarship. In the 1920s, jazz music was accused of 'turning modern men, women and children back to the stages of barbarism'; now it is the turn of heavy rock and gangsta rap to take the blame for delinquency and sociopathic behaviour. Of course it's true that music can, in the appropriate context, both heighten and quell emotions, and both for better and worse. But we should be sceptical of the idea that a sequence of notes has an intrinsic power to pervert or purify. Music alone can't make us do good or bad things.

That's not to say that music can't offer messages and suggestions. Music can be used to make explicit programmatic statements. Shostakovich's *Leningrad Symphony* was a gesture of patriotic resistance to the Nazi invasion of the Soviet Union, and was even broadcast defiantly at the German lines besieging the eponymous city. (Of course, Shostakovich suffered Stalinist condemnation when the 'message' of his works was considered ideologically unsound.) Allusions to familiar tunes or styles can prove witty or ironic, nostalgic or yearning. This, however, is the referentialist mode in which 'meaning' comes from extra-musical associations. And because context can create and transform meaning, composers can't hope to anticipate or control the way their music will be used – whether this is Wagner being enlisted by fascism or Prokofiev for televised sport. Stravinsky claimed to have been horrified by the dinosaurs frolicking to *The Rite of Spring* in Disney's *Fantasia*, though that might have been retrospective face-saving. But great art supersedes its *raison d'être* in any case – we don't really care that Handel wrote for long-forgotten royal occasions. And by conducting Wagner's works in Israel, conductor Daniel Barenboim implied that music need not be encumbered with post hoc moral or political implications.

Making shapes and telling stories

Stravinsky's dismissal of all meaning outside of the formal musical structure – the formalist position that we saw earlier championed by Eduard Hanslick – might seem discouraging, because it suggests that you need to know a thing or two about musical form before it can mean anything to you at all. At its worst, this attitude leads to the arid brand of 'music appreciation' that banishes any emotion aside from a lofty sense of formal beauty. Music is *not* about sensual pleasure, it seems to insist – and shame on you for feeling it.

Yet Hanslick's cool perspective was an understandable response to the indulgent excess that romanticism had spawned. As well as condemning the undifferentiated sentiment in which he felt audiences wallowed, Hanslick objected to how classical music had become an excuse for telling fanciful stories. Since the time of Beethoven, it had become common to think of music in narrative terms. Typically, the story involved a departure from a point of origin followed by eventual return in a transformed and enlightened state. Even within his lifetime, Beethoven's music was hemmed in by commentary and interpretation: what is he trying to *say*, audiences wanted to know. The music alone was no longer enough – it demanded words. Inevitably, those words were attached to the musical creator: Beethoven's music was deemed to be about Beethoven's experiences. It was no longer merely expressive, but *self*-expressive. That sort of thing would have been unimaginable and meaningless applied to Bach or even Mozart. Yet the view that music is 'about' something, that it is trying to express non-musical thoughts, is still prevalent today.

In trying to assess this idea, we need to be clear about one thing: it is not decided simply by the stated intentions of the composer or musician. Just because they might say that their meaning is such-and-such doesn't mean that they've communicated it. While the purpose of Western music was to depict impersonal emotional states, Mozart and Haydn would have used it as such. Since Beethoven and Schumann worked in an era that regarded music as an expression of personal thoughts, it is natural that they would indeed have attempted to express them. The question is whether anyone understood – or rather, whether the communication was clear enough to permit a *reliable* reading of intent. We've seen that in some cases a consensus might emerge about the overall emotional mood of a piece – but that's not what is generally understood by 'meaning'.

The problem is that, if you believe a composer has something more definite to say, you can generally find any message you like. And frankly, these interpretations are often repetitive and banal, like the plots of clichéd novels: Schumann's Second Symphony is read as a story of suffering leading to healing and redemption, Shostakovich's Tenth is 'a progression from dark to light or struggle to victory'. Well, so are *Star Wars* and Cinderella. Is this 'meaning' somehow expected to enhance the artistic merit of Shostakovich's Tenth? And once the readings become more complex, we plunge into a morass of unverifiable subjectivity: the 'hope' in Shostakovich's Tenth, one interpreter says, is actually 'false hope', somehow expressed through musical analogies with the 'cognitive content' of hope (as if we know what that is). Even if we had evidence that Shostakovich truly meant to convey this, are we to believe that he had such command of the cognitive emotional dynamics of music (something that, as we've seen, is barely understood even now) that he could reproduce the effect of 'false hope' in music?

The seeds of overly fanciful narrative reading of music were sown in the nineteenth century, when Richard Wagner could be found interpreting a passage from Beethoven's *Eroica* in terms of 'the loveable glad man who paces hale and hearty through the fields of Nature, looks laughingly across the meadows, and winds his merry hunting-horn from woodland heights'.* Musicologist Lawrence Kramer admits that this kind of thing is easy to mock now, but he himself offers an interpretation of Schumann's piano work *Carnaval* that includes claims like this: 'Schumann's masculine personae thus

*When Beethoven himself was asked to say what his Third Symphony meant, he is said to have sat down at the piano and begun to play it. If this is true, it shows that he knew better than his legion of 'interpreters'.

coalesce under a feminine sign. More, they coalesce under a misogynist sign, or at least a satirical one; coquettishness implies vanity, sexual teasing, triviality.'

Believe me, it can get far worse. Beethoven's Ninth Symphony, it has been asserted, has sections that are indicative of a rapist's murderous fantasy,* and his music in general (not Beethoven the man, note, but his *music*) is allegedly shot through with misogyny. Can you begin to imagine how one writes misogynistic music? Meanwhile, Schubert's Piano Sonata No. 21, D960 apparently contains a trill in the left hand that is an expression of his 'outsider' status as a homosexual. I'm reminded of Schoenberg's misgivings that 'One day the children's children of our psychologists will have deciphered the language of music . . . Woe, then, to Beethoven, Brahms, and Schumann – those men who used their human right of free speech in order to conceal their true thoughts – when they fall into such hands!' Schoenberg was worried that this 'deciphering' might happen for real; but perhaps he should have been more concerned that it would give voice to such absurd flights of arbitrary fantasy.

Yes, it's easy to mock. But there's more profit in asking what prompts people to make interpretations such as these, and why they believe them. Peter Kivy points out that music is closer to literature than to the visual arts at least insofar as both unfold over time – and this makes the narrative character of literature seem to be present in music too. Moreover, Kivy suspects that the formalist position, in which meaning can only be purely *musical*, feels to many people cold and inhuman. It could appear to imply that music is devoid of the kinds of humanistic meaning we treasure in literature and poetry, and that feels too barren a possibility to accept. So people seek comfort in the idea that music is 'telling a story' – or more grandly, that it is expressing a philosophical position.

But those reassuring notions are persuasively dismantled by Kivy. If Beethoven wished to make a philosophical statement, why on earth would he choose to do so in music, which lacks all ability to convey precise abstract ideas in a transparent (or even in an opaque) way?

*Susan McClary, who made this claim in 1987, later expressed her interpretation of the passage in question without reference to rape, perhaps acknowledging that it was problematic.

And even if some abstruse way could be concocted to 'say' in music that, for example, one must be obdurate in the face of adversity, that is not a 'philosophical position' in any case – it is a banality. Philosophy is not a sound bite, but a detailed exposition of thought. Music is no more equipped to be a vehicle for that than is football.

The same is true for 'musical narratives'. If 'stories' can be told with music alone (which I sincerely doubt), they cannot contain true plots, but at best, plot archetypes: there are no characters, no dialogue, no events. The best you might hope to glean (and this is very optimistic) is some sense of, say, 'a long journey home', without any of the abundance of riches that tells us we are reading the *Odyssey* or *The Wizard of Oz*. The problem also attaches to the notion popularized by some philosophers of music that the listener constructs a narrative by imagining some abstract persona who is experiencing the tensions, actions, emotions and sensations somehow embodied in the music. An attraction of this idea is that it leads us to demand in the music the kind of connected, coherent progression of emotions that we might reasonably expect of a real person: music that hops about between moods with no apparent relation between them seems then as implausible and unsatisfying as would be a similarly flighty, disconnected emotional journey in literature. Yet the problem remains that there is no scope for portraying nuances of narrative in a way that could be guaranteed to strike most listeners in the same way – the best one might hope for is a rather vague and humdrum series of 'happy', 'sad', 'tense' or 'serene' events.

Musicologist Susan McClary does sketch out a little detail in her narrative reading of Tchaikovsky's Fourth Symphony, which is apparently about a homosexual man who suffers under the weight of his father's conventional expectations and becomes entrapped in a relationship with a woman. Well, guess what: Tchaikovsky may have indeed been to some extent entrapped in such a relationship when he composed the symphony. He dedicated it to his patroness, the wealthy widow Nadezhda von Meck, and worked on it during the course of a short and unhappy marriage. And it is widely agreed that Tchaikovsky was a suppressed homosexual, whose father wanted him to be a civil servant, not a musician. But to comb a composition looking for musical correlates of the composer's life and circumstances is meaningless. Of course you'll find them if you're determined to, just as you can find world history prefigured in Nostradamus. Can we imagine that McClary would have proposed this reading if she knew nothing at all about Tchaikovsky? Or that anyone else would infer it without such biographical information? What exactly is the harmonic structure that expresses suppressed homosexuality, as opposed to suppressed Catholicism or frustrated ambition?

Kramer admits that interpretations of this kind are entirely unprovable, but he argues that the very fact that they can be challenged 'only enhance[s] the presence of meaning as an issue or problem at this level of utterance'. One might equally conclude that it only makes the discussion of meaning at this level of utterance profoundly pointless.

One might argue that surely composers *do* sometimes intend that their works have specific meanings, because they say so.* The private papers of Shostakovich, for example, show that he meant some of his compositions to be veiled attacks on the Stalinist state that hounded him so ruthlessly. It is no doubt valuable to know such things: when programme notes tell us that such-and-such a melody or repetition or cadence signifies some particular intent of the composer, we listen to it with heightened awareness. That's fine. But to imply that this information is the 'key' to the music is to misunderstand music altogether. Does it mean that, if we lack the inside information, we are deriving only false or faulty satisfaction from the music? Supporters of a narrative view of musical meaning often expend a lot of energy in showing that their interpretation could have been plausibly intended by the composer, forgetting that music is not a set of notations on paper but crystallizes in the mind of the listener. We have seen already that, simply because a composer has written a formal structure into the music, this does not mean it is in any sense 'present' in the musical experience. The same is true of meaning. Shostakovich may have intended those implications, but we didn't know that until we saw what he had written about his music. *They are simply not there in the music alone.*

*In fact, Tchaikovsky did this for the Fourth Symphony. At Meck's request, he wrote programme notes 'explaining' the meaning of the work. But these aren't very enlightening, and indeed succumb to mundane generalities such as 'all life is an unbroken alternation of hard reality with swiftly passing dreams and visions of happiness'. It's a reminder that often we're better advised to stick with a composer's music and ignore what they say about it.

It is sometimes said that composers suggest the tone of their message through choice of key. The *Eroica*'s E♭ major is said to be 'heroic', C minor is 'tragic', D major 'brilliant', D♭ major luxurious, A♭ minor richly sorrowful, F major pastoral, C major imbued with 'the clarity of day', and so forth. It is very hard to say how much of this is mere association – did Beethoven choose to write his Sixth 'Pastoral' Symphony in F because that was the appropriate key for the subject, or did this choice make it so? Regardless of what meaning or mood we attach to them, many musicians do feel that keys have distinctive characters – there are some, according to Geza Révész, 'who go so far as to call every transposition a rank violation of the musical feeling and an unjustified interference with the composer's intentions'.

Beethoven himself spoke of 'the absolute character of the modes' (meaning keys). But this may have been because he, like many musicians, possessed absolute pitch, and so could genuinely discriminate each key. All the same, in even temperament they should all sound identical except for a shift in pitch, so it's not obvious at first how there can really be a different 'character' to each of them. Yet these characters seem to have first been discussed during the eighteenth century, when irregular 'well' temperaments were the norm. These unequal temperaments do create differences between keys, because the pitch steps between semitones and whole tones in different scales are not the same. In 1796, for example, Francesco Galeazzi wrote that B♭ major is 'tender, soft, sweet, effeminate', and E major 'very piercing, shrill, youthful, narrow and somewhat harsh'. And even pianists who play on equal-tempered instruments will tell you that keys have different qualities that seem to go beyond the different patterns of fingering needed to play them – I would put them on a kind of smooth-to-hard scale, very evident when one switches from 'smooth' A♭ major to 'hard' A major. I have no idea where these qualities come from; it is possible they are just learnt associations, conditioned by the habits of composers. There are other distinctions that may arise even in equal temperament: open strings on stringed instruments sound different to fingered strings, for example. And the mechanics of playing other instruments also dictate key choices, and perhaps associated qualities as a result. Jazz music is full of flat keys for the convenience of the brass instruments. Rock music has a ponderous dependence on E and A, the main open-string keys of the guitar. Folk music fits itself around G and D major and E minor, the most convenient keys on the fiddle.

Are you hearing this OK?

Narratives, then, whether from composers or critics, are simply not an essential, or even an important, part of music – or at least, of Western instrumental music. People listen to it all the time in total 'ignorance' of these alleged hidden meanings – and I don't mean naïve audiences, but also profound lovers of music, expert musicians, composers and conductors of intense musical sensitivity – without having their appreciation of the art diminished one jot. Peter Kivy asks us to imagine the literary equivalent of this: a man who claims to love German poetry, but cannot read or understand a word of German. He just listens to the poems on recordings, because he loves the sound of the German tongue. Lacking any knowledge of the real content and meaning of the poems, does he in any meaningful way experience German poetry? I think we would have to say that he does not. And yet if narrative, philosophical or political 'meanings' are central to what music is 'about', do listeners who know nothing of them then fail to genuinely experience the music? Must we say, oh, they don't really *get it* at all? (And if they did *get* these interpretations, do we truly believe that their appreciation would be significantly heightened?) As Kivy puts it (though not quite in these terms), these interpretations tend to be so sketchy, so trivial, so banal, that the value of 'music plus interpretation' seems to differ imperceptibly from the value of 'music alone'. In which case, even if in some oblique sense the interpretations can be deemed 'correct', who cares?

This raises the question of whether it is possible to hear music the 'wrong' way. There is little virtue in the relativist position that says everyone's way of hearing and interpreting the music is equally valid. For one thing, the greater your musical knowledge, the more you can actually *hear* in music in the first place. And it probably requires both musical and contextual understanding to see what Handel offers that Vivaldi doesn't, or to derive from Indian ragas more than a vague impression of exoticism. The naïve listener needs at least to recognize the risk of listening in the manner that the British conductor

Thomas Beecham perceived in the British people, who, he said, don't care much for music but love the sound it makes. It's not elitist to say these things, because just about anyone can learn more about music if they care to do so. And it applies to all music – *everyone's* ear is uninformed about some traditions.

Yet this is simply to say that one can hear music incompletely, missing a great deal of what it has to offer. Can we actually get the music *wrong*? We have seen already some situations in which that could be said to happen: Westerners may fail to hear the right stresses and metres of unfamiliar Latin rhythms, and we may totally misinterpret the intended emotional quality in music of other cultures. But even here it's not obvious that we can really call these interpretations 'wrong', since they are, like all musical experience, merely structures that the mind creates to cognize and interpret the sound stimulus. From a cognitive point of view, they are valid 'interpretations'. One might say that the listener has merely transformed the music into a form that fits the mental schemas she possesses. The chances are, however, that the music achieves its greatest eloquence when there is cultural congruence between performer and listener.

And when it comes to meaning, questions of 'right' and 'wrong' become more or less void. Certainly, one cannot assert that music has a semantic meaning that the listener is obliged to ferret out. It's not uncommon for connoisseurs to claim that a great piece of classical music has a definite message, a failure to perceive which somehow signifies a lack of aesthetic sensitivity. Deryck Cooke was outraged by Hindemith's suggestion that the second movement of Beethoven's Seventh Symphony 'leads some people into a pseudo-feeling of profound melancholy, while another group takes it for a kind of scurrilous scherzo, and a third for a subdued kind of pastorale'. Each group, Hindemith said, 'is justified in judging as it does'. 'One is bound,' says Cooke, 'to regard anyone who reacts in this way as either superficial, unmusical, or unsympathetic to Beethoven' – ignorant, in any case. Now, it's true that a 'scurrilous scherzo' is a decidedly odd way to look at this piece. Yet when Cooke declares that 'The truly musical person, with a normal capacity to respond to emotion, immediately apprehends the emotional content of a piece to the degree that he can experience it', he is not only voicing the kind of snobbery that deters some people from exploring the classics, but also fundamentally misunderstanding what music does. Similarly, Roger Scruton says of the slow movement of Schubert's G major Quartet, D887 that:

> there is a tremolando passage of the kind that you would describe as foreboding. Suddenly there shoots up from the murmuring sea of anxiety a single terrified gesture, a gesture of utter hopelessness and horror . . . No one can listen to this passage without instantly sensing the object of this terror – without knowing, in some way, that death itself has risen on that unseen horizon . . . In such instances we are being led by the ears towards a knowledge of the human heart.

This is a thrilling reading of the music, offering a compelling way to listen to it. But can it really be true that 'No one can [for which read 'should'] listen to this passage without' hearing Scruton's interpretation?

It would seem a trifle perverse to insist that Holst's 'Mars' is an ode to love (the title is something of a giveaway). But excluding some interpretations is not the same as allowing only one. There are many different ways to listen to music seriously, and sometimes we *have* to do that if we are to extract all that the music offers. I can attend to the formal structures of Bach's fugues, and get delight from that; or I can let the music create an emotional affect. But I struggle to do both at once. Music that does only one thing, that offers only one way of listening, probably cannot be great.

'Emotional responses to music are neither correct nor incorrect,' says music philosopher Diana Raffman, 'typical or atypical, perhaps, but not right or wrong.' If, after all, even great composers cannot agree on what a piece of music means – Berlioz's interpretation of Beethoven's Seventh was very different from Wagner's – can the ordinary music-lover really hope to arrive at the 'right' conclusions? Besides, as Raffman points out, you can't demonstrate to someone that they have made the wrong emotional or semantic interpretation of music. You might be able to demonstrate, or at least to argue, that they've misheard the boundaries of phrasing, or missed the way in which an idea has been developed. That might alter their reading of the piece. But it might not. Interpreters like Cooke and Scruton might argue that no one will grossly misjudge the emotional content of a piece if they are a competent listener. But all the evidence can say is that there is a broad consensus on this, at least in cases where the mood or context is rather clear. To experience Barber's *Adagio for Strings* as jolly could reasonably be called eccentric or unusual; but can one meaningfully say that the listener is 'wrong' to feel that way?

I don't, therefore, think that Susan McClary's interpretations of Beethoven or Tchaikovsky are 'wrong'. She may very well discern those features or narratives in their music, and who am I to say that she doesn't? The point is that I simply don't care about such interpretations, because probably not a single other person would have arrived at them on their own. This does not mean, however, that such interpretations need be irrelevant. Some listeners may find that particular narrative readings enhance their experience. The measure of good art criticism is surely not that it tells us what we should think, but that it stimulates us to imagine possible ways of experiencing. Interpretations of meaning should be approached in this spirit. If you don't see them yourself, that may not be because you're too ignorant, but because you don't hear the music that way. And if you *do* see what the critic means, that's not because he or she has found the 'answer', but because they are a good critic.

There is another way of 'reading' music that again draws on the Romantic tradition, claiming that it places us in communion with the composer. Deryck Cooke asserted that in listening to Beethoven we feel his grief: 'The listener thus makes direct contact with the mind of a great artist.' It might be nice to think so, but this statement is pretty nigh meaningless. What does it feel like to be in 'direct contact' with the mind of absolutely anyone else? How would you know if you achieved it, especially if that person died 200 years ago? And why should Beethoven's compositions be any more an indication of his own feelings than Shakespeare's works are of his?

Such suggestions are not just facile, they undermine the value of music. They imply that we are only genuinely experiencing the *Eroica* when we achieve this sympathetic, mysterious resonance with Beethoven's mind – and worse, that *this should therefore be our objective* (an objective, moreover, that is impossible). Either it gives me a false reason to want to listen to the *Eroica*, or, more probably, it makes me think there is little point in trying. Besides, as Eduard Hanslick said, in this picture the music then becomes just the middleman, the means of connecting the mind of the composer with that of the listener.

Why in any case would I *want* to be in direct contact with Beethoven, or with Wagner? As John Blacking says, 'there are too many performers and composers whose lives are a sad contrast to the excellence of their music-making. I refer . . . to consistently mean and selfish behaviour or a startling lack of political consciousness.' Since astonishing music has been fashioned in the past several centuries by countless conceited, self-centred, bigoted, racist, misogynistic, homophobic, xenophobic, and generally unpleasant people, I don't find much comfort in the idea that the aim of music is to enable communication with the mind of the composer. It is precisely because music cannot convey semantic meaning that we do not need to worry about the people who made it. We might choose not to listen to their music, or find no pleasure in it, because of what they have said or done; but we cannot stop the music from being great.

Codes

You can represent any idea you like in music. For example, you could decide that each of the twenty-six semitones beginning on middle C represents a letter of the English alphabet, and compose melodies that spell out *King Lear* or *Look Back in Anger*, perhaps with different characters represented by different instruments. Needless to say, the result would not sound much like music, and could be 'understood' only by laboriously decoding the score. That sort of musical 'meaning' is trivial: it amounts to nothing more than making codes out of series of arbitrary perceptual symbols.*

And yet a great deal of the 'meaning' commonly attributed to Western music is prescribed by comparable codes and symbols – albeit ones that, being culturally determined and learnt since childhood, can be interpreted with minimal effort by the informed listener. It doesn't take any musical training beyond the acculturation that almost all Westerners receive to hear the martial character of Sousa's marches

* Scientists at Cornell University have developed a system for converting the chemical structures of protein-coding genes in DNA into music. This is more than a parlour trick, because it means that our finely tuned music cognition apparatus can be used to discern particular features of the protein structures which are hard to make out from simply a list of the chain of chemical building blocks.

or the dreaminess of Debussy's nocturnes. But these associations are mostly a matter of convention, nothing more.*

A confusion between convention and innateness undermines Deryck Cooke's attempts to specify meaning as well as emotion in the 'language of music' (see p. 275). Cooke claimed to have found the basic 'musical vocabulary' that composers and musicians have used to convey meaning for centuries. But he never once states that it only works *if you know the code*, and indeed he seems to imagine that the meanings are somehow universal. Since he makes no reference to music outside the Western mainstream (other than, as we saw earlier, to patronize it), one can see how he felt no great pressure to actually demonstrate this. And even in its own terms the thesis is highly questionable as a description of how anything other than the sketchiest information could be conveyed by little coded figures: the more specific it gets, the more arbitrary is Cooke's lexicon. A normal **1-3-5** triad, he claims, is 'expressive of a sense of exuberance, triumph or aspiration', as opposed to the second inversion (**5-1-3**), which connotes 'joy pure and simple'. And when Vaughan Williams' statements about his Sixth Symphony seem to contradict the code, Cooke is forced to conclude that the British composer did not understand his own music.

One common assertion that the fundamental acoustic properties of a musical structure somehow determine its symbolic associations involves the interval of the augmented fourth or tritone – the step from C to F♯, say. This, as we saw earlier, was known in the Middle Ages as the *diabolus in musica*, the devil in music, and has been used many times in Western music to signify the satanic. Although it is commonly said that the tritone has a jarring, unsettling quality that one would naturally associate with the devil and his works, in fact it sounds demonic only because we have decided to hear it so – as I explained in Chapter 4, there are worse dissonances than this, and the traditional aversion was more theoretical than acoustic. All the same, the satonic symbolism is firmly established: Liszt opens his *Mephisto Waltz* with a rapid run up a tritone span from B to F, while 'Death'

* I've chosen examples for which the mood can't so obviously be deduced from the purely 'physiognomic' aspects of the music, as described in Chapter 10. On that basis, people unfamiliar with Western music might simply suspect Sousa's music of being 'happy', and Debussy's 'sad'.

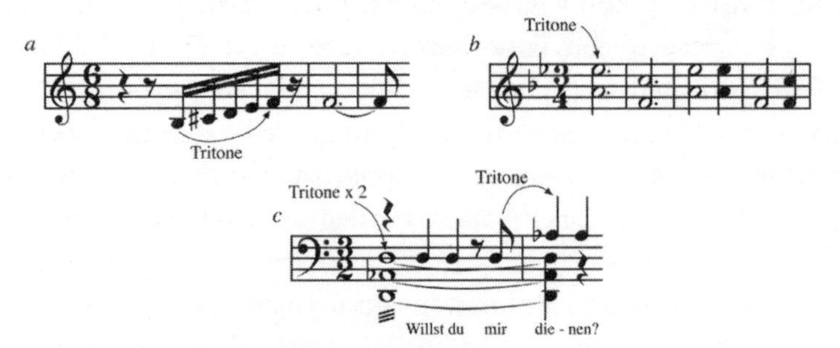

Figure 13.1 The 'demonic' tritone: Liszt's *Mephisto Waltz No. 2* (*a*), Saint-Saëns' *Danse Macabre* (*b*), and Busoni's *Doktor Faust* (*c*).

tunes his fiddle to tritone intervals in Saint-Saëns' *Danse Macabre*, and Faust uses them to summon Lucifer in Busoni's *Doktor Faust* (Figure 13.1). One has to doubt that there was any conscious attempt to join such illustrious company, but it is rather nice that the rock group Black Sabbath established this interval as the canonical sound of doom-laden heavy metal on the title track of their eponymous debut album.

A prescriptive, codified approach to expressive intent emerged in Western music in the Baroque period, when composers aimed to portray the states of the soul, such as rage, excitement, grandeur or wonder. They were not trying to tell us how *they* felt, but were offering something like symbols of ideas and feelings in a systematic language that their audiences understood. They employed stock figures and devices, often drawn from the principles of classical rhetoric, such as the *inventio* (finding a musical subject) and its *elaboratio* or exposition. No one believed that the music had some intrinsic, mystic power to evoke these things; they simply expected the audience to know the language. During the Classical era and the Age of Enlightenment, in contrast, the objective was to make music that was 'natural', that moved and entertained with its grace and lyricism – as music historian Charles Burney wrote, this made music 'the art of pleasing by the succession and combination of agreeable sounds'.

It was in the nineteenth century, when composers started to believe music had an intrinsic potential to express raw emotion without the mediation of agreed conventions, that they and their audiences lost sight of the strictly conventional assignation of meaning and started to think that music produced immediate imaginative suggestion. It is no coincidence that at the same time composers were less likely to produce works commissioned for particular patrons, audiences or events, but instead felt they were writing for eternity. Chopin and Schubert expected to be heard by select, educated groups in salons; Wagner and Mahler unveiled their grand creations to the world. The composer, like the painter, was no longer a craftsperson but a priest, prophet and genius. This is all, to a degree, part and parcel of the near-universal mystique of the musician, but it leaves observers in the West with the mistaken idea that musicians have some profound philosophical truth to tell us – which, unfortunately, they elected to express in music rather than words, thus obliging us to work out the damned puzzle for ourselves. It's hard not to suspect that some musicians *do* have a privileged window into the human soul; but all they can give us is a musical vision. If we implore them to tell us what they *mean*, we shouldn't be disappointed to receive banalities, incoherent rants, or gnomic silences.

So when composers, from Palestrina onwards, used an ascending musical figure to represent the biblical Ascension, they didn't think they were creating some motif that would be universally interpreted as an expression of ascension. They simply knew that their audience would be familiar with the symbolism. This sort of culture-specific signification of tropes seems almost inevitable in any mature musical form. Musicologist Philip Tagg believes that they exist in modern popular music too, and he calls them 'musemes', alluding to the notion of 'memes' introduced by biologist Richard Dawkins to describe ideas that propagate through a culture. Tagg argues, for example, that the conventional associations of some of the musical figures in Abba's song 'Fernando' sabotage the alleged 'revolutionary' spirit on which the lyrics try to insist.

Might the way that we seem to think about music in spatial terms be used to convey a more literal meaning connected with direction? Fred Lerdahl has examined this intriguing possibility in Wagner's *Parsifal*. He suggests that the harmonic space we encountered in Chapter 6 might have been used by Wagner as a symbolic terrain in which to plot Parsifal's spiritual journey. The various sections of the opera, says Lerdahl, follow harmonic paths either horizontally or vertically in this space, and the corresponding narrative implies that these axes have spiritual 'directions' associated with them: from evil to good (west to east) and earth to heaven (south to north) (Figure 13.2). These intersect, of course, in the form of a cross. It's not known whether Wagner knew of the theoretical map of

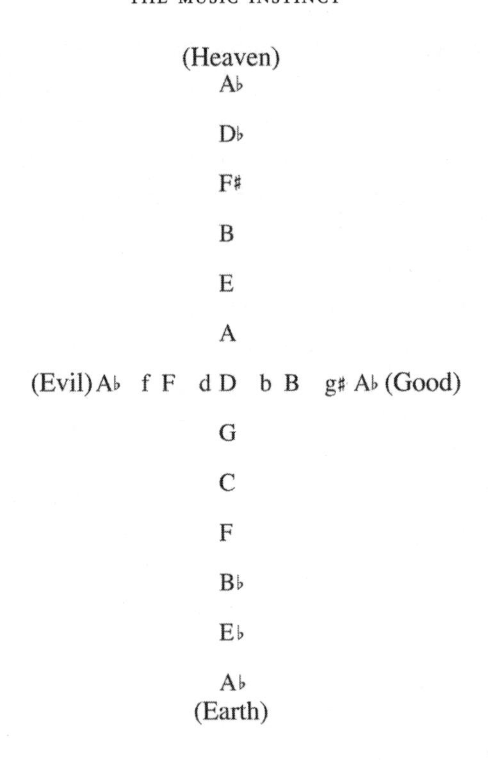

Figure 13.2 The main harmonic paths in Wagner's *Parsifal* form a cross in harmonic space, which Wagner may have used symbolically. Here lower-case letters indicate minor keys.

harmonic space, but that is possible: it appeared in a book of 1821–4 by the German theorist Gottfried Weber, which became a standard reference work in the mid-nineteenth century. Even if this was so, it seems unlikely that Wagner will have imagined this hidden narrative structure would be clearly perceived by the listener. But it seems worth considering that the informed listener might at least gain some qualitative sense of a journey (purposive or meandering, say) in pitch and harmonic space that might be created by a composer with semantic intent.

Finding the words

Must all this leave us despairing of music's intrinsic ability to communicate anything? Not by any means. It communicates a great deal. The problem, as Copland intimated, is to find words that convey *what* it is communicating. Many people suspect that the greatest music, whether by Josquin des Prez or BB King, is expressing something about the human condition. I suspect this is true. But when Cervantes or Kafka do that, the words offer concrete clues about what they are saying – not enough, perhaps, to be sure of our interpretation, but enough to open up objective debate. With music, especially non-vocal music, there is no such opportunity: we know that we have been told something by a Chopin prelude, but we don't know what it is. We may *think* we know, but I cannot prove that my interpretation is any more valid than yours.

This, however, is because we seem to feel it necessary to put the message of music in literary form, as though it were a coded book or play. That's absurd, but it is an absurdity enforced by language (or by its inadequacies). We need to talk about music, but this is hard. Very few people can do it. I suspect music critics so often review operas over purely instrumental concerts not just because they are so costly to produce but because they can be discussed and evaluated like a movie, with descriptions of plot, sets, costumes and performances, and with little obligation to talk about the music itself other than to say whether it was performed well or badly.

Music theorist Fred Maus has argued that the tendency to 'animate' and personify experience is an inevitable feature of the human mind: we make sense of events by attributing them to imaginary agents with particular intentions. We insinuate personae into the music. This impulse was demonstrated in the 1940s by psychologists Fritz Heider and Mary-Ann Simmel, who made an animated film of abstract coloured shapes – triangles and circles – that moved around the screen, apparently interacting in complex ways. When Heider and Simmel asked observers to describe what they saw, many related elaborate stories involving personalities: two shapes were in love, another tried to steal one of them away, and so forth. People identified not only narrative but also emotion in these movements. Fred Lerdahl suggests that music is like this too, with apparent attractions and repulsions between its notes and melodies and rhythms that can be anthropomorphized.

Some attempts to convey musical experience draw analogies with non-narrative *visual* art forms. Stravinsky claimed that 'One could not better define the sensation produced by music than by saying that it is identical with that evoked by the contemplation of the interplay of architectural forms.' This association is ancient, and owes much to the classical belief that both music and architecture are based on harmonious proportion. It's a useful analogy, but limited: not all music is like that. It is a helpful way to think about some of Bach's music, where plain, prosaic bricks are assembled into the most extraordinary edifices, full of allusion and symmetry and logic. But Debussy does not write architectural music – he paints with it, offering impressionistic sketches of flowing water, shimmering moonlight, soft rain.

Even here, however, we risk trying to project music on to another form of sensory input, to imply that it wants to be visual. I suspect that, while music is as beyond words as is the sensation of taste, the frustration evinced by Copland at the start of this chapter comes from the fact that it is unique in being able to *hint* at a linguistic status. Because of its modes of organization – rhythm and metre, scales and pitch relationships, hierarchies of relation and reference – it can create patterns that carry *information in a scientific sense*, as smell or taste never really can. The information is inherent in the non-randomness of the relationships between elements. We can perceive that information, but there is no meaning in it beyond that which music itself creates. This is perhaps one of the important broad lessons that the science of information theory conveys: information is not necessarily 'about' something, at least in a linguistic and semantic sense. It is a thing in itself. Perhaps one of the keys to music's power is that we are beings that have evolved to try to interpret information and to project meaning on to it. This is why we will gaze enraptured at a landscape, whether it is painted by nature, or by Turner, or by Zhu Da during the Qing dynasty. It is precisely because the meaning eludes us – because, indeed, there is none we can articulate – that we come back to it again and again.

Can we say what we cannot write?

One could easily imagine that musical notation is simply a convenient way of transmitting what composers and musicians do. But there is in fact a complex two-way interaction between how music is conceived and organized mentally, and how (if at all) it is written. The system of musical notation used in Western classical music is arguably the most developed of all such schemes, and it has allowed this music to become more polyphonic and complex than an oral tradition could sustain.

This system began, tellingly, by denoting mere contour: a series of lines called *neumes*, placed above the sung text, showed whether the melody went up (/), down (\), up and down (V), and so forth. Clearly one had to know the 'rules' to make much sense of this: it was taken for granted that the singers would take their notes from the medieval modes, and know on which to start and finish. It wasn't until around the tenth century that individual notes became notated, positioned in relation to a horizontal reference line or stave (initially just one, rather than the five used now) (Figure 13.3). A four-line stave was introduced in the eleventh century by Guido of Arezzo, whom we encountered earlier as the originator of the sol-fa mnemonic. Note durations were indicated from the tenth century, although at first these denoted only relative rather than absolute note lengths. The notation that we recognize today didn't take shape until the seventeenth century, and we'd have little difficulty interpreting the scores written by Bach in the early eighteenth century (Figure 13.4).

Notation tends to suppress what cannot be notated: gliding or microtonally pitched notes, elastic rhythms, subtle expressive gestures, and improvisation. We've seen earlier how Western notation is ill-suited to complex rhythmic structures like those of Eastern European music. And notated compositions can't evolve: their alteration comes to be seen as sacrilegious, and they stand at risk of ossifying. Reliance on notation also inhibits performance. There are extremely competent classical musicians today who cannot (or will not) play a note unless a score tells them which to use. Yet in the Baroque era, when strict notation was only just appearing, performers were invariably expected to add notes to those of the composer, and even to enliven a piece with improvised dissonances. In the early seventeenth century, Girolami Frescobaldi was happy for organists to end his toccatas wherever they felt like, or to take them apart and reassemble the material in other ways.

In the twentieth century avant-garde composers sought to find new notation systems that allowed and indeed forced the performer to find his or her own interpretations of the material. It isn't at all clear how the electronic patchworks of Ligeti and Stockhausen can be meaningfully set down on paper – Ligeti created 'listening scores'

a

b

c

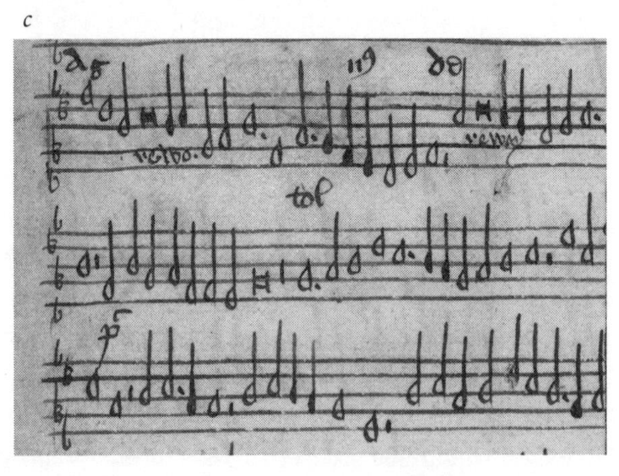

Figure 13.3 Musical notation in the eleventh (*a*), thirteenth (*b*) and fifteenth (*c*) centuries.

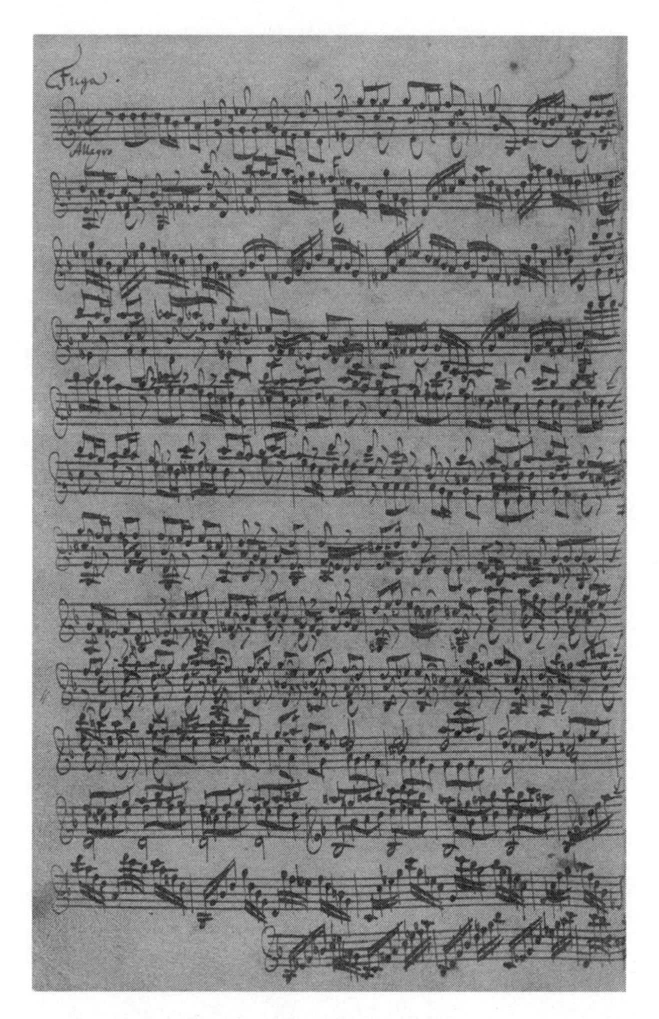

Figure 13.4 J. S. Bach's autograph transcription of his Sonata I for Unaccompanied Violin (BMV 1001), begining of second movement, writen in 1720.

that reflected what was heard rather than specifying what was performed (Figure 13.5*a*), while Stockhausen sometimes used a mixture of conventional notation and impressionistic symbols (Figures 13.5*b*). Few left quite as much to the imagination of the musician as John Cage, however (Figure 13.5*c*). In this kind of experimental music, performers are more likely to receive a set of instructions for what they must do than a set of notes they must play.

a

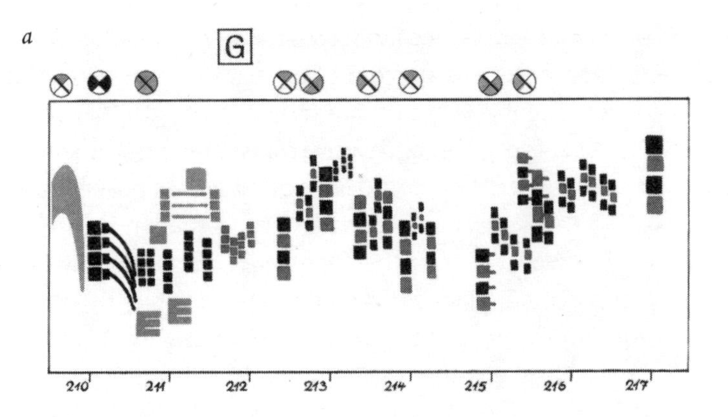

b

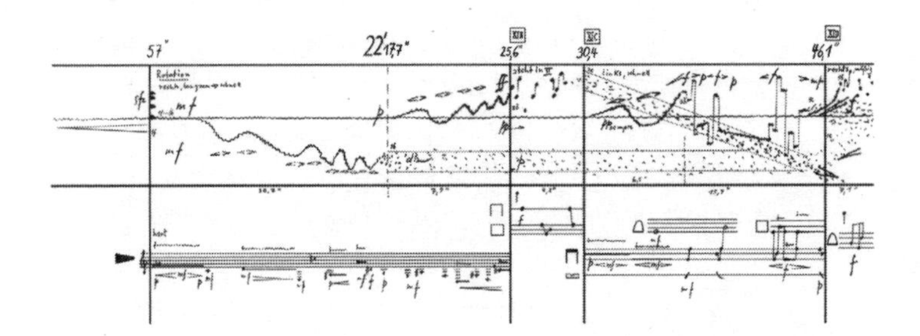

c

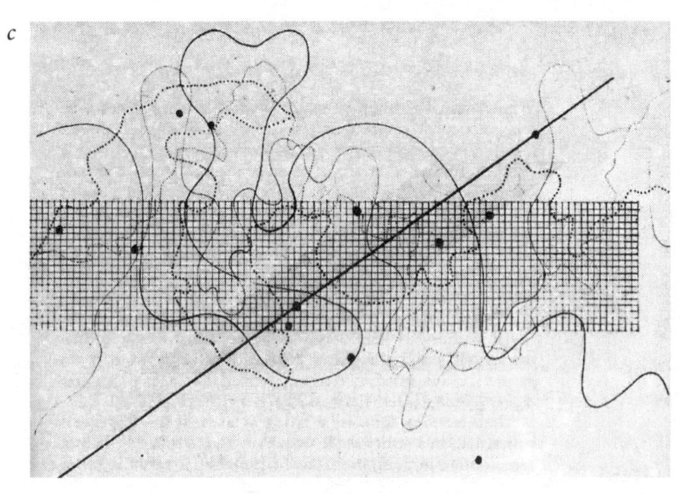

Figure 13.5 'Scores' of Ligeti's *Artikulation* (1958) (*a*); Stockhausen's *Kontakte* (1958–60) (*b*), and Cage's *Fontana Mix* (1958) (*c*).

Cage's *Fontana Mix*, for instance, is scored for any number of tracks of magnetic tape, or for any number of players using any kind and number of instruments, and is of indeterminate length. The score consists of ten sheets of paper and twelve transparencies. The sheets of paper are marked with six curved lines of different thickness and texture. Ten of the transparencies have sets of randomly distributed points, while one of the other two bears a grid and the last contains a straight line. The performer creates a 'score' by superimposing the sheets: one of the transparencies with dots is placed over one of the sheets with curved lines, and the grid goes over this. A point enclosed in the grid is then connected with a point outside using the straight-line transparency. Horizontal and vertical measurements of intersections of the straight line with the grid and the curved line somehow specify the actions to be made, and their timing. No wonder Cage often worked with specific 'performer-interpreters' who were especially adept at decoding his instructions.

One could argue that this sort of experimentation liberated the musician from notational tyranny; but in fact that tyranny has only ever really been felt in Western classical music in the first place. In British folk music, for example, conventional transcriptions of tunes are typically just skeletons which would sound deeply mundane if played as written – performers are expected to add their own elaborations (Figure 13.6). And in pop and jazz music, notation has always been superfluous: the music is all in the performance, whether live or recorded, which is improvised around an agreed format. Indeed, Steve Reich suggests that 'A more helpful distinction than classical and rock music might be nonated and non-notated music.' What we enjoy in the non-notated forms is watching how a performer can use a sketchy framework and a set of conventions as the raw material for creating novelty. In his 1949 book *Music in Java*, Dutch ethnomusicologist Jaap Kunst (who coined the very word 'ethnomusicology') wrote of the Indonesian singer that his task,

> in contrast to that of the European executant musician, is a creative one. Each time a *lagu* [the core melody of a gamelan composition] is sung, the song flowers again from the traditional melodic groundwork,

Figure 13.6 The 'skeleton' of the traditional English tune 'Speed the Plough' (*a*), and how it is played by Eliza Carthy and Nancy Kerr (*b*).

> the unalterable melodic nucleus; often to the delight of those who have learnt to esteem the native . . . style of performance.

This underlines the central tension in systems of musical notation: how to balance the composer's authorial rights and responsibilities with the fact that making music that lives and breathes is fundamentally an act of creativity.

Coda
The Condition of Music

I'd hope that no one will read this book without listening to some music along the way. Certainly I could not write it without doing so, and not just for research purposes. One of the wisest remarks about music cognition was made by John Sloboda in his book *Exploring the Musical Mind*: 'I believe that every scientist who studies music has a duty to keep his or her love of music alive.' One can't help wondering, when wading through the depths of some musicological or neurological treatise, whether that injunction is sometimes overlooked. The trouble (and the blessing) is that, with music as with a good magic trick, knowing *how* it works is no defence against wonderment at the experience itself. We can't help but suspect that the miraculous intervenes.

In any event, neither I nor anyone else can pretend to open up music's box of tricks, to show exhaustively and seamlessly how *this* leads to *that*. You should be in little doubt now how much we still don't understand, and how this sort of 'understanding' in any case has its limitations. But I hope too that you will see that music is not simply a black box into which raw notes are fed and out of which come tears and joy.

So what have we learnt?

First, that music is made in the mind. Transforming complex sound into comprehensible and meaningful music is a complicated and demanding feat, but one to which our brains are inherently attuned by the mere fact of living in the world. We are pattern seekers, clue-solvers, unravellers of sensory data, and also communicators and story-tellers. Within the auditory domain, those things are going to make us musical beings, come what may.

All the same, we have to learn these skills. From the moment of birth and in fact from some time before that, we assimilate and generalize the information we receive about our environment. We make mental maps of how stimuli relate to one another. We learn about what is more or less probable, and we use that to make predictions and form expectations – about notes and note sequences, harmonies, rhythms, timbres. We test these against the real world, and we congratulate and reward ourselves when we are proved right. We learn too the enhanced pleasures of delayed gratification and suspense. And there is something more, something that is as yet barely understood or even appreciated: a kind of delight in complex experience, in texture and sound quality, in a massaging of the auditory sense. This doesn't necessarily produce emotion in itself, but creates a susceptibility to expression, a kind of willingness to be moved.

Composers and musicians intuit these human characteristics and figure out how to toy with them. They provide sonic signposts both to heighten tension and to aid cognition: they *help us to hear.* And if they fail to do that, their music becomes marginalized, intellectualized, mathematicized. But even then, our musical sense is hard to baffle totally: we can find exciting stimuli in the most unlikely places, perhaps even in the face of the composer's intentions.

But it is performance that breathes fire into the formulas, so to speak. Certain patterns of notes work better than others, but the good performer knows how to rearrange and stretch and embody them in ways that turn a good tune into a soul-wrenching experience. This skill is hard to teach, but not impossible: it is not a mysterious gift, but requires a sensitive understanding of how music functions.

Music is a whole-brain activity. You need logic and reason, and also primitive 'gut instinct'. You need unconscious, mechanical processes for sorting pitch and classifying rhythm and metre, as well as bits of the mind that govern language and movement. Some of these functions are improved by training, but the fact is that everyone, if not hampered by physiological dysfunctions, possesses them. And everyone acquires them to some degree or another. There are, without doubt, people who have either cultivated or perhaps been born with exquisite musical sensibility, and there are many who have developed the most astonishing performance skills. And let's be grateful for that. But nearly everyone has musical ability. As John Blacking puts it, quoting from Lord Redcliffe-Maud's 1976 report to the Calouste Gulbenkian Foundation on 'support for the arts':

> The idea that educationalists must be converted 'to the belief that the arts are as important to society as reading, writing and arithmetic, not a disposable extra' will not carry conviction as long as people maintain the defeatist, elitist notion that artistic talent is [as the Redcliffe-Maud Report unhappily put it] a 'rare gift'.

And yes, music is intellectually utilitarian: it is 'good' for the brain. But that's a happy by-product, not a justification. After all, music is good for the body too, and good for the culture. In short it is, as Nietzsche said, 'something for the sake of which it is worthwhile to live on earth'.

We should never forget that music takes place in a social context. If all studies of music cognition tend to give the impression that it happens in a value-free vacuum, this danger is particularly acute when we consider the question of meaning. The American perceptual psychologist James Jerome Gibson argued in the 1960s that the cultural context of stimuli intervenes in the actual process of perception – that's to say, we don't *first* hear music and *then* interpret it within a context, but the context partially determines what we hear. This is one of the pitfalls of cross-cultural studies of music cognition, which tend to assume that if you play music to different people, they simply make different interpretations of the same sounds. All the same, music from other cultures need not be a closed book. Our minds can usually do something with it, although that is likely to happen on our own cultural terms unless we make considerable efforts to reach beyond them.

Music journalist David Stubbs has claimed that many people have a 'fear of music' that prevents them from embracing the modernity of Stockhausen even while they accept the modernity of Mark Rothko. There's something in that thesis, although one of the benefits of post-modernism is an erosion of the traditional barriers between 'difficult' art music and popular music. Some modern music is indeed somewhat fearsome in its retreat into hermetic ways of shuffling sound with no regard to cognition. But part of the responsibility for such fears must lie with the contemporary attitude towards all art, rooted in romanticism rather than modernism, which insists on 'meaning' beyond the sensory experience. People fear what they think they don't understand. And that fear is self-reinforcing, because cognition requires learning through exposure. (Both of these things are equally true for the lowbrow looking to the highbrow and vice versa, if indeed we want to insist on any such hierarchy in the first place.) This is why we should resist efforts to convert music into some kind of code, whereby Beethoven's Third Symphony becomes a narrative about his vision of heroism. A good interpretation of a piece of music is an invitation to listen to it in a certain way, and not a revelation of what the music is 'about'. This is also why we should not look for the same things in 'Hit Me With Your Rhythm Stick' as we find in the *Eroica*.

In the end we need to allow music to be music, with its own set of emotions and sensations that we have yet to name and perhaps do not need to. Music is not like other forms of art – it is *sui generis*, and therefore in some respects beyond words. And yet we cognize it using familiar neural apparatus – it is rather like an extraordinary occurrence for which our brains have accepted the need for unprecedented collaboration between departments. Wonderfully, they have decided it is worth the effort.

Credits

Figure 3.16*b*: Science Museum/Science & Society Picture Library. **Figure 3.32:** 'Do Re Mi' (Rodgers/Hammerstein). Copyright © Williamson/Hal Leonard Publishing Corp (USA). **Figure 4.8:** 'Maria' (Leonard Bernstein/Stephen Sondheim). Copyright © 1956, 1957, 1958, 1959 by Amberson Holdings LLC and Stephen Sondheim. Copyright renewed. Leonard Bernstein Music Publishing Company LLC, publisher. Reproduced by permission of Boosey & Hawkes Music Publishers Ltd. **Figure 4.12*a*:** 'Over the Rainbow' (from 'The Wizard of Oz') (Harold Arlen/E. Y. Harburg). Copyright © 1938 (Renewed) Metro-Goldwyn-Mayer Inc. Copyright © 1939 (Renewed) EMI Feist Catalog Inc. All rights controlled and administered by EMI Feist Catalog Inc. (Publishing) and Alfred Publishing Co., Inc. (Print). All rights reserved. Used by permission. **Figure 4.12*b*:** 'Alfie' (Burt Bacharach/Hal David). © Music Sales Ltd/Sony/Hal Leonard Publishing Corp (USA). **Figure 4.24:** *3 Klavierstücke* Op. 11 (Arnold Schoenberg). Copyright © 1910 Universal Edition A.G., Wien/UE 2991. **Figure 6.4*b*:** *Im Freien* (Out of Doors) (Béla Bartók). Copyright © 1927 Universal Edition A.G., Wien/UE 8892A,B. **Figure 6.19*a*:** Piano Sonata No. 5 (Sergei Prokofiev). Copyright © 1925 Hawkes & Son (London) Ltd. Reproduced by permission of Boosey & Hawkes Music Publishers Ltd. **Figure 6.19*b*:** *Cinderella* (Sergei Prokofiev). Copyright © 1948 Boosey & Hawkes Music Publishers Ltd. Reproduced by permission of Boosey & Hawkes Music Publishers Ltd. **Figure 6.27:** *Petrushka* (Igor Stravinsky). Copyright © 1912 Hawkes & Son (London) Ltd. Revised version copyright © 1948 by Hawkes & Son (London) Ltd. Reproduced by permission of Boosey & Hawkes Music Publishers Ltd. **Figure 7.12:** 'America' (Leonard Bernstein/ Stephen Sondheim). Copyright © 1956, 1957, 1958, 1959

Notes

1 *Prelude: The Harmonious Universe*

Page 2 'Music is auditory cheesecake': Pinker (1997), p. 534. • **Page 3** 'Compared with language': ibid., p. 528. 'Art, music and literature are not merely the products': Carroll (1998), p. xx. • **Page 4** 'discarded all amusements, discountenanced all story-books': Charles Dickens, *Bleak House*, Chapman & Hall, London, 1868, p. 177. • **Page 5** 'is so naturally united with us': Boethius, *Fundamentals of Music*, ed. C. V. Palisca, transl. C. M. Bower, Yale University Press, New Haven, 1989, p. 8.

2 *Overture: Why We Sing*

Page 9 'Musics can be defined as those temporally patterned': Cross (2001), p. 32. 'challenges the boundary between noise and music': Davies (2005), p. 29. • **Page 10** 'Nothing guarantees that all forms': J. Molino, in Wallin, Merker and Brown (eds) (2000), p. 169. • **Page 12** 'Music that has nothing else as its purpose': Hindemith (1961), p. xi. • **Page 13** 'the way in which music is played': M. Mead, 'Community drama, Bali and America', *American Scholar* 11, 79–88, p. 81 (194–2). 'When you are content, you sing': Merriam (1964), p. 64. 'When one shouts, he is not thinking': ibid. • **Page 14** 'Now the Spirit of the Lord had departed': 1 Samuel, 16:14–23. • **Page 15** 'more moved by the singing': St Augustine, *Confessions* X, Ch. 33, transl. J. G. Pilkington, in W. J. Oates (ed.), *Basic Writings of Saint Augustine,* Random House, New York, 1948. 'Through melody, harmony, and rhythm': Scruton (1997), p. 502. • **Page 17** 'the science that deals with the music': Merriam (1964), p. 5. 'It is a curious paradox that we probably know': Sloboda (2005), p. 320. 'rattle with wooden staves and pieces of metal': Hanslick (1891), p. 146. • **Page 19** 'As neither the enjoyment nor the capacity': Darwin (1877/2004), p. 636. • **Page 24** 'it appears probable that the progenitors of man': ibid., p. 639. • **Page 25** 'Music and language have just too many important similarities': Brown, in Wallin, Merker and Brown (eds) (2000), p. 277. 'The dance produces a condition in which the unity': A. R. Radcliffe, *The Andaman Islanders*, Free Press, Glencoe, 1948, p. 252. • **Page 26** 'The role of music in superstitions or sexual rites': J. G. Roederer, 'The search for a survival value of music', *Music Perception* 1, 350–6, p. 356 (1984). • **Page 27** 'the profound meaning of music': I. Stravinsky, *Poetics of Music*, Vintage, New York, 1947, p. 21. • **Page 28** 'were found by evolving human groups': E. Dissanayake, in Wallin, Merker and Brown (eds) (2000), p. 401. • **Page 30** 'The notion that something is either a product of biological adaptation': Patel (2008), p. 412.

3 *Staccato: The Atoms of Music*

Page 32 'I decided to call my music "organized sound"': E. Varèse, *The Liberation of Sound* (1936), quoted in A. Hugill, *The Digital Musician*, Routledge, New York, 2008, p. 68. • **Page 50** 'exceedingly coarse and tasteless': Plato, *Laws* Book II, transl. B. Jowett. • **Page 61** 'unpleasant to uncorrupted ears': H. von Helmholtz, *On the Sensations of Tone as a Physiological Basis for the Theory of Music*, transl. A. J. Ellis, Dover, Mineola, New York, 1954, p. 428. • **Page 71** 'the naïve view that by some occult process': N. Cazden, 'Musical consonance and dissonance: a cultural criterion', *Journal of Aesthetics* 4, p. 4 (1945). 'Just as little as the gothic painted arch': Helmholtz (1954), p. 236. • **Page 72** 'there are as many scales as there are gamelans': C. McPhee, *Music in Bali*, Yale University Press, New Haven, 1966. Quoted in D. Deutsch (ed.) (1982), p. 258. 'The specific size of intervals in gamelan music': J. Becker and A. Becker (1983), p. 36. • **Page 76** 'is conditioned not only by what sound is actually produced': B. Nettl (1973), p. 33. • **Page 84** 'the folk-scales in which their so-called': quoted in T. Balough, *A Musical Genius from Australia: Selected Writings by and about Percy Grainger*, Music Monographs No. 4, University of Western Australia Press, Nedlands, 1982, p. 70. 'depending on how open or how disguised': A. Hodeir (1956), p. 155.

4 *Andante: What's In a Tune?*

Page 93 'Melodies can be constructed rationally': Hindemith (1961), p. 112. • **Page 101** 'ongoing accommodation of errors': Bregman (1990), p. 475. • **Page 119** 'It is insufficient merely to hear music': Copland (1957), p. 5. • **Page 128** 'The emphasis given to a tone': A. Schoenberg, 'Composition with twelve tones

(Part 2)', 1948; reprinted in L. Stein (ed.), *Style and Idea*, Faber & Faber, London, 1975, p. 246. • **Page 130** 'it sounds as if someone has smeared the score': quoted in Rosen (1976), p. 11. • **Page 133** 'Just as our mind always recognizes': Schoenberg, in *Style and Idea*, quoted in Deutsch (ed.) (1982), p. 283. 'perceptual equivalence under transposition': D. Deutsch, in Deutsch (ed.) (1982), p. 283. 'consciously used, the motif': Schoenberg, in Strang and Stein (eds) (1982), p. 8. • **Page 135** 'as promising as attempts at avoiding the effects': Hindemith (1961), p. 64. 'will probably never understand why music': ibid., p. 65. • **Page 136** 'When the music goes everywhere': Scruton (1997), p. 303. 'in the recitative music degenerates': Hanslick (1891), p. 57–8.

5 *Legato: Keeping It Together*

Page 138 'a picture–in-sounds of the sounds of nature': C. Ives, cited in recording notes by Victor Ledin, available at http://www.medienkunstnetz.de/works/central-park-in-the-dark/ • **Page 139** 'The mind will tend to apprehend a group': Meyer (1956), p. 162. • **Page 143** 'four thin, sharp laser beams': Sacks (2007), p. 113. • **Page 154** 'It was as if someone had suddenly returned': Sloboda (1985), p. 158. • **Page 161** 'Even music of the type written by Charles Ives': Bregman (1990), p. 461. 'each enjoying its local harmony': Scruton (1997), p. 65. 'The complex polyphony of the individual parts': see, for example, S. Gerfried, G. Stocker and C. Schof (eds), *Ars Electronica 2004: Time Shift – The World in 25 Years*, Hatje Cantz, Berlin, 2004, p. 285. • **Page 162** 'perceptually precarious': Bregman (1990), p. 532.

6 *Tutti: All Together Now*

Page 163 'Now there is need, Urania, of a grander sound': Kepler, *Harmonia mundi* 441, quoted in Pesic (2005), 3.18 (http://www.sscm-jscm.org/v11/no1/pesic.html). • **Page 164** 'The universe itself is said to have been put together': Isidore of Seville, *Etymologies*, Book III.17.1, in *Isidore of Seville's Etymologies*, Vol. 1, transl. & ed. P. Throop. Lulu.com, 2006. • **Page 165** 'From Harmony, from heavenly Harmony': J. Dryden, 'Ode for St Cecilia's Day', in S. Johnson (ed.), *The Works of the English Poets, from Chaucer to Cowper*, Vol. 8, J. Johnson *et al.*, London, 1810, p. 607. • **Page 171** 'some Pairs are heard with great Delight': quoted in Fauvel, Flood and Wilson (eds) (2003), p. 84. • **Page 175** 'Chords of major and minor seconds': Blacking (1987), p. 15. 'to most ears more attractive': Rosen (1976), p. 32. • **Page 176** 'have already learned a good deal': Patel (2008), p. 382.

7 *Con Moto: Slave to the Rhythm*

Page 215 'Once a rhythm becomes established': Cooper and Meyer (1960), p. 13. • **Page 219** 'a blues lick that went round and round': quoted in M. Snow, 'The Secret Life of a Superstar', *Mojo* December 2007, p. 81–2. • **Page 222** 'It often happens among Indian Musicians': W. Sargeant and S. Lahiri, 'A study in East Indian rhythm', *Musical Quarterly* 17, 427–38, pp. 435–6 (1931).

8 *Pizzicato: The Colour of Music*

Page 229 'We do not know how to define timbre': Bregman (1990), p. 93. • **Page 233** 'Every violin has its own voice or tone': A. Gregorian, personal correspondence. • **Page 234** 'If you take each moment of each frequency component': Bregman (1990), p. 647. • **Page 238** 'spraying shower or flowing hair': ibid., p. 117.

9 *Misterioso: All In the Mind*

Page 252 'the results provide evidence of relatively modest': Schellenberg (2004), p. 513. • **Page 253** 'The development of the senses': Blacking (1987), p. 118.

10 *Apassionato: Light My Fire*

Page 254 'Suddenly [I] experienced a tremendously strong feeling': A. Gabrielsson, in Juslin and Sloboda (eds) (2001), p. 437. 'I was filled by a feeling that the music': ibid., p. 437. • **Page 225** '[There was] a chord so heart-rending and ghost-ridden': ibid., p. 439. 'music is the shorthand of emotion': in S. E. Anderson, *The Quotable Musician*, Allworth Press, New York, 2003, p. 2. 'What passion cannot Music raise and quell?': J. Dryden, 'Ode for St Cecilia's Day', in S. Johnson (ed.), *The Works of the English Poets, from Chaucer to Cowper*, Vol. 8, J. Johnson *et al.*, London, 1810, p. 607. • **Page 257** 'wrapped in a cloud of high-flown sentimentality': Hanslick (1891), p. 17. 'Definite feelings and emotions are unsusceptible': ibid., p. 33. 'The physiological process by which the *sensation* of sound': ibid, p. 116. • **Page 258** 'to a set of "effects" such as might': Sloboda (2005), p. 376. • **Page 259** 'Those who imagine that a creative artist can': quoted in J. Fisk (ed.), *Composers on Music*, Northeastern University Press, Boston, 1997, p. 157. 'the reactions which music evokes are not feelings': Hindemith (1961), p. 45. • **Page 262** 'It may be necessary to focus more strongly on terms': K. Scherer and M. Zentner, in Juslin and Sloboda (eds) (2001), p. 381. 'there exists no adequate word in any language': Copland (1957), p. 10. 'music sounds the way emotions feel': Kivy (2002), p. 40. 'The forms of human feeling': Langer (1957), p. 235 and 228. 'the dynamics and the abstract qualities of emotion': Sessions (1950), p. 23. 'a significant image of the inner flow of life': M. Tippett, 'Art, judgment and belief: towards the condition of music', in P. Abbs (ed.), *The Symbolic Order*, Falmer Press, London, 1989, p. 47. 'music mirrors the human soul': H. Schenker, *Free Composition*, Longman, London 1979; quoted in N. Cook, 'Schenker's theory of music as ethics', *Journal of Musicology* 7, 415–39 (1989). 'What we feel in art is not a simple or single': E. Cassirer, *An Essay on Man: An Introduction to a Philosophy of Human Culture*, Meiner, Hamburg, 2006, p. 161. • **Page 265** 'Music operates on our emotional faculty': Hanslick (1891), p. 107. 'My intuition is that musical emotions typically occur': I. Peretz, in Juslin and Sloboda (eds) (2001), p. 126. • **Page 267** 'Of the harmonies I know nothing': Plato, *The Republic*, Book III, transl. B. Jowett. • **Page 270** 'That which for the unguarded feelings of so many lovers of music': Hanslick (1891), p. 123. 'If music be the food of love': W. Shakespeare, *Twelfth Night*, Act I, i. • **Page 271** 'many people feel they are being told': in Juslin and Sloboda (eds) (2001), p. 458. • **Page 272** 'Taking advantage of the fact that I still was not alone': M. Proust, *Remembrance of Things Past*, Vol. 2, transl. C. K. S. Moncrieff and S. Hudson, Wordsworth, Ware, 2006, p. 576. • **Page 273** 'One answer to the question': in Juslin and Sloboda (eds) (2001), p. 98. • **Page 277** 'In 1971 Ravi Shankar, the Indian virtuoso': D. Hajdu, 'Fascinating rhythm', *New York Review of Books* 20 July 2000, p. 41. 'the more closely the external musical impression approaches': Hindemith (1961), p. 20. • **Page 278** 'the essential possibility of foreseeing and anticipating': ibid., p. 23. 'music goes astray, disappears in chaos': ibid., p. 23. 'the building material cannot be removed very far away': ibid., p. 24. • **Page 279** 'To the scientist our method': ibid, p. ix. • **Page 280** 'In music the state of suspense involves an awareness': Meyer (1956), p. 29. • **Page 281** 'emotions are felt to be pleasant, even exhilarating': L. B. Meyer, in Juslin and Sloboda (eds) (2001), p. 359. • **Page 283** 'Nature's tendency to overreact provides a golden opportunity': Huron (2006), p. 6. • **Page 285** 'properly curb the violent impulses of his imagination': Grout (1960), p. 205. • **Page 286** 'The delight of intelligent mental play': L. B. Meyer, *Explaining Music*, University of California Press, Berkeley, 1973, p. 213. 'will immediately conclude that the whole concert': quoted in Meyer (1956), p. 208. 'Certain purposeful violations of the beat': C. P. E. Bach, *An Essay on the True Art of Playing Keyboard Instruments*, transl. W. J. Mitchell, W. W. Norton, New York, 1949, p. 150. • **Page 293** 'Rubato must emerge spontaneously from the music': M. Pollini, interviewed by Carsten Dürer for Deutsche Grammophon; see http://www2.deutschegrammophon.com/special/insights.htms?ID=pollini-nocturnes. • **Page 295** 'always represents the extraordinary': E. E. Lowinsky, *Secret Chromatic Art in the Netherlands Motet*, transl. C. Buchman, Columbia University Press, New York, 1946, p. 79. • **Page 296** 'frustration of conventional expectations': Rifkin (2006), p. 146. • **Page 302** 'The unlimited resources for vocal and instrumental expression': C. E. Seashore, introduction to M. Metfessel, *Phono-photography in Folk Music*, University of North Carolina Press, Chapel Hill, 1928, p. 11. 'must not be considered faulty, off-pitch singing': Bartók and Lord (1951), p. 4. • **Page 303** 'is not necessarily worse than [that of]': Meyer (1956), p. 204. 'Every now and then there was a little deviation': J. Kunst, *Music in Java*, Martinus Nijhoff, The Hague, 1949, p. 59.'When an expressive intention shows up in the blues': Hodeir (1956), p. 227. • **Page 304** 'Probably no one has ever doubted': quoted in Grout (1960), p. 455. 'A melody without ornament is like': quoted in A. Danielou, *Northern Indian Music*, Christopher Johnson, London, 1949, p. 102. 'may well be due to the contrast between the essential simplicity': G. Herzog, in the introduction to Bartók and Lord (1951), p. xiii. • **Page 306** 'toward the emotion, and illuminate the sense': J. Mattheson, *Der vollkommene Capellmeister* (1739), transl. E. C. Harriss, University of Michigan Research Press, Ann Arbor, Michigan, 1981, p. 370. • **Page 307** 'One will say 'love'. He may be right': Hanslick (1891), p. 44. • **Page 311** 'to calculate all the beauties in *one* symphony': H. C. Oerstedt, *Der Geist in der Natur*, Vol. III, p. 32 (Hanslick's translation). • **Page 314** 'What is profound about the experience of a listener': Meyer (1996), p. 462. • **Page 315** 'there would hardly be a measure of music': Sloboda (2005), pp. 229–30. • **Page 316** 'the entire interaction between the attractional field': Lerdahl (2001), p. 190. • **Page 317** 'The interplay of tension, release, surprise, and confirmation': Juslin and Sloboda (eds) (2001), p. 93. • **Page 318** 'aggressive force implies no future goal': L. B. Meyer, ibid., p. 357. • **Page 319** 'we should resist cognitive theories of expression': Scruton (1997), p. 359. 'the great triumphs of music . . . involve this synthesis': ibid. 'Many people are moved by music': ibid., p. 371.

11 *Capriccioso: Going In and Out of Style*

Page 325 'Always remember that a theme is, after all, only a succession of notes': Copland (1957), pp. 19–20. • **Page 329** 'A classical composer did not always need themes': quoted in Rothstein (1996), p. 96. • **Page 333** 'After sufficient training, GenJam's playing': Biles (1994), p. 136. 'it does not please *us* particularly well': Spector and Alpern (1994), p. 7. • **Page 335** 'a decline in morals': Scruton (1997), p. 502. • **Page 336** 'it has become increasingly clear that the coherence systems': Becker and Becker (1979/1983), pp. 34–5. • **Page 338** 'traditional modes of deviation are exaggerated to extremes': Meyer (1956), p. 71. 'Modulations, cadences, intervals and harmonious progressions': Hanslick (1891), p. 81. 'unifies all elements so that their succession and relation': A. Schoenberg, *Style and Idea*, ed. L. Stein, Faber & Faber, London, 1975, p. 279. • **Page 339** 'What would remain of the art of painting': Scruton (1997), p. 291. • **Page 341** 'makes ever heavier demands upon the training': Babbitt (1958), available at http://www.palestrant.com/babbitt.html. 'I dare suggest that the composer would do himself': ibid. • **Page 343** 'the best music utilizes the full potential of our cognitive resources': Lerdahl, in Sloboda (ed.) (1988), p. 255.

12 *Parlando: Why Music Talks to Us*

Page 355 'alone in the world with his more or less unintelligible language': quoted in Cooke (1959), p. ix. • **Page 358** 'Whenever someone spoke to me': quoted in M. Zemanová, *Janáček: A Composer's Life*, John Murray, London, 2002, p. 75. • **Page 359** 'Nations creates music, composers only arrange it': quoted in F. Bowers, *Scriabin: A Biography*, 2nd revised edn, Dover, Mineola, New York, 1996, p. 20. • **Page 362** 'an intensified version of the actual sounds': *Grove's Dictionary of Music and Musicians*, Vol. 5, Macmillan, New York, 1935, p. 605. • **Page 363** 'If I took a Czech, or an English, French or other folk-song': in Z. E. Fischmann (ed.), *Janáček-Newmarch Correspondence*, Kabel, Rockville, 1989, p. 123. • **Page 373** 'closer to the surface': Lerdahl (2001), p. 143. • **Page 374** 'a brilliant creative work': quoted in Sloboda (1985), p. 260. • **Page 380** 'can prime representations of meaningful concepts': Koelsch *et al.* (2004), p. 306.

13 *Serioso: The Meaning of Music*

Page 381 'It is easier to pin a meaning-word on a Tschaikovsky': Copland (1957), pp. 10–11. 'Is there meaning in music?': ibid., p. 9. 'If a composer could say what he had to say in words': attributed to Mahler in a letter to Max Marschall, 26 March 1896. • **Page 382** 'People usually complain that music is too many-sided': quoted in L. Botstein, 'Recreating the career of Felix Mendelssohn', in *Mendelssohn and His World*, ed. R. L. Todd, Princeton University Press, Princeton, 1991, p. 60. 'Language is a common and practical element': P. Valéry, *Pure Poetry: Notes for a Lecture*, quoted in *The Creative Vision*, ed. H. M. Black and Salinger, Grove Press, New York, 1960, pp. 25–6.

'all art constantly aspires towards the condition of music ': W. Pater, *The Renaissance: Studies in Art and Poetry*, Macmillan, London, 1873, p. 111. 'If, as is nearly always the case': I. Stravinsky, *Igor Stravinsky: An Autobiography*, W. W. Norton, New York, 1962, p. 53. 'Strictly speaking you cannot write about music': quoted in Critchley and Henson (1977), p. 217. • **Page 383** 'Music expresses, at different moments': Copland (1957), p. 10. 'music which always says the same thing': ibid., p. 11. • **Page 384** 'when men hear imitations, even apart from the rhythms': Aristotle, *Politics* 8, part V, transl. B. Jowett. 'has the power either to improve or to debase': quoted in Hindemith (1961), p. 8. 'Let me make the songs of a nation': quoted in Grout (1960), p. 8. 'when there are no words, it is very difficult': Plato, *Laws* Book II, transl. B. Jowett. 'music, like humus in a garden soil': Hindemith (1961), p. 6. • **Page 385** 'The riches of music are so excellent': M. Luther, Foreword to G. Rhau, *Symphoniae iucundae* (1538), quoted in H. Lockyer, Jr., *All the Music of the Bible*, Hendrickson, Peabody, 2004, p. 144. 'a discipline, and a mistress of order and good manners': quoted in J. E. Tarry, 'Music in the educational philosophy of Martin Luther', *Journal of Research in Music Education* 21, 355–65 (1973). 'turning modern men, women and children back to the stages': Anon., 'Condemns age of jazz', *New York Times* 27 January 1925, p. 8; quoted in Merriam (1964), p. 242. • **Page 387** 'the loveable glad man who paces hale and hearty': R. Wagner, *Judaism in Music and Other Essays*, transl. W. A. Ellis, London, 1894, pp. 222–3. 'Schumann's masculine personae thus coalesce under a feminine sign': Kramer (2002), p. 113. • **Page 388** 'One day the children's children of our psychologists': A. Schoenberg, in *Style and Idea*, ed. L. Stein. Faber & Faber, London, 1975; quoted in Cooke (1959), p. 273. • **Page 390** 'only enhance[s] the presence of meaning as an issue or problem': Kramer (2002), p. 16. • **Page 391** 'who go so far as to call every transposition a rank violation': Révész (2001), p. 113. 'tender, soft, sweet, effeminate': quoted in Duffin (2007), p. 44. • **Page 393** 'leads some people into a pseudo feeling of profound melancholy': Hindemith (1961), p. 47. 'One is bound to regard anyone who reacts in this way': Cooke (1959), p. 23. 'The truly musical person, with a normal capacity': ibid., p. 22. • **Page 394** 'there is a tremolando passage of the kind': Scruton (1997), p. 43. 'Emotional responses to music are neither correct nor incorrect': Raffman (1993), p. 59. • **Page 395** 'The listener thus makes direct contact': Cooke (1959), p. 19. • **Page 396** 'there are too many performers and composers whose lives': Blacking (1987), p. 40. • **Page 398** 'the art of pleasing by the succession and combination': C. Burney, 'Essay on musical criticism', introduction to his *General History of Music*, Book III, 1789; quoted in Grout (1960), p. 451. • **Page 401** 'One could not better define the sensation produced by music': I. Stravinsky, *Chronicle of My Life*, Gollancz, London, 1936, p. 93. • **Page 407** 'A more helpful distinction than classical and rock music': quoted in *Guardian* Review 27 June 2009, p. 12. 'in contrast to that of the European executant musician': J. Kunst, *Music in Java*, Martinus Nijhoff, The Hague, 1949, p. 401.

Coda: *The Condition of Music*

Page 409 'I believe that every scientist who studies music': Sloboda (2005), p. 175. • **Page 411** 'The idea that educationalists must be converted': Blacking (1987), p. 120. 'something for the sake of which it is worthwhile': F. Nietzsche, *Beyond Good and Evil*, transl. R. J. Hollingdale, Penguin, London, 2003, p. 111.

Bibliography

Adler, D. 'The earliest musical tradition', *Nature* 460, 695–6 (2009).

Adorno, T. *Prisms*, transl. S. and S. Weber. MIT Press, Cambridge, Ma., 1981.

Adorno, T. *Philosophy of Modern Music*, transl. A. G. Mitchell and W. V. Blomster. Continuum, New York, 2003.

Apel, W. *The Notation of Polyphonic Music 900–1600*. Medieval Academy of America, 1953.

Aristotle, *Politics*, transl. T. A. Sinclair. Penguin, Harmondsworth, 1981.

Assmann, J., Stroumsa, G. G. and Stroumsa, G. A. G. *Transformations of the Inner Self in Ancient Religions*. Brill, Leiden, 1999.

Avanzini, G., Lopez, L. and Majno, M. (eds). 'The Neurosciences and Music', *Annals of the New York Academy of Sciences* 999 (2003).

Avanzini, G., Koelsch, S. and Majno, M. (eds). 'The Neurosciences and Music II: From Perception to Performance', *Annals of the New York Academy of Sciences* 1060 (2006).

Ayari, M. and McAdams, S. 'Aural analysis of Arabic improvised instrumental music (Taqsim)', *Music Perception* 21, 159–216 (2003).

Babbitt, M. 'Who cares if you listen?', *High Fidelity* February 1958.

Bailes, F. 'Timbre as an elusive component of imagery for music', *Empirical Musicology Review* 2, 21–34 (2007).

Balkwill, L. L. and Thompson, W. F. 'A cross-cultural investigation of the perception of emotion in music: psychophysical and cultural clues', *Music Perception* 17, 43–64 (1999).

Ball, P. 'Mingle bells', *New Scientist* 13 December 2003, pp. 40–3.

Ball, P. 'Facing the music', *Nature* 453, 160–2 (2008).

Balter, M. 'Seeking the key to music', *Science* 306, 1120–2 (2004).

Balzano, G. 'The group-theoretic description of 12-fold and microtonal pitch systems', *Computer Music Journal* 4, 66–84 (1980).

Barbour, J. M. *Tuning and Temperament*. Michigan State College Press, East Lansing, 1951.

Bartók, B. and Lord, A. B. *Serbo-Croatian Folk Songs*. Columbia University Press, New York, 1951.

Becker, A. L. and Becker, J. 'A grammar of the musical genre srepegan', *Journal of Music Theory* 23, 1–43 (1979); reprinted in *Asian Music* 14, 30–73 (1983).

Becker, A. L. and Becker, J. 'Reflections on srepegan: a reconsideration in the form of a dialogue', *Asian Music* 14, 9–16 (1983).

Benamou, M. 'Comparing musical affect: Java and the West', *The World of Music* 45, 57–76 (2003).

Bergeson, T. R. and Trehub, S. E. 'Infants' perception of rhythmic patterns', *Music Perception* 23, 245–60 (2006).

Bernstein, L. *The Unanswered Question*. Harvard University Press, Cambridge, Ma., 1976.

Bharucha, J. J. 'Anchoring effects in music: the resolution of dissonance', *Cognitive Psychology* 16, 485–518 (1984).

Bharucha, J. J. 'Music cognition and perceptual facilitation', *Music Perception* 5, 1–30 (1987).

Bigand, E. and Parncutt, R. 'Perceiving musical tension in long chord sequences', *Psychological Research* 62, 237–54 (1999).

Bigand, E., Poulin, B., Tillmann, B., D'Adamo, D. A. and Madurell, F. 'Sensory versus cognitive components in harmonic priming', *Journal of Experimental Psychology: Human Perception and Performance* 29, 159–71 (2003).

Biles, J. A. 'GenJam: A generative algorithm for generating jazz solos', in *Proceedings of the 1994 International Computer Music Conference*, International Computer Music Association, San Francisco, 1994. Available at http://www.it.rit.edu/~jab/GenJam94/Paper.html.

Bischoff Renninger, L., Wilson, M. P. and Donchin, E. 'The processing of pitch and scale: an ERP study of musicians trained outside of the Western musical system', *Empirical Musicology Review* 1 (4), 185–97 (2006).

Blacking, J. *How Musical is Man?* Faber & Faber, London, 1976.

Blacking, J. *A Commonsense View of All Music*. Cambridge University Press, Cambridge, 1987.

Blood, A., Zatorre, R. J., Bermudez, P. and Evans, A. C. 'Emotional responses to pleasant and unpleasant music correlate with activity in paralimbic brain regions', *Nature Neuroscience* 2, 382–87 (1999).

Blood, A. J. and Zatorre, R. J. 'Intensely pleasurable responses to music correlate with activity in brain regions implicated in reward and emotion', *Proceedings of the National Academy of Sciences USA* 98, 11818–23 (2001).

Boulez, P. 'Timbre and composition – timbre and language', *Contemporary Music Review* 2, 161–72 (1987).

Braun, A., McArdle, J., Jones, J., Nechaev, V., Zalewski, C., Brewer, C. and Drayna, D. 'Tune deafness: processing melodic errors outside of conscious awareness as reflected by components of the auditory ERP', *PLoS One* 3 (6), e2349 (2008).

Bregman, A. S. *Auditory Scene Analysis: The Perceptual Organization of Sound*. MIT Press, Cambridge, Ma., 1990.

Burkeman, O. 'How many hits?', *Guardian Weekend* 11 November 2006, 55–61.

Callender, C., Quinn, I. and Tymoczko, D. 'Generalized voice-leading spaces', *Science* 320, 346–8 (2008).

Carroll, J. 'Steven Pinker's Cheesecake for the Mind', *Philosophy and Literature* 22, 478–85 (1998).

Castellano, M. A., Bharucha, J. J. and Krumhansl, C. L. 'Tonal hierarchies in the music of North India', *Journal of Experimental Psychology: General* 113, 394–41 (1984).

Chen, J. L., Perhune, V. B. and Zatorre, R. J. 'Listening to musical rhythms recruits motor regions of the brain', *Cerebral Cortex* 18, 2844–54 (2008).

Chen, J. L., Zatorre, R. J. and Penhune, V. B. 'Interactions between auditory and dorsal premotor cortex during synchronization to musical rhythms', *Neuroimage* 32, 1771–81 (2006).

Clark, S. and Rehding, A. (eds). *Music Theory and Natural Order from the Renaissance to the Early Twentieth Century*. Cambridge University Press, Cambridge, 2001.

Clarke, E. F. 'Subject-position and the specification of invariants in music by Frank Zappa and P. J. Harvey', *Music Analysis* 18, 347–74 (1999).

Clarke, E. F. 'Structure and expression in rhythmic performance', in *Musical Structure and Cognition*, ed. P. Howell, I. Cross and R. West, 209–36. Academic Press, London, 1985.

Cogan, R. *New Images of Musical Sound*. Harvard University Press, Cambridge, Ma., 1984.

Cohen, D. 'Palestrina counterpoint: A musical expression of unexcited speech', *Journal of Music Theory* 15, 85–111 (1971).

Cole, H. *Sounds and Signs: Aspects of Musical Notation*. Oxford University Press, Oxford, 1974.

Conard, N. J., Malina, M. and Münzel, S. C. 'New flutes document the earliest musical tradition in southwestern Germany', *Nature* 460, 737–40 (2009).

Cook, N. 'The perception of large-scale tonal structure', *Music Perception* 5, 197–206 (1987).

Cook, N. and Everist, M. *Rethinking Music*. Oxford University Press, Oxford, 1999.

Cooke, D. *The Language of Music*. Oxford University Press, Oxford, 1959.

Cooper, G. W. and Meyer, L. B. *The Rhythmic Structure of Music*. University of Chicago Press, Chicago, 1960.

Copland, A. *What To Listen For in Music*. McGraw-Hill, New York, 1957.

Critchley, M. and Henson, R. A. (eds). *Music and the Brain*. Heinemann, London, 1977.

Cross, I. 'Music, cognition, culture and evolution', *Annals of the New York Academy of Sciences* 930, 28–42 (2001).

Cross, I. 'Bach in mind', *Understanding Bach* 2; available online only at http://www.bachnetwork.co.uk/ub2_contents.html (2007).

Cuddy, L. L. and Lunney, C. A. 'Expectancies generated by melodic intervals: perceptual judgements of melodic continuity', *Perception & Psychophysics* 57, 451–62 (1995).

Dalla Bella, S. and Peretz, I. 'Differentiation of classical music requires little learning but rhythm', *Cognition* 96, B65–B78 (2005).

Darwin, C. *The Descent of Man*, ed. J. Moore and A. Desmond. Penguin, London, 2004.

Davies, J. B. *The Psychology of Music*. Hutchinson, London, 1978.

Davies, S. *Musical Meaning and Expression*. Cornell University Press, New York, 1994.

Davies, S. 'The expression of emotion in music', *Mind* 89, 67–86 (1980).

Davies, S. 'Profundity in instrumental music', *British Journal of Aesthetics* 42, 343–56 (2002).

Davies, S. *Themes in the Philosophy of Music*. Clarendon Press, Oxford, 2005.

Deliège, I. 'A perceptual approach to contemporary musical forms', *Contemporary Music Review* 4, 213–30 (1989).

Deliège, I. and El Ahmadi, A. 'Mechanism of cue extraction in musical groupings: a study of perception on *Sequenza VI* for viola solo by Luciano Berio', *Psychology of Music* 18 (1), 18–44 (1990).

Deliège, I., Mélen, M., Stammers, D. and Cross, I. 'Music schemata in real-time listening to a piece of music', *Music Perception* 14, 117–60 (1996).

Deliège, I. and Sloboda, J. A. (eds). *Perception and Cognition of Music*. Psychology Press, Hove, 1997.

Deutsch, D. and Feroe, J. 'The internal representation of pitch sequences in tonal music', *Psychology Review* 86, 503–22 (1981).

Deutsch, D. (ed.) *The Psychology of Music*. Academic Press, London, 1982.

Deutsch, D., Henthorn, T., Marvin, E. and Xu, H.-S., 'Absolute pitch among American and Chinese conservatory students: prevalence differences, and evidence for a speech-related critical period', *Journal of the Acoustical Society of America* 119, 719–22 (2006).

Deutsch, D., Dooley, K., Henthorn, T. and Head, B. 'Absolute pitch among students in an American conservatory: association with tone language fluency', *Journal of the Acoustical Society of America* 125, 2398–403 (2009).

Dibben, N. 'The cognitive reality of hierarchic structure in tonal and atonal music', *Music Perception* 12, 1–25 (1994).

Donington, R. *The Interpretation of Early Music*. Faber & Faber, London, 1963.

Dowling, W. J. 'Scale and contour: two components of a theory of memory for melodies', *Psychological Review* 85, 342–54 (1978).

Dowling, W. J. and Harwood, D. L. *Musical Cognition*. Academic Press, London, 1986.

Duffin, R. W. *How Equal Temperament Ruined Harmony (and Why You Should Care)*. W. W. Norton, New York, 2007.

Edwards, P. 'A suggestion for simplified musical notation', *Journal of the Acoustical Society of America* 11, 323 (1940).

Eerola, T. and North, A. C. 'Expectancy-based model of melodic complexity', in *Proceedings of the 6th International Conference of Music Perception and Cognition* (Keele University, August 2000), ed. Woods, C., Luck, G., Brochard, R., Seddon, F. and Sloboda, J. Keele University, 2000.

Fauvel, J., Flood, R. and Wilson, R. *Music and Mathematics: From Pythagoras to Fractals*. Oxford University Press, Oxford, 2003.

Fedorenko, E., Patel, A.D., Casasanto, D., Winawer, J. and Gibson, E. 'Structural integration in language and music: evidence for a shared

system', *Memory & Cognition* 37, 1–9 (2009).
Fitch, W. T. 'On the biology and evolution of music', *Music Cognition* 10, 85–8 (2006).
Fitch, W. T. and Rosenfeld, A. J. 'Perception and production of syncopated rhythms', *Music Perception* 25, 43–58 (2007).
Forte, A. *Tonal Harmony in Concept and Practice*. Holt, Rinehart & Winston, New York, 1962.
Forte, A. *The Structure of Atonal Music*. Yale University Press, New Haven, 1973.
Forte, A. and Gilbert, S. E. *Introduction to Schenkerian Analysis*. W. W. Norton, New York, 1982.
Francès, R. *La Perception de la Musique*. Vrin, Paris, 1958.
Fritz, T. *et al.* 'Universal recognition of three basic emotions in music', *Current Biology* 19, 1–4 (2009).
Gabriel, C. 'An experimental study of Deryck Cooke's theory of music and meaning', *Psychology of Music* 6 (1), 13–20 (1978).
Gaser, C. and Schlaug, G. 'Brain structures differ between musicians and non-musicians', *Journal of Neuroscience* 23, 9240–5 (2003).
Gjerdingen, R. 'The psychology of music', in *The Cambridge History of Western Music*, ed. T. Christensen, 956–81. Cambridge University Press, Cambridge, 2002.
Gjerdingen, R. 'Leonard B. Meyer', *Empirical Musicology Review* 3, 2–3 (2008).
Gosselin, N. *et al.* 'Impaired recognition of scary music following unilateral temporal lobe excision', *Brain* 128, 628–40 (2005).
Grant, M. J. *Serial Music, Serial Aesthetics*. Cambridge University Press, Cambridge, 2001.
Gregory, A. H. and Varney, N. 'Cross-cultural comparisons in the affective response to music', *Psychology of Music* 24, 47–52 (1996).
Grey, J. An Exploration of Musical Timbre. PhD thesis, Stanford University, California, 1975.
Grey, J. 'Multidimensional perceptual scaling of musical timbres', *Journal of the Acoustical Society of America* 61, 1270–7 (1977).
Grout, D. J. *A History of Western Music*, revised edn. J. M. Dent & Sons, London, 1960.
Hanslick, E. *The Beautiful in Music*, transl. G. Cohen. Novello, Ewer & Co., London, 1891.
Hargreaves, J. R. 'The effects of repetition on liking for music', *Music Education* 32, 35–47 (1984).

Hatten, R. *Musical Meaning in Beethoven*. Indiana University Press, Bloomington, 1994.

Hindemith, P. *A Composer's World*. Anchor, Garden City, New York, 1961.

Ho, L.-T. 'On Chinese scales and national modes', *Asian Music* 14, 132–54 (1983).

Hodeir, A. *Jazz: Its Evolution and Essence*, transl. D. Noakes. Secker & Warburg, London, 1956.

Howell, P., Cross, I. and West, R. (eds). *Musical Structure and Cognition*. Academic Press, London, 1985.

Huron, D. 'The melodic arch in Western folksongs', *Computing in Musicology* 10, 3–23 (1996).

Huron, D. 'The avoidance of part-crossing in polyphonic music: perceptual evidence and musical practice', *Music Perception* 9, 93–104 (1991).

Huron, D. 'Tonal consonance versus tonal fusion in polyphonic sonorities', *Music Perception* 9, 135–54 (1991).

Huron, D. 'Tone and voice: a derivation for the rules of voice-leading from perceptual principles', *Music Perception* 19, 1–64 (2001).

Huron, D. *Sweet Anticipation: Music and the Psychology of Expectation*. MIT Press, Cambridge, Ma., 2006.

Huron, D. 'Lost in music', *Nature* 453, 456–7 (2008).

Huron, D. 'Asynchronous preparation of tonally fused intervals in polyphonic music', *Empirical Musicology Review* 3, 11–21 (2008).

Huron, D. and Veltman, J. 'A cognitive approach to medieval mode: evidence for an historical antecedent to the major/minor system', *Empirical Musicology Review* 1, 33–55 (2006).

Iverson, P. and Krumhansl, C. L. 'Isolating the dynamic attributes of musical timbre', *Journal of the Acoustical Society of America* 94, 2595–603 (1993).

Iversen, J. R., Patel, A. D. and Ohgushi, K. 'Perception of rhythmic grouping depends on auditory experience', *Journal of the Acoustical Society of America* 124, 2263–71 (2008).

Jackendoff, R. Review of Leonard Bernstein's 'The Unanswered Question', *Language* 53, 883–94 (1977).

Jackendoff, R. and Lerdahl, F. 'The capacity for music: what is it, and what's special about it?', *Cognition* 100, 33–72 (2006).

Jairazbhoy, N. A. *The ragas of North Indian Music: Their Structure and Evolution*. Faber & Faber, London, 1971.

Janata, P. 'Brain electrical activity evoked by mental formation of auditory expectations and images', *Brain Topography* 13, 169–93 (2001).

Janata, P., Birk, J. L., Van Horn, J. D., Leman, M., Tillmann, B. and Bharucha, J. J. 'The cortical topography of tonal structures underlying Western music', *Science* 298, 2167–70 (2003).

Jennings, H. D., Ivanov, P. Ch., Martins, A. M., da Silva, P. C. and Viswanathan, G. M. 'Variance fluctuations in nonstationary time series: a comparative study of music genres', *Physica A* 336, 585–94 (2004).

Jones, M. R. and Holleran, S. (eds). *Cognitive Bases of Musical Communication*. American Psychological Association, Washington, DC, 1992.

Juslin, P. N. and Laukka, P. 'Communication of emotions in vocal expression and music performance: Different channels, same code?', *Psychological Bulletin* 129, 770–814 (2003).

Juslin, P. N. and Laukka, P. 'Expression, perception, and induction of musical emotions: a review and questionnaire study of everyday listening', *Journal of New Musical Research* 33, 217–38 (2004).

Juslin, P. N. and Sloboda, J. A. (eds). *Music and Emotion*. Oxford University Press, Oxford, 2001.

Juslin, P. N. and Västfjäll, D. 'Emotional responses to music: the need to consider underlying mechanisms', *Behavioral and Brain Sciences* 31, 559–75 (2008).

Karno, M. and Konecni, V. 'The effects of structural interventions in the first movement of Mozart's Symphony in G Minor K550 on aesthetic preference', *Music Perception* 10, 63–72 (1992).

Kessler, E. J., Hansen, C. and Shepard, R. N. 'Tonal schemata in the perception of music in Bali and the West', *Music Perception* 2, 131–65 (1984).

Kivy, P. *Introduction to a Philosophy of Music*. Oxford University Press, Oxford, 2002.

Kivy, P. *The Corded Shell: Reflections on Musical Expression*. Princeton University Press, Princeton, 1980.

Kivy, P. *New Essays on Musical Understanding*. Oxford University Press, Oxford, 2001.

Koelsch, S. and Mulder, J. 'Electric brain responses to inappropriate harmonies during listening to expressive music', *Clinical Neurophysiology* 113, 862–69 (2002).

Koelsch, S., Kasper, E., Sammler, D., Schulze, K., Gunter, T. and Friederici,

A. D. 'Music, language and meaning: brain signatures of semantic meaning', *Nature Neuroscience* 7, 302–7 (2004).

Koelsch, S. and Siebel, W. A. 'Towards a neural basis of music perception', *Trends in Cognitive Sciences* 9, 578–84 (2005).

Koelsch, S., Gunter, T. C., Wittfoth, M. and Sammler, D. 'Interaction between syntax processing in language and in music: an ERP study', *Journal of Cognitive Neuroscience* 17, 1565–77 (2005).

Koelsch, S., Fritz, T., Yves von Cramon, D., Müller, K. and Friederici, A. D. 'Investigating emotion with music: an fMRI study', *Human Brain Mapping* 27, 239–50 (2006).

Koelsch, S., Kilches, S., Steinbeis, N. and Schelinski, S. 'Effects of unexpected chords and of performer's expression on brain responses and electrodermal activity', *PLoS One* e2631 (2008).

Koelsch, S., Fritz, T. and Schlaug, G. 'Amygdala activity can be modulated by unexpected chord functions during music listening', *NeuroReport* 19, 1815–19 (2008).

Koopman, C. and Davies, S. 'Musical meaning in a broader perspective', *Journal of Aesthetics and Art Criticism* 59, 261–73 (2001).

Kramer, J. D. *The Time of Music: New Meanings, New Temporalities, New Listening Strategies*. Schirmer, New York, 1988.

Kramer, L. *Musical Meaning: Towards a Critical History*. University of California Press, Berkeley, 2002.

Kreutz, G., Bongard, S., Rohrmann, S., Hodapp, V. and Grebe, D. 'Effects of choir singing and listening on secretory immunoglobulin A, cortisol, and emotional state', *Journal of Behavioral Medicine* 27, 623–35 (2004).

Krumhansl, C. L. *Cognitive Foundations of Musical Pitch*. Oxford University Press, New York, 1990.

Krumhansl, C. L. 'Music psychology: tonal structures in perception and memory', *Annual Reviews of Psychology* 42, 277 (1991).

Krumhansl, C. L. 'A perceptual analysis of Mozart's piano sonata, K282: segmentation, tension, and musical ideas', *Music Perception* 13, 401–32 (1996).

Krumhansl, C. L. and Keil, F. C. 'Acquisition of the hierarchy of tonal functions in music', *Memory & Cognition* 10, 243–51 (1982).

Krumhansl, C. L. and Kessler, E. J. 'Tracing the dynamic changes in perceived tonal organization in a spatial representation of musical keys', *Psychological Review* 89, 334–68 (1982).

Krumhansl, C. L. and Lerdahl, F. 'Modeling tonal tension', *Music*

Perception 24, 329–66 (2007).

Krumhansl, C. L., Louhivuori, J., Toiviainen, P., Järvinen, T. and Eerola, T. 'Melodic expectation in Finnish folk hymns: convergence of statistical, behavioral, and computational approaches', *Music Perception* 17, 151–97 (1999).

Krumhansl, C. L., Sandell, G. J. and Sergeant, D. C. 'The perception of tone hierarchies and mirror forms in twelve-tone serial music', *Music Perception* 5, 31–78 (1987).

Krumhansl, C. L. and Schmuckler, M. A. 'The Petroushka chord', *Music Perception* 4, 153–84 (1986).

Krumhansl, C. L., Toivanen, P., Eerola, T., Toiviainen, P., Järvinen, T. and Louhivuori, J. 'Cross-cultural music cognition: cognitive methodology applied to North Sami yoiks', *Cognition* 76, 13–58 (2000).

Krumhansl, C. L. 'An exploratory study of musical emotions and psychophysiology', *Canadian Journal of Experimental Psychology* 51, 336–53 (1997).

Langer, S. K. *Philosophy in a New Key: A Study in the Symbolism of Reason, Rite, and Art*, 3rd edn. Harvard University Press, Cambridge, Ma., 1957.

Large, E. W. and Palmer, C. 'Perceiving temporal regularity in music', *Cognitive Science* 26, 1–37 (2002)

Lerdahl, F. and Jackendoff, R. *A Generative Theory of Tonal Music*, 2nd edn. MIT Press, Cambridge, Ma., 1996.

Lerdahl, F. 'Tonal pitch space', *Music Perception* 5, 315–50 (1988).

Lerdahl, F. 'Cognitive constraints on compositional systems', in J. Sloboda (ed.) *Generative Processes in Music*. Oxford University Press, Oxford, 1988.

Lerdahl, F. *Tonal Pitch Space*. Oxford University Press, New York, 2001.

Lerdahl, F. 'Calculating tonal tension', *Music Perception* 13, 319–63 (1996).

Lerdahl, F. and Krumhansl, C. L. 'Modeling tonal tension', *Music Perception* 24, 329–66 (2007).

Levitin, D. J. *This is Your Brain on Music: The Science of a Human Obsession*. Plume, New York, 2007.

Lippman, E. *A History of Western Musical Aesthetics*. University of Nebraska Press, Lincoln, 1992.

Lomax, A. 'Universals in song', *World of Music* 19 (1/2), 117–29 (1977).

Longuet-Higgins, H. C. 'Making sense of music', *Proceedings of the Royal Institution of Great Britain* 45, 87–105 (1972).

Longuet-Higgins, H. C. and Lee, C. S. 'The rhythmic interpretation of monophonic music', in H. C. Longuet-Higgins (ed.), *Mental*

Processes, MIT Press, Cambridge, Ma., 1987, pp. 150–68.

Longuet-Higgins, H. C. and Lee, C. S., 'The perception of musical rhythms', *Perception* 11, 115–28 (1982).

Lu, T.-C. 'Music and salivary immunoglobulin A (sIgA): a critical review of the research literature'. Thesis, Drexel University, 2003.

Maess, B., Koelsch, S., Ganter, T. C. and Friederici, A. D. 'Musical syntax is processed in Broca's area: an MEG study', *Nature Neuroscience* 4, 540–5 (2001).

Manning, J. *The Finger Ratio*. Faber & Faber, London, 2008.

Marvin, E. W. and Brinkman, A. 'The effect of modulation and formal manipulation on perception of tonic closure by expert listeners', *Music Perception* 16, 389–408 (1999).

Masataka, N. 'Preference for consonance over dissonance by hearing newborns of deaf parents and of hearing parents', *Developmental Science* 9, 46–50 (2006).

Maus, F. E. 'Music as drama', *Music Theory Spectrum* 10, 56–73 (1988).

Maus, F. 'Music as narrative', *Indiana Theory Review* 12, 1–34 (1991).

May, E. (ed.) *Music of Many Cultures: An Introduction*. University of California Press, Berkeley, 1980.

McAdams, S. and Bigand, E. (eds). *Thinking in Sound: The Cognitive Psychology of Human Audition*. Oxford University Press, Oxford, 1993.

McClary, S. *Feminine Endings: Music, Gender and Sexuality*. University of Minnesota Press, Minneapolis, 1991.

McDermott, J. 'The evolution of music', *Nature* 453, 287–8 (2008).

McDermott, J. and Hauser, M. 'Are consonant intervals music to their ears? Spontaneous acoustic preferences in a nonhuman primate', *Cognition* 94, B11–B21 (2004).

McDermott, J. and Hauser, M. 'The origins of music: innateness, uniqueness and evolution', *Music Perception* 23, 29–59 (2005).

McDermott, J. and Hauser, M. D. 'Thoughts on an empirical approach to evolutionary origins of music', *Music Perception* 24, 111–6 (2006).

McDermott, J. H., Lehr, A. J. and Oxenham, A. J. 'Is relative pitch specific to pitch?', *Psychological Science* 19, 1263–71 (2008).

McDonald, C. and Stewart, L. 'Uses and functions of music in congenital amusia', *Music Perception* 25, 345–55 (2008).

Merriam, A. P. *The Anthropology of Music*. Northwestern University Press, Evanston, 1964.

Meyer, L. B. *Emotion and Meaning in Music*. University of Chicago Press, Chicago, 1956.

Meyer, L. B. 'Commentary', *Music Perception* 13, 455–83 (1996).

Mithen, S. *The Singing Neanderthals*. Weidenfeld & Nicolson, London, 2005.

Moore, A. F. *Rock: The Primary Text*. Ashgate, Aldershot, 2001.

Moore, A. 'The so-called "flattened seventh" in rock', *Popular Music* 14, 185–201 (1995).

Musacchia, G., Sams, M., Skoe, E. and Kraus, N. 'Musicians have enhanced subcortical and audiovisual processing of speech and music', *Proceedings of the National Academy of Sciences USA* 104, 15894–8 (2007).

Musicae Scientiae special issue on emotions and music (2001/2002).

Narmour, E. *The Analysis and Cognition of Basic Melodic Structures*. University of Chicago Press, Chicago, 1990.

Narmour, E. 'The top-down and bottom-up systems of musical implication: building on Meyer's theory of emotional syntax', *Music Perception* 9, 1–26 (1991).

Narmour, E. 'Analysing form and measuring perceptual content in Mozart's sonata K282: a new theory of parametric analogues', *Music Perception* 13, 265–318 (1996).

Nasr, S. 'Audio software for the moody listener', *Technology Review* online, 19 July 2006; available at http://www.technologyreview.com/read_article.aspx?id=17183&ch=infotech.

Nattiez, J. J. *Music and Discourse: Toward a Semiology of Music*. Princeton University Press, Princeton, 1990.

Nettl, B. *Folk and Traditional Music of the Western Continents*, 2nd edn. Prentice-Hall, Eaglewood Cliffs, NJ, 1973.

Newcomb, A. 'Once more "between absolute and program music": Schumann's Second Symphony', *19th Century Music* 7, 233–50 (1984).

Oram, N. and Cuddy, L. L. 'Responsiveness of Western adults to pitch-distributional information in melodic sequences', *Psychological Research* 57, 103–18 (1995).

Page, M. F. 'Perfect harmony: a mathematical analysis of four historical tunings', *Journal of the Acoustical Society of America* 116, 2416–26 (2004).

Palmer, C. and Kelly, M. H. 'Linguistic prosody and musical meter in song', *Journal of Memory and Language* 31, 525–42 (1992).

Parncutt, R. *Harmony: A Psychoacoustical Approach*. Springer, Berlin, 1989.

Partch, H. *Genesis of a Music*. Da Capo, New York, 1974.

Patel, A. D., Gibson, E., Ratner, J., Besson, M. and Holcomb, P. J.

'Processing syntactic relations in language and music: an event-related potential study', *Journal of Cognitive Neuroscience* 10, 717–33 (1998).
Patel, A. D. 'Syntactic processing in language and music: different cognitive operations, similar neural resources?', *Music Perception* 16, 27–42 (1998).
Patel, A. D. 'Language, music, syntax and the brain', *Nature Neuroscience* 6, 674–81 (2003).
Patel, A. D. 'Musical rhythm, linguistic rhythm, and human evolution', *Music Perception* 24, 99–104 (2006).
Patel, A. D. 'Talk of the tone', *Nature* 453, 726–7 (2008).
Patel, A. D., Iversen, J. R. and Rosenberg, J. C. 'Comparing the rhythm and melody of speech and music: the case of British English and French', *Journal of the Acoustical Society of America* 119, 3034–47 (2006).
Patel, A. D. and Iversen, J. R. 'The linguistic benefits of musical abilities', *Trends in Cognitive Sciences* 11, 369–72 (2007).
Patel, A. D. *Music, Language, and the Brain*. Oxford University Press, New York, 2008.
Patel, A. D., Iversen, J. R., Bregman, M. R. and Schultz, I. 'Studying synchronization to a musical beat in nonhuman animals', *Annals of the New York Academy of Sciences* (in press).
Pederson, P. 'The perception of octave equivalence in twelve-tone rows', *Psychology of Music* 3, 3–8 (1975).
Peretz, I., Gagnon, L. and Bouchard, B. 'Music and emotion: perceptual determinants, immediacy, and isolation after brain damage', *Cognition* 68, 111–41 (1998).
Peretz, I. and Gagnon, L. 'Dissociation between recognition and emotional judgements for melodies', *Neurocase* 5, 21–30 (1999).
Peretz, I. and Hébert, S. 'Towards a biological account of musical experience', *Brain & Cognition* 42, 131–4 (2000).
Peretz, I. and Zatorre, R. (eds). *The Cognitive Neuroscience of Music*. Oxford University Press, Oxford, 2003.
Peretz, I., Radeau, M. and Arguin, M. 'Two-way interactions between music and language: evidence from priming recognition of tune and lyrics in familiar songs', *Memory and Cognition* 32, 142–52 (2004).
Perttu, D. 'A quantitative study of chromaticism: changes observed in historical eras and individual composers', *Empirical Musicology Review* 2 (2), 47–54 (2007).

Pesic, P. 'Earthly music and cosmic harmony: Johannes Kepler's interest in practical music, especially Orlando di Lasso', *Journal of Seventeenth Century Music* 11 (2005).

Plantinga, J. and Trainor, L. J. 'Melody recognition by two-month-old infants', *Journal of the Acoustical Society of America Express Letters* 125, E58–E62 (2009).

Plomp, R. and Levelt, W. J. M. 'Tonal consonance and critical band width', *Journal of the American Acoustical Society* 38, 548–60 (1965).

Poulin-Charronnat, B., Bigand, E. and Koelsch, S. 'Processing of musical syntax tonic versus subdominant: an event-related potential study', *Journal of Cognitive Science* 18, 1545–54 (2006).

Raffman, D. *Language, Music and Mind*. MIT Press, Cambridge, Ma., 1993.

Ratner, L. G. *Classic Music: Expression, Form and Style*. Schirmer, New York, 1980.

Rauscher, F. H., Shaw, G. L. and Ky, K. N. 'Music and spatial task performance', *Nature* 365, 611 (1993).

Rauscher, F. H., Shaw, G. L. and Ky, K. N. 'Listening to Mozart enhances spatial-temporal reasoning: towards a neurophysiological basis', *Neuroscience Letters* 185, 44–7 (1995).

Read, G. *Musical Notation*. Gollancz, London, 1974.

Révész, G. *Introduction to the Psychology of Music*. Dover, New York, 2001.

Rifkin, D. 'Making it modern: chromaticism and phrase structure in 20th-century tonal music', *Theory and Practice* 31, 133–58 (2006).

Rifkin, D. 'A theory of motives for Prokofiev's music', *Music Theory Spectrum* 26, 265–90 (2004).

Robinson, J. (ed.) *Music and Meaning*. Cornell University Press, Ithaca, NY, 1997.

Rohrmeier, M. and Cross, I. 'Statistical properties of harmony in Bach's chorales', in *Proceedings of the 10th International Conference on Music Perception and Cognition (ICMPC 2008)*. Sapporo, Japan, 2008.

Rosen, C. *Schoenberg*. Fontana, London, 1976.

Ross, A. *The Rest Is Noise*. Fourth Estate, London, 2008.

Rothstein, E. *Emblems of the Mind*. Harper, New York, 1996.

Sachs, C. *Our Musical Heritage*. Prentice-Hall, New York, 1948.

Sacks, O. *Musicophilia*. Picador, London, 2007.

Saffran, J. R., Johnson, E. K., Aslin, R. N. and Newport, E. L. 'Statistical learning of tone sequences by human infants and adults', *Cognition* 70, 27–52 (1999).

Sakai, K., Hikosaka, O., Miyauchi, S., Takino, R., Tamada, T., Kobayashi Iwata, N., and Nielsen, M., 'Neural representation of a rhythm depends on its interval ratio', *Journal of Neuroscience* 19, 10074–81 (1999).

Samplaski, A. 'Comment on Daniel Perttu's "A quantitative study of chromaticism"', *Empirical Musicology Review* 2 (2), 55–60 (2007).

Schellenberg, E. G. 'Music lessons enhance IQ', *Psychological Science* 15, 511–14 (2004).

Schellenberg, E. G. 'Music and cognitive abilities', *Current Directions in Psychological Science* 14, 317–20 (2005).

Schellenberg, E. G. 'Long-term positive associations between music lessons and IQ', *Journal of Educational Psychology* 98, 457–68 (2006).

Schellenberg, E. G., Bigand, E., Poulin, B., Garnier, C. and Stevens, C. 'Children's implicit knowledge of harmony in Western music', *Developmental Science* 8, 551–6 (2005).

Schellenberg, E. G. and Hallam, S. 'Music Listening and cognitive abilities in 10- and 11-year-olds: the Blur effect', *Annals of the New York Academy of Sciences* 1060, 202–9 (2005).

Schellenberg, E. G. and Trehub, S. E. 'Natural musical intervals: evidence from infant listeners', *Psychological Science* 7, 272–7 (1996).

Schoenberg, A. *Fundamentals of Musical Composition*, ed. G. Strang and L. Stein. Faber & Faber, London, 1982.

Scruton, R. *The Aesthetics of Music*. Clarendon Press, Oxford, 1997.

Sessions, R. *The Musical Experience of Composer, Performer, Listener*. Princeton University Press, Princeton, 1950.

Sethares, W. A. 'Adaptive tunings for musical scales', *Journal of the Acoustical Society of America* 96, 10–18 (1994).

Sethares, W. A. 'Real-time adaptive tunings using Max', *Journal of New Music Research* 31, 347–55 (2002).

Sethares, W. A. *Tuning Timbre Spectrum Scale*, 2nd edn. Springer, Berlin, 2004.

Shepherd, J., Virden, P., Vulliamy, G. and Wishart, T. (eds). *Whose Music? A Sociology of Musical Languages*. Latimer, London, 1977.

Slevc, L. R., Rosenberg, J. C. and Patel, A. D. 'Making psycholinguistics musical: self-paced reading time evidence for shared processing of linguistic and musical syntax', *Psychonomic Bulletin and Review* 16, 374–81(2009).

Sloboda, J. A. 'The uses of space in music notation', *Visible Language* 25, 86–110 (1981).

Sloboda, J. A. *The Musical Mind: The Cognitive Psychology of Music.* Clarendon Press, Oxford, 1985.

Sloboda, J. A. (ed.) *Generative Processes in Music,* Oxford University Press, Oxford, 1988.

Sloboda, J. A. and Lehmann, A. C. 'Tracking performance correlates of changes in perceived intensity of emotion during different interpretations of a Chopin piano prelude', *Music Perception* 19, 87–120 (2001).

Sloboda, J. A. 'Music structure and emotional response: some empirical findings', *Psychology of Music* 19, 110–20 (1991).

Sloboda, J. *Exploring the Musical Mind.* Oxford University Press, Oxford, 2005.

Sluming, V. A. and Manning, J. T. 'Second to fourth digit ratio in elite musicians: evidence for musical ability as an honest signal of male fitness', *Evolution and Human Behavior* 21, 1–9 (2000).

Smith, J. and Wolfe, J. 'Vowel-pitch matching in Wagner's operas: implications for intelligibility and ease of singing', *Journal of the Acoustical Society of America* 125, EL196–EL201 (2009).

Smith, N. A. and Cuddy, L. L. 'Perceptions of musical dimensions in Beethoven's Waldstein sonata: an application of tonal pitch space theory', *Musicae Scientiae* 7.1, 7–34 (2003).

Spector, L. and Alpern, A. 'Criticism, culture, and the automatic generation of artworks', in *Proceedings of the Twelfth National Conference on Artificial Intelligence, AAAI-94,* pp. 3–8. AAAI Press/MIT Press, Menlo Park, Ca. and Cambridge, Ma. 1994.

Steedman, M. J. 'The perception of musical rhythm and metre', *Perception* 6, 555–69 (1977).

Stewart, L., von Kriegstein, K., Warren, J. D. and Griffiths, T. D. 'Music and the brain: disorders of musical listening', *Brain* 129, 2533–53 (2006).

Storr, A. *Music and the Mind.* Collins, London, 1992.

Stravinsky, I. *Poetics of Music in the Form of Six Lessons,* transl. A. Knodel and I. Dahl. Harvard University Press, Cambridge, Ma., 1947.

Strong, G. (ed.) *Fundamentals of Music Composition.* St Martin's Press, London, 1967.

Stubbs, D. *Fear of Music.* Zero Books, Ropley Hants., 2009.

Sundberg, J. (ed.) *Gluing Tones: Grouping in Music Composition, Performance and Listening.* Royal Swedish Academy of Music, Taberg, 1992.

Sundberg, J. and Lindblom, B. 'Generative theories in language and music descriptions', *Cognition* 4, 99–122 (1976).

Tagg, P. 'Analysing popular music: theory, method and practice', *Popular Music* 2, 37–65 (1982).

Temperley, D. 'Syncopation in rock: a perceptual perspective', *Popular Music* 18, 19–40 (1999).

Temperley, D. *The Cognition of Basic Musical Structures*. MIT Press, Cambridge, Ma., 2001.

Thompson, W. 'From sounds to music: the contextualizations of pitch', *Music Perception* 21, 431–56 (2004).

Tillmann, B., Janata, P. and Bharucha, J. J. 'Activation of the inferior frontal cortex in musical priming', *Cognitive Brain Research* 16, 145–61 (2003).

Tillmann, B. and Bigand, E. 'Does formal musical structure affect perception of musical expressiveness?', *Psychology of Music* 24, 3–17 (1996).

Timmers, R. and Ashley, R. 'Emotional ornamentation in performances of a Handel sonata', *Music Perception* 25, 117–34 (2007).

Trainor, L. J. 'The neural roots of music', *Nature* 453, 598–99 (2008).

Trainor, L. J. and Heinmiller, B. M. 'The development of evaluative responses to music: infants prefer to listen to consonance over dissonance', *Infant Behavior and Development* 21, 77–88 (1998).

Trainor, L. J. and Trehub, S. E. 'A comparison of infants' and adults' sensitivity to Western musical structure', *Journal of Experimental Psychology: Human Perception and Performance* 18, 394–402 (1992).

Trainor, L. J. and Trehub, S. E. 'Musical context effects in infants and adults: key distance', *Journal of Experimental Psychology: Human Perception and Performance* 19, 615–26 (1993).

Trainor, L. J. and Trehub, S. E. 'Key membership and implied harmony in Western tonal music: developmental perspectives', *Perception and Psychophysics* 56, 125–32 (1994).

Trainor, L. J., Tsang, C. D. and Cheung, V. H. W. 'Preference for consonance in 2-month-old infants', *Music Perception* 20, 185–192 (2002).

Trehub, S. E., Thorpe, L. A. and Trainor, L. J. 'Infants' perception of *good* and *bad* melodies', *Psychomusicology* 9, 5–19 (1990).

Trehub, S. E., Schellenberg, E. G. and Kamenetsky, S. B. 'Infants' and adults' perception of scale structure', *Journal of Experimental Psychology: Human Perception and Performance* 25, 965–75 (1999).

Trehub, S. E. 'The developmental origins of musicality', *Nature*

Neuroscience 6, 669 (2003).

Van den Toorn, P. C. *The Music of Igor Stravinsky.* Yale University Press, New Haven, 1983.

Wallin, N. L., Merker, B. and Brown, S. (eds). *The Origins of Music.* MIT Press, Cambridge, Ma., 2000.

Walser, R. *Running with the Devil: Power, Gender and Madness in Heavy Metal Music.* Wesleyan University Press, Hanover, NH, 1993.

Warren, J., 'How does the brain process music?', *Clinical Medicine* 8, 32–6 (2008).

Waterman, M. 'Emotional responses to music: implicit and explicit effects in listeners and performers', *Psychology of Music* 24, 53–67 (1996).

Weber, M. *The Rational and Social Foundations of Music*, transl. D. Martindale, G. Neuwirth and J. Riedel. Southern Illinois University Press, Carbondale, Ill., 1958.

White, B. 'Recognition of distorted melodies', *American Journal of Psychology* 73, 100–7 (1960).

Winkler, I., Haden, G., Ladinig, O., Sziller, I. and Honing, H. 'Newborn infants detect the beat in music', *Proceedings of the National Academy of Sciences USA* 106, 2468–71 (2009).

Zajonc, R. B. 'On the primacy of affect', *American Psychologist* 39, 117–23 (1984).

Zatorre, R. J. and Peretz, I. (eds), 'The Biological Foundations of Music', *Annals of the New York Academy of Sciences* 930 (2001).

Zatorre, R. J., Chen, J. L. and Penhune, V. B. 'When the brain plays music: auditory-motor interactions in music perception and production', *Nature Reviews Neuroscience* 8, 547–58 (2007).

Zbikowski, L. M. *Conceptualizing Music: Cognitive Structure, Theory and Analysis*. Oxford University Press, Oxford, 2002.

Zentner, M. and Kagan, J. 'Perception of music by infants', *Nature* 383, 29 (1996)

Index